DOREEN MASSEY: CRITICAL DIALOGUES

Edited by

MARION WERNER
JAMIE PECK
REBECCA LAVE
BRETT CHRISTOPHERS

agenda
publishing

First published in 2018 by Agenda Publishing

Agenda Publishing Limited
The Core
Science Central
Bath Lane
Newcastle upon Tyne
NE4 5TF
www.agendapub.com

ISBN 978-1-911116-85-1 (hardcover)
ISBN 978-1-911116-86-8 (paperback)

British Library Cataloguing-in-Publication Data
A catalogue record for this book is available from the British Library

Typeset by Out of House Publishing
Printed and bound in the UK by TJ International

CONTENTS

ACKNOWLEDGEMENTS

This volume and its companion, *The Doreen Massey Reader*, were spurred by Doreen Massey's untimely passing in March 2016. As we reflected upon her contributions, we lamented that so many of Massey's works were relatively difficult to access and thus began the process of re-reading her works and making the invidious decisions about what could be included in the *Reader*. Because Massey was in the middle of so many conversations at the moment of her ill-timed death, we also sought a way to continue those discussions among her long-time collaborators and friends and to extend them to other scholars who, while not directly engaged with Massey, were interested in thinking with the many wonderful conceptual tools that she developed over her career. When we sent out invitations to potential authors, we were overwhelmed by the enthusiastic response and the willingness of the contributors to conform to a relatively short timeline. We would like to thank the 39 authors for their insightful and generous contributions to the *Doreen Massey: Critical Dialogues* volume. The process of creating both volumes benefited immensely from the insights and guidance of John Allen, who serves as the literary executor of Doreen's estate. We are also grateful for the support of Agenda Publishing, its managing director, Steven Gerrard, and especially, Alison Howson, who has worked closely with us on the numerous moving parts of the project, as well as the editorial assistance of Rachel Brydolf-Horwitz at UBC. We are indebted to Doreen's sister, Hilary Corton, who graciously granted us permission to reprint the work included in *The Doreen Massey Reader*. All royalties from both volumes will be donated to charities designated by the Massey Estate.

Marion Werner
Jamie Peck
Rebecca Lave
Brett Christophers

CONTRIBUTORS

Abel Albet is Associate Professor in the Department of Geography at the Universitat Autònoma de Barcelona.

Trevor Barnes is Professor and University Distinguished Scholar in the Department of Geography, University of British Columbia.

Núria Benach is Associate Professor in the Department of Geography at the Universitat de Barcelona.

Christian Berndt is Professor of Economic Geography at the University of Zurich, Switzerland.

Huw Beynon is Emeritus Professor at the Wales Institute of Social and Economic Research, Data and Methods at Cardiff University.

John Clarke is Emeritus Professor of Social Policy at the Open University and a Visiting Professor in the Department of Sociology.

Allan Cochrane is Emeritus Professor of Urban Studies at the Open University.

Brett Christophers is Professor in the Department of Social and Economic Geography at Uppsala University.

Michael Dear is Emeritus Professor of City and Regional Planning in the College of Environmental Design at the University of California, Berkeley, and Honorary Professor in the Bartlett School of Planning, University College London.

Sarah Elwood is Professor of Geography at the University of Washington and co-founder of the Relational Poverty Network with Victoria Lawson.

Coleen Fox teaches in the Department of Geography and the Environmental Studies Program at Dartmouth College.

Vinay Gidwani is Professor of Geography and Global Studies at the University of Minnesota.

Gillian Hart is Professor of the Graduate School at the University of California, Berkeley, and Distinguished Professor at the University of the Witwatersrand.

Nik Heynen is Professor in the Department of Geography at the University of Georgia.

Ray Hudson is Emeritus Professor of Geography at Durham University.

Jennifer Hyndman is Professor in Social Science and Geography at York University in Toronto, where she is also Director of the Centre for Refugee Studies.

Jessica Jacobs is a geographer-filmmaker currently based at Queen Mary University of London.

Caroline Keegan is a PhD student in the Department of Geography at the University of Georgia.

Rebecca Lave is Associate Professor in Geography at Indiana University.

Victoria Lawson is Professor of Geography at the University of Washington, Director of the University of Washington Honors Program and co-founder of the Relational Poverty Network with Sarah Elwood.

Helga Leitner is Professor of Geography at the University of California, Los Angeles.

Nikki Luke is a PhD student in the Department of Geography at the University of Georgia.

Frank Magilligan is Professor in the Geography Department at Dartmouth College and was recently awarded the Frank J. Reagan '09 Chair in Policy Studies.

Linda McDowell worked with Doreen Massey at the Open University between 1983 and 1992. After leaving the OU she held posts at Cambridge and London Universities, ending up as Professor of Human Geography at Oxford University.

Richard Meegan first worked with Doreen Massey at the government-funded Centre for Environmental Studies in London. He is an Honorary Research Fellow in the Department of Geography and Planning at the University of Liverpool.

Katharyne Mitchell is Dean of the Social Sciences and Professor of Sociology at the University of California, Santa Cruz.

Alison Mountz is Professor of Geography and Canada Research Chair in Global Migration at the Balsillie School of International Affairs at Wilfrid Laurier University.

Jamie Peck is Canada Research Chair in Urban and Regional Political Economy, Distinguished University Scholar and Professor of Geography at the University of British Columbia, Vancouver.

John Pickles is the Patterson Distinguished Professor of International Studies and Geography and adjunct Distinguished Professor in Cultural Studies and Anthropology at the University of North Carolina, Chapel Hill.

Geraldine Pratt is Professor of Geography and Canada Research Chair in Transnationalism and Precarious Labour at the University of British Columbia.

Priti Ramamurthy is Professor of Gender, Women and Sexuality Studies at the University of Washington, Seattle.

Susan M. Roberts is Professor of Geography and Associate Provost for Internationalization at the University of Kentucky.

Michael Rustin is Professor of Sociology at the University of East London, and a Visiting Professor at the Tavistock Clinic and at the University of Essex.

Andrew Sayer is Professor of Social Theory and Political Economy at Lancaster University.

Erica Schoenberger is Professor in the Department of Environmental Health and Engineering at the Johns Hopkins University, with a joint appointment in Anthropology.

Eric Sheppard is the Alexander von Humboldt Chair of Geography at UCLA.

Christopher Sneddon is Professor in the Department of Geography and the Environmental Studies Program at Dartmouth College.

Matthew Sparke is Professor of Politics at the University of California, Santa Cruz.

Kendra Strauss is Associate Professor and Director of the Labour Studies Program at Simon Fraser University, and an Associate Member in the Department of Geography.

Hilary Wainwright is co-editor of the magazine *Red Pepper* and Fellow of the Transnational Institute.

Richard Walker is Professor Emeritus of Geography at the University of California, Berkeley, where he taught from 1975 to 2012 and served at various times as Chair of Geography, Global Metropolitan Studies and California Studies.

Marion Werner is Associate Professor in the Department of Geography at the University of Buffalo, State University of New York.

Perla Zusman is Professor at the University of Buenos Aires and Researcher at the National Research Council (Argentina).

CHAPTER 1

OUT OF PLACE: DOREEN MASSEY, RADICAL GEOGRAPHER

Jamie Peck, Marion Werner, Rebecca Lave and
Brett Christophers

Doreen Massey changed geography. As a creative scholar, an inspiring teacher and a restless activist, she initiated new ways of seeing, understanding and indeed changing the world. She launched critiques, both in the relatively small world of economic geography and the much bigger worlds of social theory and progressive politics, which would prove to be truly transformative; she developed arguments against a host of establishment and orthodox positions that left something better and more productive in their place; she confronted structurally embedded power relations, most notably of class and gender, while steadfastly resisting political and analytical foreclosure; and she started conversations that continue to resonate and reverberate, not least those around the protean potential of place, even in these challenging times.

"There is no point of departure" is a line that Massey liked to quote from Louis Althusser (1971: 85; Massey 1995b: 351; Featherstone & Painter 2013). For her, it meant that socially made historical geographies really matter, always and everywhere, and that futures are neither singular nor pre-given. Her own life was a case in point. The product of an "ordinary place" (Massey 2001b: 459), a public-housing estate just south of Manchester, Doreen Massey knew where she came from and for that matter, which side she was on. "I'm from the North West [of England] and have lived with, through and kind of in combat with regional inequality [since] my childhood", Massey once explained (Massey with HGRG 2009: 405). Out of the conformity of postwar Britain, Massey fashioned a transformative trajectory not least, she later reflected, by participating in political movements "in the late 60s and the 70s with the emergence of Marxism, feminism, sexual liberation, being part of the GLC [Greater London Council] in the 1980s, or the kind of stuff that has happened more recently", from Chavismo in Venezuela to the Occupy movement in London (*ibid.*: 403, 405; Featherstone *et al.* 2013: 253, 257).

From her adopted home of Kilburn, in North London, she would sometimes commute to work at the Open University with her longtime friend and collaborator, Stuart Hall. The quotidian experience of driving to campus and back served as a reminder to both that one "can never 'go home' ... You can't go back", since neither of them came from "this tract of southeastern England", nor was it possible for them to return to the Jamaica or Manchester of their youth, which of course were not "the same as when we left" (Massey 2000b: 230). Walks down Kilburn High Street would evoke for Massey, indelibly, "a global sense of place", quite the opposite of its sometimes parochial, introspective, singular, or static meaning, but instead an open-ended, processual, intersectional and dynamic sense of place, always in the (re)making (Massey 1991a).

This understanding of place as an emergent constellation, or moving configuration, of social trajectories echoed the way in which Massey sought to problematize connection and difference – not separately but in the same time-space. As she would reflect in relation to her travels (and conversations) with Stuart Hall:

> What the simultaneity of space really consists in, then, is absolutely not a surface, a continuous material landscape, but a momentary coexistence of trajectories, a configuration of a multiplicity of histories all in the process of being made. This is ... part of the delight, and the potential, of space.   (Massey 2000b: 229)

That Massey's journey should take her here, all the way from regional science and industrial geography, is a story in its own right. Since there can be no singular point of departure, nor any final moment of closure, our purpose in this chapter is to trace some of the contours and milestones of Doreen Massey's transformative intellectual and political journey. Reassembling the story will require, inevitably, some measure of chronology. But as David Featherstone and Joe Painter wrote in an earlier collection on Massey's career-long contributions, "Any attempt to fit her work into a neat sequential account of geography's recent past ... would be doomed to fail" (Featherstone & Painter 2013: 2). What follows then should be understood less as a sequence of steps or stages, and more as a collection of episodes in the formation of an intellectual and political biography. We begin in Manchester and end in Kilburn. In between, we arc selectively through Massey's paradigm-making interventions in political-economic geography and late-neoliberal politics, through her distinctive interventions in Marxist and feminist theory, seeking to trace along the way some of the contours of her foundational contributions to the understanding of space and spatiality. And it still feels like we are barely scratching the surface ...

OUT OF MANCHESTER

Doreen Massey grew up in a working-class family in the Wythenshawe council estate in south Manchester, a public housing development that was for a while the largest in Europe (see Massey 2001b). Along with many of their neighbours, the Massey family relied upon the state for subsidized housing, free schooling and healthcare. This would be especially important for their eldest daughter, who was born with a calcium disorder that made her bones fragile and subject to breaks throughout her life. "[H]ad there not been a welfare state and the hospitals", Massey later reflected, "I would probably not have survived so well. I really feel in a kind of physical, personal way the need for a welfare state, not as a 'safety net', but just for ordinary people simply to provide a decent life" (Freytag & Holyer 1999: 85). She would take a hardly typical path for a working-class girl from the North, going to Oxford University in the mid-1960s, where not for the first nor indeed last time in her life she would sometimes feel like a "space invader" (see Telegraph 2016: 33; cf. Massey 1994: 185). Somewhat ironically perhaps, given the discipline's overwhelmingly conservative cast at the time, it was through Geography that she discovered her way out, even as she retained an abiding anti-establishment sensibility.

Despite being awarded a First at Oxford, Massey initially rejected the academic path and instead went to work in the computing department of a market research firm. As she described her thinking later, "The reason I left Oxford not wanting to be an academic was that I'd seen what I thought it meant to be an academic. And I didn't want to be that. So I went into industry – and hated it!" (Freytag & Hoyler 1999: 84). Abandoning the private sector, she began her research career in earnest at the Centre for Environmental Studies (CES), an independent research institute founded by Harold Wilson's Labour Government in 1967 with a remit in spatial planning. Here she would make some of the first moves in what would amount to a radical rethinking of industrial and class restructuring. In 1970, Massey returned briefly to Oxford to participate in the UK's first Women's Liberation Movement conference of that era. "From then on", she later reflected, "I have always been involved in feminist movements" (Albet & Benach 2012: 53, editors' translation). Massey would shape and channel her anti-establishment sensibility in productive tension with feminism in the ensuing decades, wary of currents within feminism that tended towards essentialism and narrow identity politics, and always emphasizing the imbrications of class and gender (Albet & Benach 2012: 54). She insisted that "[C]lass and feminism ... [affect] what kind of voice you have; what kind of role you can play, and want to play" (Featherstone *et al*. 2013: 261).

3

In 1971, Massey was granted a leave from CES to undertake a Master's degree in regional science and "mathematical economics" at the University of Pennsylvania in Philadelphia, then a citadel of neoclassical location theory, which she viewed as a "'you ought to know your enemy' kind of thing" (Massey with HGRG 2009: 404). It was here that an extracurricular visit to the French Department provided a quite unexpected introduction to Louis Althusser and to a distinctive (and generative) interpretation of Althusserianism. After Penn, it was not the mathematics of location theory that she pursued, but instead an altogether more radical path. For a while, CES would be an accommodating home for what would prove to be formative work on the political economy of Britain's "regional problem", some of this in collaboration with Richard Meegan (see Meegan 2017). But in 1979, when the incoming Thatcher administration abruptly withdrew funding for the organization, Massey found herself at a crossroads of sorts. Fortunately, at least for the short term, she had a research grant enabling her to work at the London School of Economics, and to make what would prove to be a remarkably catalytic visit to the University of California at Berkeley (Peck & Barnes 2019). It was here, in what was otherwise an especially lean spell for British universities, that an unexpected opportunity arose:

> While I was in California the advertisement came up that offered a post at the Open University and that seemed to me to be a place where it might be possible to be an intellectual, a teacher, a researcher without being at a more formal university, and I applied for that job and I got it. There was a short time to go before I would have been unemployed. So it was either a chair or the dole. (Freytag & Hoyler 1999: 85)

Doreen Massey would spend the next 27 years of her academic career teaching at what many would consider, rather ironically, the most placeless of British universities, the Open University, a distance-learning institution in the infamously anonymous new town of Milton Keynes, where in the context of limited face-to-face contact with students she pioneered innovative ways of "teaching at a distance" (Clarke 2016: 360), not to say engaging across distance.

Massey was on the frontline in some of the signature struggles against Thatcherism, including the miners' strike of 1984–85 and the municipal-socialist project of the Greater London Enterprise Board, events that in retrospect marked an historical inflection point between a regionalist model of labour organization and the ascendancy of the "new urban left". She was also heavily involved in a wide range of intellectual projects, as an early editorial board member of *Capital & Class* and *Red Pepper*, as a co-founder of *Soundings* with Stuart Hall and Michael Rustin, and as the key mover in an extended series of remarkably influential Open University course texts. For many years, Massey was engaged in political struggles in Latin America and South Africa too, as a researcher and activist. And she would devote the later part of her career to a creative and politically inspiring analysis of why the global financial crisis of 2008 had not led to the collapse of neoliberalism, culminating in a final project in collaboration with Hall and Rustin. Characteristically, what was known as the Kilburn Manifesto could be considered both a product of their (shared) place, as a model of conjuncturally situated political, economic, and cultural analysis, and a contribution that resonated and reached significantly beyond that place (see Hall *et al.* 2012; Peck *et al.* 2014). In these, as in so many of her other endeavours, Doreen Massey consistently gave the lie to the idea that the price of theoretical sophistication had to be paid in political irrelevance or incomprehensibility outside (and sometimes even inside!) academia. Similarly, she refused to accept that there should be a dividing line between political and intellectual work.

These were principles that she quite literally embodied. One of the most striking things about Doreen Massey was the contrast between her very small physical stature and her very large personality. Possessed of a radiant smile, ready laughter and boundless curiosity, she had a notable ability to connect personally and intellectually with those around her. Throughout her life, she moved in academic and political circles dominated mostly by men, and not only rejected but actively challenged the masculinist cultures of both those worlds. "This is a political position, not an essentialization around masculinity, femininity or whatever", she once explained. "But I do find myself amazed by and wary of the ease with which writers make Olympian statements about the age … [while] standing outside society and describing it and forgetting that we're also within it" (Featherstone *et al.* 2013: 261). Massey never forgot that, despite the enormous influence that her own work would have in geography, in feminism, in social theory, and in left scholarship more generally. She handled this, as she noted in amusement in an interview with graduate students at the University of Kentucky in the early 1990s, by "just carry[ing] on being different!"

> I can't speak like some of these big guys do. If you are five foot one, and you are fair-haired, and you are female, and quite often you can barely see over the podium, then just physically and materially you cannot be imposing in the same way that you can when you are six foot five and have a big male deep voice. The very physicality and materiality of it, as well as the fact that they just take those people

more seriously than they take us, starts you off in a different situation. So what I've tried to do is just carry on being different.

(Ijams *et al.* 1994: 101)

In what follows, we sketch some of the many generative and inspiring ways in which Doreen Massey carried on by being different, beginning where her scholarly career began, with a transformative critique of industrial location theory.

INDUSTRIAL DISLOCATIONS

Doreen Massey made the first of many field-shaping interventions in 1973, with her practically terminal critique of the science of location theory. "Towards a Critique of Industrial Location Theory", published in the recently launched radical journal, *Antipode*, marked her uninvited arrival to the male-dominated, white-bread field known at the time as industrial geography. Hers was a nominally "tentative" critique from which the field would never really recover. The journey implied by the article's title did not in the end result in a "march … into a newly-formulated industrial location theory" (Massey 1973: 38, 33), but instead would take a more circuitous route to an entirely different paradigm. Not unlike British industry itself, the field of industrial geography was in a parlous state at the time, dogged as it was by unreflexive strains of empiricism and economism, and falling short in its attempts to account for disorienting patterns of path-disrupting, radical, and often divergent change (see Williams & Thomas 1983; Martin *et al.* 1993).

Critical not only of the epistemological but also the ideological affinities between location theory and neoclassical economics, Massey challenged the prevailing conception of abstract firms operating in abstract space on the grounds both of analytical insufficiency and an evident estrangement from real-world conditions:

> What are emerging as "locational problems", whether intra-urban, interregional or international, are the spatial manifestations of the contradictions of capitalism … spatial development can only be seen as part of the overall development of capitalism. However, it is also true that many of the emerging contradictions of the economic system both take on a specifically spatial form, and are exacerbated by the existence of the spatial dimension. To this end, consideration of "the spatial element" is essential to all effective economic analysis. (Massey 1973: 38–9)

Here, Massey was not only calling into question the plausibility of location theory's claim to a "separate existence", as a project of closed-system theorizing premised on the principles of rational action and general equilibrium, she was also anticipating an entirely different ontology of the economic, together with an understanding of its constitutive spatiality.

The course (and cause) of Massey's work during the 1970s was taking shape. There was a recognition that the disorderly economic conditions of the time – oil crises, deindustrialization, stagflation, industrial-relations strife, anti-immigrant backlash, rising unemployment, the International Monetary Fund's "bailout" of the UK economy, the failure of both Conservative and Labour governments on the altar of economic policy – more than amply confirmed the redundancy of the timid orthodoxy of regional science. More than this, though, they demanded a radical alternative, one that offered analytical purchase on the real-time restructuring of (regional) economies in crisis. A further indicator of Massey's thinking at the time was a book review commissioned by her former boss at CES, Alan Wilson, just a few months after the publication of the *Antipode* article. In a broadly sympathetic but exacting review of David Harvey's *Social Justice and the City*, Massey credited the book for raising questions and problematizing issues, "which quite simply *cannot* arise within the normal framework of Anglo-Saxon regional science". The book offered a necessary critique, she continued, albeit an insufficient alternative; it built suggestively on Marxian notions of rent, but less persuasively on the accompanying apparatus of a narrowly defined class analysis. Overall, she concluded, Harvey's intervention would surely "shake a good many economist-geographers out of their implicit and tautological assumptions" (Massey 1974: 235). Thus, more or less in parallel, Massey and Harvey spurned the trappings and pretensions of positivist geography – with its "trivial notions of causality, [its] idea that a scientific 'law' was something that could be spotted simply through empirical regularity [and with] the mathematics (or the problems in the mathematics) leading the direction of enquiry rather than questions which arose from real world processes themselves" (Massey 1985: 10). The next steps that they would take, however, would be different ones.

Convinced that geography mattered, in a material, political, social, and constitutive sense, but deeply sceptical of positivist claims concerning supposedly spatial laws, Massey found in more abstract forms of Marxism both an affinity and an irreconcilable limit. As a member of the editorial board of *Capital & Class* and elsewhere, she had for years been engaged in debates around value theory, dialectics, uneven development and more, but often found these base categories of analysis to be bloodless, "byzantine entanglements" inadequate in political as well as explanatory terms (1995b: 307). As she later reflected, "the way in which I was thinking was definitely influenced, utterly influenced, by Marx", but at the same time, because of its orthodox remit, she "found it very, very difficult to count myself as a Marxist" (Massey with the HGRG 2009: 403, 405). Some of these critical reservations stemmed from Massey's feminism, which shaped her orientations to theory and politics for some time before it was explicitly incorporated into her analytical schema. Initially, then, feminist influences manifested in her reaction against macrotheoretical abstractions that presumed, rather than interrogated, their impact on the world. Neither value, as an abstract category, nor class, understood primarily in economic terms, could explain the lived realities of restructuring (as it would come to be known) observed in particular places and industries. Gender blindness was part of the problem, but there was a wider failure to account for *who* it was that filled the "empty spaces" of so many abstract Marxian categories (cf. Hartmann 1979), and for that matter the different ways in which those empty spaces were filled in different places.

Massey's twin convictions that geography mattered and that Marxian categories must be rendered concrete if they were to furnish analytical and political value were on clear display in her first monograph, *Capital and Land*, co-authored with her CES colleague Alejandrina Catalano (Massey & Catalano 1978). The product of several years of collaborative labour, the book was a trenchant critique of a central pillar of Britain's radically unequal political economy (then, but also now): private landownership. It seamlessly combined theoretical sophistication and empirical exposition, marrying a clear explication of that most abstruse of Marxian demesnes – rent theory – with the first major investigation of patterns of landownership in Britain for over a century, those ownership patterns having remained essentially undocumented since 1873's "Modern Domesday" book, *The Return of Owners of Land*. The book became a touchstone for scholars of the land question in Britain in the ensuing decades. Its main message – that the social institution of landed property profoundly colours British capitalism – was one that stayed with Massey for the rest of her career (indeed, it was to be explicitly reiterated in the Kilburn Manifesto), although she would not work on land issues again. Instead, her attention turned to what were seen at the time as more pressingly urgent questions – labour, (un)employment and the political economy of capitalist restructuring.

In collaborative work with Richard Meegan, Massey had been tracking patterns of employment change across dozens of UK industries, in the process uncovering a heterogeneous tangle of sectoral dynamics that belied simplified narratives of the causes of manufacturing job losses and the phenomenon known as deindustrialization. Unpacking what they would call the "anatomy" of job loss meant taking account of a repertoire of causally distinct processes at the sectoral level: in some cases, production systems and employment regimes were being rationalized; in others, they were being reorganized through new waves of technological investment; while elsewhere, the imperative was to drive improvements in productivity by way of intensification (Massey & Meegan 1982). To isolate, analyse, and document these distinctive strategies was not to tune out the supposedly steady signal of structural change in favour of cacophonic "noise", Massey and Meegan insisted, nor was it to be diverted by local details or confounding contingencies; instead, it was to theorize generatively with and across difference, to illuminate varied configurations and pathways, and to point to conjuncturally specific stress points and sites of intervention. Crucially, this was not just about wading into the empirical undergrowth and then insisting upon the need for a more granular account; neither was it simply a matter of adding texture while continuing to colour between the lines of big-picture accounts of industrial transformation and uneven development. Instead, these were early steps on the path towards a more deeply relational form of geographical political economy.

When Massey first introduced what was to become one of her signature concepts, the spatial division of labour, she did so "in order to make a point" (1979: 234): the geographically differentiated conditions of production, including for instance the availability and cost of labour, should not be seen as some inert surface across which profit-seeking firms maximized returns. Rather, the relationship between the dynamics of accumulation and the shifting geographies of work and production was one of mutual interaction and adaptation. Beginning, in effect, with just two dimensions – industrial sectors and employment geographies – Massey conjured a vividly three-dimensional understanding of capitalist spatiality, displaying a characteristic combination of complex reasoning and unvarnished exposition. Granting that the orthodox observation that economic activities are distributed systematically in space according to the principle of profit maximization was "correct [but] also trivial", she set out in the space of two paragraphs an alternative (and demanding) remit for political-economic geography:

> What [the orthodox account] ignores is the variation in *the way in which different forms of economic activity incorporate or use the fact of spatial inequality* in order to maximise profits. This manner of response to geographical unevenness will vary both between sectors and, for any given sector, with changing conditions of production … [There is an] interaction between, on the one hand the existing characteristics of spatial differentiation, and on the other hand the requirements at that time of the particular process of production. Moreover, if it is the case that different industries will use spatial variation in different ways, it

9

is also true that these different modes of use will subsequently produce/contribute to different forms of geographical inequality. Different modes of response by industry, implying different spatial divisions of labour within its overall process of production, may thus generate different forms of "regional problem."

One schematic way of approaching this as a historical process is to conceive of it as a series of "rounds" of new investment, in each of which a new form of spatial division of labour is evolved. In fact, of course, the process of change is much more diversified and incremental (though certainly there are periods of radical redirection) ... In any empirical work, therefore, it is necessary both to analyse this complexity and to isolate and identify those particular divisions which are dominant in reshaping the spatial structure. The geographical distribution of economic activity which results from the evolution of a new form of division of labour will be overlaid on, and combined with, the pattern produced in previous periods by different forms of division of labour. *This combination of successive layers will produce effects which themselves vary over space*, thus giving rise to a new form and spatial distribution of inequality in the conditions of production, as a basis for the next "round" of investment. "The economy" of any given local area will thus be a complex result of the combination of its succession of roles within the series of wider, national and international, spatial divisions of labour.

(Massey 1979: 234–5, emphasis added)

The implications of this remarkably succinct formulation would be far reaching, intellectually and politically. It presaged a style of relational theorizing that did more than transcend the rapidly fading orthodoxy of location theory; it challenged Marxian political economy to engage *and work with* the patterned specificities of industrial restructuring, locality effects and the complex recombination of class and gender relations, rather than to override them, or to subsume them within reductionist or all-encompassing categories of analysis.

Relational understandings of space and spatiality, in short, were central to Massey's formulations, rather than being secondary to (or derivative of) social processes. This concern with how social processes *take place*, as it were, implied nothing less than a relational ontology. "'Spatial outcomes' are not simply the 'outfall' of restructuring", Richard Meegan (2017: 1287, 1288) has explained, but are constitutively active in shaping successive rounds of investment. And while this is "[o]ften misinterpreted as a geological metaphor of the layering of strata, the notion of the historical layering of rounds of investment spatially is more accurately a metaphor of interaction and articulation". Subsequently, "articulation" would become a hallmark of Massey's approach, initially by way of an Althusserian treatment of the combination of economic, political, and ideological forms and practices within regional conjunctures, and then much more expansively, in her influential notion of "relational space" (Massey 1995b: 315–23; Featherstone & Painter 2013: 4–7).

LOCALITY EFFECTS

Massey's foundational arguments concerning the spatiality of capitalism were worked out in extended form, and operationalized too, in *Spatial Divisions of Labour*, a book on which she worked for several years and to which she would return – taking the opportunity to append substantial methodological elaborations – just over a decade later (Massey 1984b; 1995b). In its first edition, the book crystallized what was taking shape as the restructuring approach, although in retrospect it may also have represented its zenith (see Warde 1985; Lovering 1989). It carried the burdens of complexity and specificity alongside its mandate for creative conceptualization and active theorization, the explanatory incision and persuasive power of which was not to be matched in the wider research programmes that followed. Most conspicuously, the somewhat ill-starred "localities" research initiative, a national project funded by the Economic and Social Research Council in the UK and directly inspired by Massey's framework, largely failed to deliver, at least in its own terms. Controversial practically from the outset, the localities research programme became the focus for a series of proxy debates around the status of Marxism and postmodernism, the politics of scale, the methodological potential (and limits) of critical realism and structuration theory, and more, not to mention the prosaic, organizational and scientific challenges of managing an expressly polycentric study comprising multiple case-study sites and research teams (see Cochrane 1987; Smith 1987; Bagguley *et al.* 1990).

While initially choosing to remain one step removed from the debates that roiled around the localities initiative – which her work had been instrumental in launching, but in which she had no direct role – Massey continued to support its programmatic rationale. Nationally oriented political debates were only rarely taking account of the diversity of regional and local experiences, she maintained, often asserting the existence of trends and (causal) connections that simply did not hold across scale and space; there were significant limits to explanations of (diverse) local transformations that were crudely pegged to "capitalism in general"; and a host of new social movements and municipal-socialist projects had been seeking to harness "the local" for progressive ends.

11

Empirically, "[s]omething that might be called 'restructuring' was clearly going on, but its implications both for everyday life and for the mode and potential of political organising were clearly highly differentiated and we needed to know how" (Massey 1991c: 269). Theoretically and methodologically, Massey was challenging the conflation of the global scale with supposedly "general" theory claims, or abstraction itself, as well as the false equivalence between the local and the concrete. With a debt to the *Grundrisse*, she insisted on an understanding of the concrete as the synthesis of multiple determinations, pointing out that abstract analysis might just as well focus on small objects as large ones, as it might on the local as well as the global.

Notwithstanding their operational limitations, the locality projects had not been originally conceived as idiosyncratic case studies, detached from broader explanatory frameworks; by design, they had been concerned to explore the spatial constitution of social processes, and the implications – both causal and political – of different, localized configurations of social relations. From the perspective of the localities research programme,

> not only was the character of a particular place [understood to be] a product of its position in relation to wider forces (the more general social and economic restructuring, for instance), but also that that character in turn stamped its own imprint *on* those wider processes ... The facts of distance, betweenness, uneven-ness, nucleation, copresence, time-space distantiation, settings, mobility and differential mobility, all of these affect how specified social relations work; they may even be necessary for their exist-ence or prevent their operation ... [The] fact of spatial variation itself, and of interdependence – of uneven development – has major implications.   (Massey 1991c: 271–2)

Massey's approach would become synonymous with the slogan "geography matters" (Massey & Allen 1984). While she went as far as to say that "the unique is back on the agenda" (Massey 1985: 19), Massey never displayed any interest in the unmoored pursuit of idiographic detail as an end in itself. To the contrary, her position was that "places [may be] unique but that does not make them inimical to theory" (Graham 1998: 942).

Yet there were influential (mis)readings of *Spatial Divisions of Labour*, and of the rationale for localities research, in which an alleged departure from Marxian theory was equated with a turn not just towards empirics but empiricism (Smith 1986, 1987; Harvey 1987; cf. Scott 2000). David Harvey contended that, "Massey is so anxious to deny structuralist leanings or that the 'logic of capitalist development' has any explanatory power in local settings that all theorising disappears between a mass of contingent labour-management relations in place", asserting that her approach had become "laden down with a rhetoric of contingency, place, and the specificity of history", to the point that the "guiding thread of Marxian argument is reduced to a set of echoes and reverberations of inert Marxian categories" (Harvey 1987: 369, 373). However, there is little in Harvey's critique to suggest (or indeed to recognize) a two-way interplay between the social and the spatial, or for that matter an effort to hold together the general and the particular. Instead, the project was interpreted as a reversal into the cul-de-sac of empirical specificity.

There is no disputing the fact, of course, that *Spatial Divisions of Labour* explicitly grappled with concrete complexity. As Ann Markusen reflected in a symposium convened to mark its enduring contribution, "students sometimes find it difficult to master Massey's book, not because it is densely written – on the contrary it has a light and colloquial tone – but because the analysis is so complex and multifaceted". Massey's response to Harvey's charge that the book was weighed down with contingency and complexity was characteristically forthright: "it certainly was and … I meant it to be!" (Martin *et al.* 1993: 70–1). Far from a repudiation, or retreat from, the concerns of Marxian political economy, Massey had in fact been continuing to employ the classic entry points of the capitalist labour process, the social relations of production and the problematic of industrial transformation. Operating from – but reaching beyond – this known analytical territory, she had not just illustrated but *elaborated* the working out of processes of uneven development, specifying their (somewhat divergent) sectoral dynamics, and breaking down tendential historical claims in favour of a sharper focus on "layers" of (dis)investment. Not least, her approach was intended to open up and then occupy the space for much less deterministic accounts of place-based conditions and local change. She deftly reworked an Althusserian sense of overdetermination into an alternative conception of spatiotemporal relations, in dialogue with feminist theory and critical realism, by evoking notions such as the "combination of layers", the variable intersection of class and gender relations, the specificities of capitalist class divides and allegiances, and the vagaries of localized politics. Massey's framework put flesh on the bones of sparse formulations like core–periphery and "see-sawing" capital movements, insisting on the irreducibly sociospatial content of relations that had too often been portrayed mechanistically.

Massey's project in this period, through *Spatial Divisions of Labour* and the remit for locality studies, thus is best understood as a dexterous elaboration of Marxian political economy rather than some radical departure from it. She insisted that theory claims had to be *read through* (and across) conjunctural specificities; theoretical abstractions were not somehow floating above the particular in a "general" sense. Massey's intention had never been to abandon received conceptions of the forces and relations of production or the dialectics of uneven geographical development, but instead to operationalize a framework for mobilizing these relatively abstract formulations *through* the structurally necessary mediations of (industrial) sector and (regional) space. While she did not conceive this work as an escape route from industrial geography, it nevertheless opened the door to quite different ways of thinking – ways of thinking that would soon carry Massey's own work beyond the relatively restrictive problematic of capitalist restructuring.

A VIEW FROM SOMEWHERE

By the end of the 1980s, with the fall of Communism and triumphalist "end of history" narratives ringing out from the heartlands of capitalism, geographers and the broader social sciences became immersed in often-fervent disputes over the meaning and status of the watchwords of the time: postmodernism and globalization. Massey had spent the previous decade developing the apparatus of spatial divisions of labour, at least in part, to historicize the restructuring present, to reanimate and contextualize received concepts, and to specify more or less familiar forms of recombination. But as the objective grounds were shifting, Massey now parlayed these conceptual tools derived from her interventions in the far narrower field of industrial geography, along with her poststructuralist and feminist sensibilities, to intervene in these larger debates. Her contributions would unsettle prevailing claims that were often shared by the early boosters and sceptics of globalization about the death of geography, time-space compression, and the global (market) as a scale of "out there" forces and economic predetermination.

These debates benefited immensely from Massey's way of doing theory and being in the world. It is not simply that she critiqued the way that certain academics were validating the global in their work, sometimes echoing the prevailing formulations of corporate and governmental elites, even as they endeavoured to critique capitalism. She also called out the manner in which these universalized arguments were so often decontextualized by jetsetter scholars who seemed to survey (and write about) the world from 30,000 feet, as if to project their own placelessness. Massey argued instead for a more grounded engagement, literally and figuratively. In stark contrast to the "hype and hyperbole" that had become so typical of writing on globalization and postmodernism, she wrote clearly and accessibly, with wry awareness that, "Much of life for many people ... still consists of waiting in a bus-shelter with your shopping for a bus that never comes" (Massey 1992a: 8). She would often share anecdotes about where and how her ideas emerged and how they worked in practice. They might be stories about walking down Kilburn High Road, or sharing a commute with Stuart Hall, or about rounding Lake Windermere in the Lake District. This was how she explained the idea of the global sense of place, for example, and the notion of places as trajectories and multiplicities. Hers was a notion of globalization inseparable from deep commitments to place, but not in some nostalgic or parochial sense; instead, this was a variously more intimate, positional and political sense of place, imbued always with the recognition of difference. Place was a site of encounter, of engagements with and across difference, globalization being experienced, as a far-from-universal condition, "in here" just as much as "out there".

This kind of view from somewhere had always been a compelling characteristic of Massey's writing and indeed her conceptual outlook, but it would be enriched in new and distinctive ways through her engagements with feminist theory and situated epistemologies, notably the work of Donna Haraway and Sandra Harding, whom she credited with a profound influence on her thinking (Massey 1995b; Albet & Benach 2012). Massey went on to provide object lessons in the power of feminist epistemology and critical spatial thinking in her key interventions of the early 1990s. She put in sharp relief the quite particular privileged positionality of scholars of globalization peddling seemingly universal and generalized claims about the dissolution of the subject, boundaries, and certainties, observing pointedly that

> Those who today worry about a sense of disorientation and a loss of control must once have felt they knew exactly where they were, and that they had control ... The assumption that runs through much of this literature is that this openness, this penetrability of boundaries is a recent phenomenon. [And yet for the colonized periphery] the security of boundaries of the place one called home must have dissolved long ago, and the coherence of one's local culture must long ago have been under threat.
>
> (Massey 1992a: 9–10)

In her critique of the masculinist gaze exhibited by David Harvey's *Condition of Postmodernity* and Ed Soja's *Postmodern Geographies* in particular, Massey (1991b) exposed the occlusions, exclusions and elisions typical of "unreconstructed" variants of Marxism in human geography. Her critique echoed and built upon key feminist interventions intended to overturn masculinist Marxism across disciplines, including Deutsche (1991), Scott (1988/1999), Hartsock (1987) and Christopherson (1989).

Declaring her sympathy with the overall projects underpinning Soja's and Harvey's books, she challenged their exclusions (not least of swathes of feminist literature) and their apparently unthinking recirculation of supposed universals which "are so often in fact quite particular; not universals at all but white, male, Western, heterosexual, what have you", formulations that also carried broad and deep implications for academic politics, for political representation, and for the practice of political economy:

> Harvey has produced a fascinating, if arguably economistic, exploration of the relation between the definition, production, and experience of space, on the one hand, and modes of production and class formation on the other. But it completely misses other ways, other power-relations, in which space is also structured and experienced ... This leads to an unnecessarily monolithic view of the modernist period; it shifts the definition of what it was and, by missing out the voices on the margins and in the interstices of what was accepted, it also misses the full force of the critique which those voices, among them feminists, were making of the modernism he does discuss ... After all the feminist debate about representation [including] the directly political critique of modernist representation, it is surely inadequate to put the whole crisis down to time-space compression and flexible accumulation.
>
> (Massey 1991b: 52, 53, 51)

Recognizing that "neither of the authors would want to be thought of as antifeminist", Massey took Harvey and Soja to task for reducing culture and politics to epiphenomena of the mode of production; for positioning class relations "above" other sources and sites of social difference, ascribing overarching or covering status to the former while subordinating or localizing the latter; for normalizing an essentially masculinist analytical gaze; and for universalizing sites and subjects that should always be situated.

ARTICULATING DIFFERENCE

If her intervention into postmodernism was spurred by a principled rejection of masculinist and Western-centric thought, Massey's provocation to imagine "a global sense place" was a timely, and still remarkably relevant, response to the horrors of ethnic cleansing in the former Yugoslavia and the eruptions of nationalist violence across the former Soviet Union (Massey with the HGRG 2009: 417). The equally appalling options of exultant globalism and reactionary ethnonationalism had fuelled a tendency within progressive and liberal thinking to brand place-based identities as de facto reactionary. Confronting this foreboding conjuncture, Massey countered that local, place-based negotiations of difference were very often fundamental to progressive politics and to democratic struggles more generally. After all, place is one of the arenas in which we "learn to negotiate with others – to learn to form this thing called society", she argued, not least because "a healthy democracy requires, not pacification into conformity, but an open recognition of difference, and an ability to negotiate it with mutual respect" (Massey 2002b: 294).

In contrast to readings of the local as a regressive site of restructuring, reactionary elements, and enforced conformity, this is a vision of place as a space of open-ended intersections, of journeys anything but complete:

> [W]hat gives a place its specificity is not some long internalised history but a particular constellation of social relations, meeting and weaving together at a particular locus. If one moves in from [a] satellite towards the globe, holding all those networks of social relations and movements and communications in one's head, then each "place" can be seen as a particular, unique point of their intersection. It is, indeed, a *meeting* place. Instead, then, of thinking of places with boundaries around, they can be imagined as articulated moments in networks of social relations and understandings, but where a large proportion of those relations, experiences and understandings are constructed on a far larger scale than what we happen to define for that moment as the place itself … And this in turn allows for a sense of place which is extroverted, which includes a consciousness of its links with the wider world, which integrates in a positive way the global and the local.
> (Massey 1991a: 28, original emphasis)

Against a static, bounded, and introverted sense of place, Massey argued for a processual understanding of socially structured but always becoming spaces – against deterministic readings of the global as a scale of imperative forces, she insisted on the local as a scale of political vitality and creative potential; against all-encompassing visions of time-space compression, Massey made the case for a more intricate conception of sociospatial difference. Building from the recognition of multiple identities in (and between) places, her approach was sensitive to the ways in which "power-geometries" variously include, exclude, connect, divide, empower, and disempower different social groups. This is a reading of place and place identities "constituted *out of* social relations, social interactions, and for [this] reason always and everywhere an expression and a medium of power", implying a conception of place as a "particular articulation [of] power-filled social relations" (Massey 1995c: 284, emphasis added).

The work of constructing an "open", progressive, and processual sense of place was for Massey both philosophical and political. It meant supplanting received conceptions of space as inert and empty in favour of a much more disruptive understanding of time-space as a zone of endemic frictions, of both order and chaos (Massey 1992b). Massey sharpened these arguments in dialogue with leading lights of the New Left in the UK, especially Ernesto Laclau, Chantal Mouffe and Stuart Hall. In the early 1980s, they had founded the Hegemony Group, a monthly reading collective that poured over texts by Gramsci, Althusser, Poulantzas and others. It was from her interactions and debates with these figures that Massey became convinced of the necessity for an anti-essentialist politics that dispensed with the dualism of space (as passive) and time (as active), rejecting also the structuralist conceit of grafting social relations onto space as a static, apolitical container.

In her Hettner Lectures in Heidelberg in 1998, Massey proposed a threefold reconceptualization of space carrying a series of political implications: first, the view of space as "a product of interrelations [and therefore] constituted through interactions" is complementary with an anti-essentialist politics in which identities are always in relational (re)construction; second, since space is a domain of multiplicity and emergence, then it follows that (as feminist and postcolonial theorists have long maintained) stories and visions of the world are multiple too, not singular and universal; and third, because space is "always being made ... never finished [and] never closed", political futures are neither known nor should they be foreclosed (Massey 1999d: 2–3). This amounts to a conceptualization of space that is both theoretically principled and politically open.

Concerned to deconstruct the binaries of orthodox theories, frameworks, and formulations in political, conceptual, and empirical terms, Massey's work in the 1990s and beyond pushed out in many directions, in part by virtue of an increasing engagement with feminist, postcolonial, post-structural, and queer theories (see Massey 1992b, 1994, 1995a, 1999d; Henry & Massey 1995). But it would be quite wrong to represent this as a rupture with the reflexive, heterodox Marxism of *Spatial Divisions of Labour* and the restructuring approach (cf. Saldanha 2013). With the benefit of hindsight, Julie Graham would re-read the book as "a founding text in the emerging tradition of poststructuralist economic geography" (Graham 1998: 942; cf. Soja 1987). Massey's (self)-positioning on this score was careful but also quite explicit, plainly stating in the book's second edition that "I do not see Marxist categories as inert", while classifying her approach as "clearly related to historical materialism" (1995b: 312). She acknowledged the continuing salience of Marxist categories while distancing herself from their off-the-shelf, inflexible deployment, (re)stating the case for a non-essentialist analysis of capitalist restructuring. This would seek to cut a path between totalizing metanarratives and the restrictive remit of "local theory", mobilizing a dexterous interpretation of critical realism's layered conception of social reality over "flaccid" acceptance of free-form indeterminacy.

A distinctive handling of articulation was therefore a longstanding methodological feature of Massey's work. While in her work during the 1980s she had been predominantly concerned with circumstances that were "specifically capitalist", neither then nor later did Massey have any time for "deterministic laws, predetermined outcomes or ineluctable stages of history" (1995b: 301–2). After all, tendencies in capitalist restructuring could be immanent even as their realization was – inescapably – contingently conditioned, situationally mediated, and conjuncturally framed, just as local circumstances were emergent in both a causal and a political sense. In other words,

> *Connection, as well as differentiation*, are what it is all about ... It is not necessary, having rejected metanarratives, to reject any notion of broad (but non-totalising) structures, most especially if at the same time their multiplicity and complexity is recognised ... Thus, *Spatial Divisions of Labour* rejects metanarratives ... but it does not therefore adopt a position which restricts itself only to local structures; broad structures, but which are assumed to be multiple, non-totalising and without pregiven narratives are acceptable, indeed necessary to the approach ... None of the structures which are identified need be assumed to have any inexorability in their unfolding. And in the analysis here it is resolutely assumed that they do not; outcomes are always uncertain, history – and geography – have to be made.
>
> (Massey 1995b: 303–4, emphasis in the original)

Convinced that "the usual categories of economic geography", including its privileged object of analysis (industry), were "simply not good enough", Massey's approach to theory was based on "rigorous conceptualisation", which would draw upon, without being unnecessarily constrained by, "previously-achieved understandings of the phenomena in question". Her methodological preference was to "wrestle with theoretical problems in the laboratory of an empirical case, rather than simply 'in the abstract', as they say" (Massey 1995b: 309, 304, 315; Massey & Meegan 1985). Social scientists, she argued, are routinely confronted by conditions of endemic complexity but this need not imply unprincipled indeterminacy. Political futures, like social systems themselves, are open but this must not be confused with the claim that every outcome is equally likely, or that tangential patterns are beyond specification.

Articulation, in this sense, meant theorizing both with and across difference; it meant theorizing through relations, connections, and interdependencies, while refusing to freeze (or reify) particular combinations. "In Massey's world", Julie Graham (1998: 942) wrote, "things are related to each other not primarily through replication or reflection (sameness) but through articulation – the transformative intellectual and social process of creating connections and generating in the process unique beings, situations, and possibilities". At the heart of *Spatial Divisions of Labour* had been an effort to work a conception of capitalist spatiality from the abstract through to the concrete, and back again, as a set of spatiotemporal interdependencies and connections. This was achieved by interrogating a particular "spatial structure", that of intrafirm relations. That left a great deal out, of course, including the relatively familiar territory of *inter*firm relations as they were being progressively (re)articulated through markets and networks. As Michael Storper (1986: 456) had said of the first edition of the book, to focus on the "locational processes that are part of the intracorporate hierarchy … is to underestimate the complexity of the spatial division of labour". Once attention had shifted, during the 1990s, towards notionally "post-Fordist" dynamics like vertical disintegration, subcontracting and outsourcing, agglomeration economies, interfirm clusters and industrial districts, many of the substantive arguments in the book were to be overtaken by the restless process of restructuring itself (see Martin *et al.* 1993; Scott 2000). Such is the fate, of course, of so many arguments that take seriously the facts of conjunctural specificity.

In her subsequent reflections, Massey argued for extending the concept of spatial divisions of labour to these novel geographies of industrial districts and interfirm networks (1995b: 335–9). To do so, however, would have meant foregrounding the chop and churn of uneven development as a principal spatial domain (albeit a contingent and overdetermined one) in the analysis, yet the drift of subsequent work in the field – including in economic geography – augured against such an approach. Indeed, while subsequent work on geographies of agglomeration and global production networks often cited Massey's foundational text, it generally relegated its core concerns to the conceptual background or neglected them altogether (Peck 2016; Werner 2016). Massey opted not to pursue these possibilities herself either, at least not in an explicit way. The same cannot however be said for a second area that she thought offered the possibility, and indeed the necessity, for extending the concept of the spatial division of labour: unpaid labour, or more accurately, "those variable processes through which the paid/unpaid boundary is constructed" (Massey 1995b: 334). Observing ascendant patterns of economic transformation in the global North, not least through her research on gendered divisions of labour in the much-vaunted field of R&D (see Massey *et al.* 1992), and the experiences of structural adjustment in the global South, which was transforming paid jobs into unpaid burdens shouldered primarily by women in the home, Massey argued that any analysis of paid work must incorporate the dynamics of households, gender ideologies and state practices that shape the moving boundaries between remunerated and unremunerated work in a given conjuncture (1995b: 334–5). This was just one of the ways, Massey later reflected, in which feminist thought influenced the substance and overall project of *Spatial Divisions of Labour*. She maintained that while "gender is essential to the empirical story" in the book, it had not been explicitly positioned "within what was at the time defined as the field of feminist geography"; nevertheless, its style of (relational) theorizing warranted (re)consideration from a feminist standpoint, not least as a contribution to the ongoing project of "re-thinking the way we think" (Massey 1995b: 345, 341, 354).

RELATIONAL SPACE

Massey's response to the first wave of globalization debates in the early 1990s had been to counter with what would eventually become a mainstream critical position: that global space was no less "concrete", no less constructed, than local places; that the global is therefore socially and politically made, not least in *and from* certain places; that simplified and self-fulfilling "impact" models of globalization, which assume vertical relations of scalar domination, are not only misleading but will also tend to constrain and distort political strategies and imaginaries; and that conventional narratives of (neoliberal) globalization represented just one vision of global and transnational politics, against which alternative models could be developed based on principles of egalitarianism, respect, responsibility and democratic deepening. Many of these contentions would be echoed in other strands of critical scholarship and commentary.

Arguably more distinctive, indeed radical, was Massey's parallel rethinking of "the local" in relational terms as a corollary of her critical reading of globalization. Against romanticized or static conceptions of the local, as the scale of essentialized and fixed identities, Massey repeatedly made the case for more processual understandings of space and place as *always* in the making, never closed and never finished. "The character of place is not somehow a product only of what goes on within it, but results too from the juxtaposition and intermixing there of flows, relations, connections from 'beyond' [some of which] may, indeed, go around the world" (Massey 2004b: 98). Places, moreover, were anything but a singular category; they always had to be situated in relation to *other* places and scales, and within relations of uneven and combined development. In this context, the political vision that there are other globalizations to be imagined, to be had, not only "prevents us slipping into easy oppositions such as global = bad, local = good", but also "prevents us facing up to neoliberal globalisation simply by retreating into the defensive laager of local place" (Massey 2004b: 98).

This formulation meant that Massey was no less wary of across-the-board versions of horizontalism and the "flat ontologies" that pervaded many nominally poststructuralist accounts (Routledge & Cumbers 2013), than she was of top-down visions of globalism advanced by some orthodox Marxists. Once again, her insistence on places as intersections of multiple relations stemmed from both political and analytical commitments.

> [W]hile we are indeed all discursively subject to a disempowering discourse of the inevitability and omnipotence of globalisation, materially the local identities created through globalisation vary substantially. Not all local places are simply "subject to" globalisation. The nature of the resubjectivation required, and of the responsibility implied, in consequence also varies between places.
>
> (Massey 2004a: 10)

Massey would go on to explore the relational constitution of identity formation both within and between places. She was to become increasingly convinced that "rethinking identity [could be] a crucial complement to a politics which is suspicious of foundational essentialism; a politics which, rather than claiming 'rights' for pre-given identities ... based on assumptions of authenticity, argues that it is at least as important to challenge the identities themselves and thus – *a fortiori* – the relations through which those identities have been established" (2004a: 5). It followed that identities and subjectivities are constitutively relational, that they do not precede (social) interactions in some fixed sense, but are malleable outcomes of those ongoing interactions (Massey 2006a: 92–3). Massey's decisive intervention was to point out that *spatial* identities were constituted in relational ways as well, such that they must be understood as "internally complex, essentially unboundable in any absolute sense, and inevitably historically changing" (Massey 2004a: 5). As Chantal Mouffe would later reflect, this conceptualization necessitated a spatialized understanding of hegemony; more than this, it opened up new potential for thinking through counter-hegemonic politics. Writing against immanent conceptions of "the multitude" as formulated by Hardt and Negri, Mouffe argued that "to acknowledge the ineradicability of antagonism implies recognising that every form of order is necessarily a spatialized, hegemonic one, that it constitutes a 'geometry of power'" (2013: 29).

Massey would make her most extensive attempt to elaborate these arguments for an audience beyond geography in the monograph, *For*

Space (2005). The book drew upon an extraordinarily wide range of texts – including French poststructuralism, phenomenology, art criticism, post-colonial studies, anthropology, geography, feminist theory and more – to tease out a series of problems associated with understandings of space as an inert, static category, developing in its stead an alternative reading of space as a necessary component of multiplicity and becoming, along with a more open approach to theoretical and political practice. As Matt Sparke wrote, the book's title referenced a parallel problematic in Althusser's *For Marx*, even as it constructed a critique of structuralism's spatial stasis and its entrenched time-space dualisms (Massey 2005: 40–1; Sparke 2007; see also Saldanha 2013). Rather than proclaim a proper way to be "for space", Massey instead illustrated the many dead ends and limits to thought enabled by a category of stasis that permeates Western philosophy which is convention-ally given meaning and circulated as "space". Massey revisited the threefold proposition on space (as relational, a multiplicity, and in-process) that she had originally presented in the Hettner lectures, insisting that this should serve neither as a model nor as a formula but instead as a provocation to critical, progressive theory and politics – jogging them out of the deeply engrained habits of Western thinking that tended to mobilize space towards essentialist, identitarian ends. Her deconstruction of space, here and else-where, sought a redefinition and reclamation of the term as necessary to the project of shaping a truly radical politics of identity and place.

In particular, Massey revisited and rejected two common but "evasive imaginations", both of which failed to confront the "challenges of space". Firstly, there was the long-established habit of turning space into time, and geography into history, where "spatial difference [is] convened into tem-poral sequence" (Massey 2005: 68). Against Rostovian, sequentialist, and developmentalist visions of a convergent (global) future, Massey refuted the contention that there is but one way for places to "catch up", since this meant that those positioned "behind" are effectively denied equal standing, quite lit-erally having no space to imagine (let alone take) alternative pathways (Massey 2006a: 90). Secondly, there was the practice of thinking of space as a surface, *over* which journeys are made. Against this colonizing imaginary, which also tends to compress multiple histories and geographies into a master narrative-cum-trajectory of modernity, Massey insisted on seeing space and place in terms of constellations, conjunctures, crystallizations and shifting patterns of coexistence, in which space-time exists as a "simultaneity of unfinished, ongoing, trajectories" or "stories-so-far" (Massey 2006a: 92, 2005).

These philosophical explorations of space-time were to lead Massey in an additional, somewhat surprising, direction. In collaboration primarily with two physical geographers, Stephan Harrison and Keith Richards, Massey convened a series of conversations to probe the moral, conceptual and empirical possibilities for bridging the longstanding divide between physical and human geography (Massey 1999g; Harrison *et al.* 2004, 2006, 2008). She had earlier contended that, for all their differences, human and physical geographers shared an imagination of physics as the model science, which had saddled the discipline as a whole with unhelpful and flawed conceptualizations of time, space and space-time. Against static understandings of space and teleological conceptualizations of time, Massey (1999g: 262) argued that space-time should instead be understood as "relative (defined in terms of the entities 'within' it), relational (as constituted through the operation of social relations, through which the 'entities' are also constituted), and integral to the constitution of the entities themselves (the entities are local time-spaces)". So conceived, this alternative understanding of space-time provided an admittedly abstract but nevertheless potentially generative means to bridge between physical and human geography, building upon the shared concern to explain dynamism, complexity, emergence and process. In *For Space*, for example, Massey put this relational view of the physical and social to work in the empirical analysis of the landscape of Skidaw, in the Lake District in England, which she saw to be constituted as much by the dynamic physicality of immigrant rocks as of migrant humans (Hinchcliffe 2013). And many other scholars, such as Diane Rocheleau (Harcourt *et al.* 2013), were to find Massey's work an inspiration for the integrated analysis of relations among physical and social systems, grounded in particular places. Though Massey did concede that it can "make your head hurt to think in this way", she argued that it was worth it because, quoting Raper and Livingstone (1995: 364), "the way that spatio-temporal processes are studied is strongly influenced by the model of space and time that is adopted", or as she chose to underscore the point, characteristically, "it matters; it makes a difference" (Massey 1999g: 262).

Massey later offered a unique and highly personal justification for this portion of her intellectual project, and her relational view of human and physical geography. Already suffering from advanced osteoporosis, she described her lifelong struggle with a debilitating bone disease. Constant therapy and numerous broken bones had blighted her childhood, a burden shouldered by her parents, especially her mother, who shuttled her in and out of doctor's offices and hospitals. Growing up with a "disability" had shaped a very particular, indeed personal, sense of marginality. As Massey recounted in an interview with Catalan geographers Núria Benach and Abel Albet,

> some of my bones have metal pieces, my feet are full of fragments. All told, my body is a major disaster but this disaster has always formed an important part of me, in the sense that each day it poses a small grand challenge. I cannot be exclusively cerebral ...

because my body is here each day and, well, it is a very important part of me.　　　(Albet & Benach 2012: 62, editors' translation)

These reflections brought her to the question of norms – of gender and the body – that she had always challenged in her life and in her work. When her physical therapists refused the label "normal", or rejected her qualifications of what she saw as her "good" versus "bad" leg, she felt subjected to what in Foucauldian terms amounted to a strong form of social constructionism. "And it's true, in one way, but in another … it hurts … It's pure materiality!" (Albet & Benach 2012: 63). In a characteristic manner, she used the contrast between the lived experience of her aching body and her therapists' descriptions of it as a springboard to reflect on how false dichotomies such as these, between the discursive and the material, continue to mark the divisions between human and physical geography: "we need both at the same time", she concluded.

Massey's life experiences would also be echoed in reflections on the evolution of her theoretical commitments and the tensions between the personal and the political. If Althusser had been Massey's awakening to anti-foundationalism, not least in her invocations of the idea that "there is no point of departure", in her later writing, especially in *For Space*, there were to be stronger affinities to the work of Deleuze and Guattari (1994). Massey (2005) would draw upon the engagements between Deleuze and Guattari and the philosopher Henri Bergson in her explorations of the concepts of multiplicity and becoming – even as she roundly critiqued Bergson for predicating his reading of open ontology on a shallow, fixed treatment of space. Indeed, Massey's former student and Deleuze scholar, Arun Saldanha, argues that it is Deleuze and Guattari's notion of geophilosophy that "comes closest to Massey's philosophical-geographical project" (Saldanha 2013: 50). Yet there were also tensions between Massey's position and the Deleuze–Guattari programme. While Massey would embrace the notion of assemblage, especially for its role in facilitating the theorization of more-than-human geographies, she would become increasingly "wary of the 'all is process' view of the world, in so far as it's been translated from a reconceptualization to almost a denial of the existence of 'things' … The body is all process, certainly, but we still have bodies" (Featherstone *et al.* 2013: 255). She expressed similar reservations concerning the widespread use of the notion of "relationality", which too often would be evacuated of the political conundrum that the term had been meant to index:

The degree to which that conceptual questioning emerged from and was developed within political struggle – including, very importantly, feminist struggles – seems to me to be under-recognised. Many of the explorations of the construction of identity [and] … relational thinking more generally, were importantly first explored in debates within political constituencies – anti-racism, feminism and sexual liberation. It is a great pity when those roots are forgotten; conceptual debate can then lose its urgency and real meaning. The real two-way movement between the conceptual and the political has always, for me, been very important.

(Featherstone *et al.* 2013: 260)

Massey steadfastly refused to separate, on the one hand, ontological and epistemological theories of space, and on the other, the political possibilities that were always being opened and foreclosed by contested understandings of space. More than a geophilosophy, then, hers was first and foremost a geography of praxis.

LOCATING RESPONSIBILITY

Massey's commitment to praxis was evident in her work during the miners' strike of 1984–85 and with the Greater London Council; in wide-ranging activities with *Soundings*, as a journal of political commentary and debate; and in projects and collaborations in Nicaragua, South Africa, Venezuela and elsewhere. It was also manifested in her effort to learn Spanish and French, not simply to translate across languages, but instead to comprehend the different idioms and forms of academic and political debate outside the Anglo sphere (Albet & Benach 2012: 39; see also Vaiou 2017). Never one for the international conference circuit or the trappings of academic celebrity, Massey remained sceptical of scholarly exchange as a detached practice, as an end in itself. This led her to find new ways to interweave theoretical practice with political engagement, becoming a public intellectual even as she bristled at the label. The public meetings convened to debate the collaborative project that was the Kilburn Manifesto – an impassioned plea for a new politics, involving new languages and new alliances – represented Massey in her element, despite her sometimes-faltering health. And in a host of other projects, involving all manner of interlocutors and collaborators, she would continue to explore generative connections between relational thinking, questions of identity and the politics of responsibility. These would span the sciences, the arts and the humanities, and they would take her from the corridors of the Open University to Turbine Hall at the Tate Modern, and to Caracas, Venezuela, where the notion of "power-geometries" was adopted by the Bolivarian movement under Hugo Chávez (Massey 2000a, 2009b; Eliasson 2013).

The openness that Massey accorded to space, and of place as a space of potential, was never divorced from her preoccupation with its unevenly distributed (and contested) potential, itself continuously reshaped in (mutual) relation to the global and the more-than-local. To the extent to which the generalized dynamics of capitalism, and not only in their globalized and neoliberalized variants, are associated with systematic tendencies towards sociospatial inequality, it follows that "the local relation to the global will … vary [along with] the coordinates of any potential local politics of challenging that globalization" (Massey 2004a: 13).

> For in this [alternative] imagination "places" are criss-crossings in the wider power-geometries which constitute both themselves and "the global." In this view local places are not simply always the victims of the global; nor are they always politically defensible redoubts *against* the global. For places are also the moments through which the global is constituted, invented, coordinated, produced. They are "agents" *in* globalisation. There are two immediate implications. First this fact of the inevitably local production of the global means that there is potentially some purchase through "local" politics on wider global mechanisms. Not merely defending the local against the global, but seeking to alter the very mechanisms of the global itself. A local politics with a wider reach [this represents a different] basis for the recognition of the potential agency of the local … The second implication of this line of reasoning [is that, if] the identities of places are indeed the product of relations which spread way beyond them … then what should be the political relationship to those wider geographies of construction? (Massey 2004a: 11)

Massey would go on to explore some of the distinctive power-geometries of this "local production of the global" in her adopted home city of London. In her final monograph, *World City,* she offered a book not so much "about" the place, in the more conventional sense, but rather an extended meditation from, through, and *out of* London, a search for alternative, disruptive and progressive political-geographical imaginaries, beyond the entrenched hegemony of neoliberal globalism. *World City* positions London as "a crucial node in the production of an increasingly unequal world" (Massey 2007: 8), as a crucible of uneven development, both locally and far beyond the city limits. It is a plea for this epicentre of financialized capitalism to face up to its responsibilities, at home and abroad. The book presents a trenchant critique of New Labour's contradictory embrace of economic competitiveness and social amelioration – epitomized by the Blairite conceit that "while poverty matters, wealth does not" – calling attention to London's simultaneous role as a wealth machine and generator of poverty. The horizons of the book, however, extend far beyond the questions of global-city polarization. Contra the neoliberal construction of cities as competitive agents, engaged in a war of all against all for mobile resources and "assets", Massey develops a relational account of the more-than-local production of economic "success". Yet this was never merely some conventional critique of neoliberalized, global-city urbanism. Beyond some of the established formulations of localism and horizontalism on the left, she reiterates her critique of such defensive formulations of place politics – leaning as they tend to do on received constructions of authenticity and community, and on the idea that localized havens can somehow be constructed, sheltered from out-there forms of globalization – in favour of an "outward-looking politics that seeks to address [the] wider geography of place and to ponder what might be thought of as [its] global responsibilities" (Massey 2007: 176). There is a domestic strain to this argument, for instance where Massey demonstrates the absurdity of the proposition that London-style growth should (or even could) be exported to Britain's provinces, as a solution to a longstanding "regional problem" (see also Allen *et al.* 1998; Amin *et al.* 2000). To the contrary, London-centric growth is seen to be constitutive of this very problem, with the capital itself arguably being the country's *real* "problem region" (John *et al.* 2002; cf. Massey 1979). Transnational extensions of this argument include the suggestion that alternative "geographies of allegiance" might be forged through initiatives providing restitution or reciprocation to those developing countries that effectively subsidize the capital's health service through asymmetrical migration flows of nurses and doctors trained at public expense in countries of the global South.

In 2007, just a few days before the release of *World City*, Massey received an invitation to Venezuela to give a series of lectures and to write a set of essays, including a widely-distributed popular pamphlet, that would revisit her notion of power-geometries in the context of the real-time implementation of "21st century socialism" (Massey 2009b; Massey with HGRG 2009). She would spend time in dialogue with activists working to form communal councils as part of a broader effort to decentralize state power and to remedy the deep-seated political, economic, and social imbalances that produced and reinforced profound regional inequalities in the country. "My appreciation for the forms and nuances of power was vastly enriched", she reflected, "the concept of power-geometries itself was dynamised – precisely because it was being used in a political *process*" (Featherstone *et al.* 2013: 263, emphasis in the original; see also Massey 2015a). Although Massey had always been politically engaged, the publication of *World City* and her work in Venezuela marked a shift away from what she increasingly saw as abstruse debates within her discipline. She would retire from the Open University in 2009, after almost three decades, but in some respects her intellectual and political activities intensified during her remaining years, especially in the tumultuous context of the global financial crisis and its lingering aftermath.

In what would be her last collaboration with *Soundings* colleagues Michael Rustin and Stuart Hall, the Kilburn Manifesto was an effort to unpack the unfolding crisis and its apparent failure to bring down an increasingly sclerotic neoliberalism. As Massey explained, in the wake of the financial crash of 2008,

> today we are sitting here with [UK Prime Minister] Cameron saying that the big problem is the public deficit, and the big state. The economic crisis is partly being solved, at least for the time being, and that is seen as the only problem. The implosion of neoliberal ideology is no longer on the agenda. It's as though they've separated those two instances [the economic and the philosophical] again. (Hall & Massey 2012: 58)

The Kilburn Manifesto project involved a series of political events and collaborative writing projects targeted on what were seen, in Gramscian terms, as "antagonisms" in the reigning conjuncture. It included initiatives focused on feminism, generational politics and race/ethnicity, bringing together social activists, trade unionists and academics from around the country. The goal was to engage in a process of debate, collaboration and strategic (re)thinking in order to expose "cracks" or "fissures" in the stubbornly hegemonic ideology of neoliberalism. Amongst other things, this would involve a sustained effort to take apart the notorious "there is no alternative" formulation that since the 1980s had been entrenched as a cornerstone of the neoliberal common sense in Britain, not least to demonstrate that, in Massey's words, this "was a triumph that was engineered" (Peck *et al.* 2014: 2036). In explicating how this common sense was laboriously cobbled together from contradictory elements, the Manifesto and related publications, including columns in the *Guardian* newspaper, sought to develop an alternative narrative aimed at prizing apart political spaces in order to enable "the possibility of thinking a radically different future" (Peck *et al.* 2014: 2039; see also Massey 2013b).

Recounting the story of the Kilburn Manifesto project to a packed house at the 2014 annual geography meetings in the United States, a country that she chose rarely to visit, Massey powerfully wove together her enduring commitments to place-based theorizing, on the one hand, and to forging progressive coalitions for political change on the other. She reminded her audience of the necessary *political* work of relational thinking – the need for a politics of articulation. One of her problematics on this occasion was to grapple with the meaning of "social settlement", understood in the Gramscian sense of how it is that "a glue of ideas, a common sense" is able to articulate, or hold together, seemingly contradictory elements into a conjuncture (Peck *et al.* 2014: 2036). Invoking the tradition of British cultural studies, she insisted that the cultural never "floats free" of the economic, just as the economy, even in its globalized and financialized form, is never disengaged from the cultural and the political, despite attempts to present it otherwise. The task at hand – inseparably analytical and political – is precisely to combine these cultural and economic "instances", and others too, in order to understand the formation of a hegemonic ideology, which after Raymond Williams and Stuart Hall is never fixed or final, but always in the making, always in need of renewal, always contested. When confronting the fracturing hegemony of neoliberalism, she emphasized, it would be necessary to work at the intersection of "constituent power" – which had been on display in movements like Occupy in the wake of the financial crisis – and "constituted power" – efforts to transform the prevailing rules of the game within representative democracy. The latter had been evidenced by the rise of parties like Syriza in Greece, and by the left turn in Latin America. There were even progressive disruptions to politics-as-usual closer to home. In fact, the last blog post that Massey wrote, for *Soundings*, called on the left to mobilize around the leadership of Jeremy Corbyn, who had recently been elected leader of the Labour Party in the UK (Massey 2015b). When Massey died, the world – and the British political and academic world in particular – lost a unique chronicler of, commentator on and *participant in* progressive disruptions such as these. She would have been surprised neither by the powerful counter-mobilizations against these breaches of the neoliberal edifice, nor by their resultant mixed fortunes, and would doubtless have carried on tilting at social injustice and its underlying power-geometries in her own inimitable, and different way.

Doreen Massey, we have argued here, changed geography – and in much more than a disciplinary sense. She developed and popularized entirely new ways of thinking and acting geographically, leaving changed worlds behind her. As Roger Lee wrote in one of many obituaries,

> She was never prepared to separate theory from the complexity of experience and the possibilities of politics and, in this, Doreen was the epitome of a geographer – wholly aware of the need to understand the multiplicity of relations that interact to shape social and environmental life and of the possibilities for progressive change.
> (Lee 2016: 311)

From her early efforts to rethink and revitalize Marxist categories and her engagements with critical realism, and in her subsequent elaboration of the implications of relational thinking, space as multiplicity, global sense of place, and so many more, Massey always sought to demonstrate the materiality of sociospatial processes. Throughout her career, she worked to explore the dialectical relations between the material, the social and the spatial in human geography and far beyond in ways that deftly balanced the structural and the conjunctural, embracing openness and multiplicity while resolutely confronting dominant forms of space and power. Doreen Massey's sense of place was never bounded, restrictive or limiting; it was where new things could happen, where new alliances could be fashioned, where change could be made. Thankfully, it was one that she shared.

THE BOOK AHEAD

The 25 original essays in this volume highlight the generative potential of Doreen Massey's remarkably wide-ranging oeuvre, not least as a means to confront the major social challenges of our time. Contributors include some of Massey's long-time interlocutors and collaborators, together with other prominent scholars drawn from the fields of geography, political economy, gender and feminist studies, labour studies, and more. These short, provocative essays provide readers with a multi-sited assessment of the political and social circumstances that gave rise to many of Massey's key ideas and contributions, and of how those contributions subsequently travelled, and were translated and transformed, both within and outside of academia. Looking forward and outward, rather than merely backward, the collection also highlights some of the diverse ways in which Massey's formulations and frameworks provide a basis for new interventions into critical contemporary debates.

All of the essays present conversations or dialogues *with* Massey and her work. The conversation is not one-way. The authors listen attentively to Massey, utilizing her concepts, formulations and provocations to help illuminate their own fields of study, but they also speak back to her, pressing her on certain premises and postulates, and wondering, too, how she might respond to this or that question were she still with us. Readers are invited to listen in on the conversation and, looking ahead, to engage with it and build upon it.

The eight essays in Part I all engage in various ways with the specific personal *contexts* in which Doreen Massey lived and worked. While they are not strictly biographical, many of them do have strong – and sometimes quite intimate – biographical elements. The essays examine Massey's ideas in the unique context of their situated germination – a particular stage of her life, her institutional affiliation, the local ideational milieu, and so forth. And they reflect in turn on how Massey's ideas shaped the scholarly and political environments within which they were produced and circulated. Part I begins with the three chapters that are in a sense the most biographical. In Chapter 2, Linda McDowell accounts for some of the ways in which "place mattered" in the life that Doreen Massey lived, from her upbringing in postwar Manchester, through her education in Oxford and Philadelphia, to her adopted home in Kilburn, North London, in an essay that is simultaneously sociological, geographical and personal. This is followed in Chapter 3 by Trevor Barnes's revelatory discussion of Massey's "dark past", which included a brief stint in the private sector and work as an especially restless regional scientist, which is not presented here as a search for "roots" but an exploration of "routes" towards what would be transformative interventions in economic geography. Following this, Michael Dear offers a personal account of many of these same events, beginning with the journey that he shared with Doreen Massey from London to Philadelphia in 1971, where they studied regional science together, to their shared paths to different places but always with continuing connections (Chapter 4).

Chapter 5, by Gillian Hart, is biographical in a rather different way, recounting as it does Massey's influence on her own path from Economics to Geography via the locality debates of the late 1980s and their subsequent journeys – sometimes together, sometimes diverging – as they engaged with various strands of Marxism and radical social theory. Michael Rustin, another long-time friend and collaborator of Doreen Massey, and a co-founder of *Soundings* along with Stuart Hall, presents his own perspective on Massey's conception of space and how, for him, she was to breathe new life into Geography (Chapter 6). These are themes that Andrew Sayer also picks up in Chapter 7, where he reflects on both the ontological and the political status of space in Massey's thinking, drawing on their long conversations together over many years on sociospatial theory, critical realism, geology, bird-watching, and more. A powerful sense of the reach and impact of these ideas is provided by Núria Benach and Abel Albet in Chapter 8, who describe how Massey lived, with one foot in academia and the other in politics, and the far-reaching impacts both of her work and her example. For Susan Roberts, too, it is difficult to separate Doreen Massey the thinker from Doreen Massey the person, or the feminist commitments that shaped both her scholarly programme and her way of being in the world (Chapter 9). It is Massey's advice to graduate students at the University of Kentucky that provides the title for this chapter, "Just carry on being different."

In the eight essays in Part II, the optic is expanded from particular personal contexts to wider historical, geographical and political *conjunctures*, the idea of *conjunctural analysis* being one that Massey herself helped to refine and popularize. Here, Massey and the details of her own personal trajectory move to the background, while broader stories and experiences – the narratives of places – come to the fore. Many of these narratives concern the nations, regions and cities of the United Kingdom, which was where Massey always lived, the place she knew best, and the place about which she most often wrote. (The United Kingdom in the final decades of the twentieth century and the first of the twenty-first was, if you like, *Massey's* conjuncture.) In Chapter 10, John Pickles explores Massey's contributions to British Cultural Studies and her insistence that the open and contingent relations that everywhere constitute the conditions of life are necessarily spatial as well as social and historical. In the process, Pickles argues, Massey developed a "conjunctural theory of place", the fundaments of which he fleshes out and reflects critically upon. Chapter 11, by Richard Walker and Erica Schoenberger, is the first of four that explicitly examine either Massey's work on the United Kingdom, the United Kingdom in the light of Massey's work, or, indeed, both of these. Walker and Schoenberger do so by revisiting Massey's first (co-authored) monograph, 1978's *Capital and Land*, and using it to reflect both on the particular political economy of the postwar conjuncture in Britain (and how much has changed in the neoliberal decades since) and on the key conceptual issues addressed in the book such as rent theory, class fractions, land development, and financialization. In Chapter 12, Huw Beynon and Ray Hudson trace the conjunctural shifts that coincided with the publication of *Spatial Divisions of Labour* in the United Kingdom, published the year of the historic miners' strike, whose critical defeat would consolidate Thatcher's power. Massey's involvement in support committees for the strikers served as an object lesson in the progressive defence of a place, the coalfields, which would become a bastion of Brexit support after more than three decades of neglect by the main political parties. Chapter 13, by Richard Meegan, offers a companion narrative to Beynon and Hudson's focus on deindustrialized areas, arguing for the acute continuing relevance of spatial divisions of labour (the book and the concept) to understanding patterns of uneven development in contemporary Britain, not least the privileged position of London. And the latter is Allan Cochrane's focus in Chapter 14, which asks the question of what – or more precisely, where – *is* London. Drawing on Massey's work on London as a "world city" and on geographies of responsibility as well as spatial divisions of labour, Cochrane situates London in the context of multiple interlocking sociospatial scales, all characterized by pronounced inequality, and reads Brexit as a manifestation of entrenched and worsening patterns of uneven development very much centred on and in the capital city.

The final three chapters of Part II move beyond the borders of the United Kingdom, but remain with the concept of conjuncture. In Chapter 15, John

Clarke theorizes "our" current conjuncture in terms of states falling back on protectionist nationalisms – focused on the UK but in dialogue with processes elsewhere – as a regressive solution to the social fallout from neoliberalism, and suggests that Massey's open notion of space and multiple trajectories can help us conceive of progressive disruptions to these resurgent, closed identitarian claims. Such nationalisms are equally evident, and the need for progressive disruptions equally urgent, in the German city of Hamburg, where, as Matthew Sparke and Katharyne Mitchell show in Chapter 16, citizenship is today subject to acute contestation in the context of a refugee crisis. Describing this local conjuncture as an example of what Massey termed "throwntogetherness", Sparke and Mitchell draw on her work to conceptualize progressive forms of global city citizenship. Lastly, Chapter 17, by Victoria Lawson and Sarah Elwood, undertakes a conjunctural analysis of poverty, distinguishing between "thinkable" and "unthinkable" poverty politics. Where the former is individualizing, sanctioned by the state, and governed through established programs, the latter excavates breaches in this common sense to encourage solidarities within and across class lines. Their discussion engages with numerous examples of contemporary Left social movements in the United States that are effectively unsettling the hegemonic poverty discourse, from Black Lives Matter to domestic-worker organizing.

Part III, containing nine essays, shifts the focus of the collection from conjunctures to *connections*. In some of the essays, the connections in question are sociospatial, for example involving extra-local versions of the solidarities alluded to by Lawson and Elwood. In other essays, meanwhile, intellectual connections are the primary concern, with the authors connecting Massey's work to novel or "other" spheres such as human–environment relations, architecture and film. The latter types of connection are in focus first. In Chapter 18, for example, Nik Heynen, Nikki Luke and Caroline Keegan urge urban political ecologists to engage Massey's ideas, pointing not only to her substantive discussions of "nature" (for example in *Capital and Land*) and her attempts to bridge physical and human geography (discussed earlier in this chapter), but also to her unique model for marrying theory and praxis. Chapter 19, by Frank Magilligan, Christopher Sneddon and Coleen Fox, treads related ground, positing river restoration, and in particular dam removal in the northeastern United States, as an appropriate topic for the kind of integrated physical-social research that Massey championed. Chapters 20 and 21 see Massey and her ideas taken to very different intellectual terrain. In the former, Geraldine Pratt and Jessica Jacobs explore Massey's relationship with film, visual media and the screen, connections that dated back to the distance-learning methods of the Open University but which would flourish in collaborations with artists, activists and filmmakers later in her career. In the latter, Kendra Strauss examines campaigns for institutional pension plans to divest from fossil-fuel investments, showing, in dialogue with Massey, how discursive construction of the "responsible investor" mobilizes a politics of responsibility for distant others linked to the causes and consequences of climate change.

The final five chapters of Part III take the discussion beyond sites of interrogation in the global North. First, in Chapter 22, Perla Zusman considers Massey's engagement with Latin America, particularly Nicaragua and Venezuela, emphasizing how her work – including the concept of power-geometries and the idea of space as a zone of coexistence with others – speaks to and resonates with distinctively Latin American conceptions of space and time. Then, in Chapter 23, Helga Leitner and Eric Sheppard use studies of urban commoning and struggles for rights to the city in Los Angeles and Jakarta respectively to reflect on and extend Massey's conceptualizations of place, space-time and power-geometries. In a sympathetic critique, they argue that these concepts remain overly state-centric and tend to focus on the politics of spatiality to the exclusion on the spatialities of politics. Chapter 24, by Priti Ramamurthy and Vinay Gidwani, seeks to advance earlier feminist theorizations of social reproduction by conceiving of the latter as a space of heterogeneous spatiotemporalities, drawing in the process on Massey's relational conception of space in conjunction with queer phenomenology. These arguments are developed through engagement with the oral histories of rural-to-urban migrants working in the urban informal economies of two Indian cities. In Chapter 25, Christian Berndt problematizes mainstream idealist narratives of globalization's geographies and seeks to install in their place an alternative, progressive Left imaginary. His arguments are illustrated through a case study of a Mexican border city and its maquiladora economy, and in critical conversation with Massey's work on globalization, in particular its understanding of the latter as a renewal of longstanding dynamics of uneven development and differentiated place-making. And in Chapter 26, Jennifer Hyndman and Alison Mountz consider how Massey's enduring notion of power-geometries can shed critical light on the new geographies of migration and detention shaping refugee and migrant trajectories today. Policies at multiple scales and the proliferation of island detention facilities have intensified power-geometries and expanded their dimensions, while migrants themselves are remaking the economies and meanings of these carceral spaces.

And so the dialogues continue … We began this chapter by noting the practical impossibility of the task of attempting to encompass all of Doreen Massey's far-reaching contributions. The essays that follow provide a sense of just how generative these contributions have been – and continue to be. We close this collection with a remembrance of Doreen Massey written by her close friend, Hilary Wainwright, who pays tribute to this "socialist feminist, engaged geographer and influential public intellectual [who] radiated polit-ical energy and humanity, sparkling with a cheeky wit" (Epilogue: 367). With Doreen Massey's passing, on 11 March 2016, the profound sense of loss was felt far and wide, but so has been the recognition of just how much was left behind. In this respect, just as there was no single point of departure, quite appropriately there is also no end. The dialogues will continue, with Massey's contributions remaining an essential guide to the open horizon.

REFERENCES

For works authored and co-authored by Doreen Massey, please see Select Bibliography of Doreen Massey, beginning on p. 371.

Albet, A. & N. Benach 2012. *Doreen Massey: Un Sentido Global del Lugar*. Barcelona: Icaria.

Althusser, L. 1971. *On Ideology*. London: New Left Books.

Allen J., D. Massey & A. Cochrane 1998. *Rethinking the Region*. London: Routledge.

Amin, A., D. Massey & N. Thrift 2000. *Cities for the Many Not the Few*. Bristol: Policy Press.

Bagguley, P., J. Mark-Lawson, D. Shapiro, *et al.* 1990. *Restructuring: Place, Class, and Gender*. London: Sage Publications.

Christopherson, S. 1989. "On Being Outside 'The Project'". *Antipode* 21 (2): 83–9.

Clarke, J. 2016. "Doreen Massey (1944–1966): Making Geography Matter". *Cultural Studies* 30(3): 357–61.

Cochrane, A. 1987. "What a Difference the Place Makes: The New Structuralism of Locality". *Antipode* 19 (3): 354–63.

Deleuze, G. & F. Guattari 1994. *What is Philosophy*? London: Verso.

Deutsche, R. 1991. "Boys Town". *Environment and Planning D: Society and Space* 9 (1): 5–30.

Eliasson, O. 2013. "Your Gravitational Now". In D. Featherstone & J. Painter (eds) *Spatial Politics: Essays for Doreen Massey*. Oxford: Wiley-Blackwell, 125–32.

Featherstone, D. & J. Painter 2013. "'There Is No Point of Departure': The Many Trajectories of Doreen Massey". In D. Featherstone & J. Painter (eds) *Spatial Politics: Essays for Doreen Massey*. Oxford: Wiley-Blackwell, 1–18.

Featherstone, D., S. Bond & J. Painter 2013. "'Stories So Far.' A Conversation with Doreen Massey". In D. Featherstone & J. Painter (eds) *Spatial Politics: Essays for Doreen Massey*. Oxford: Wiley-Blackwell, 253–66.

Freytag, T. & M. Hoyler 1999. "'I Feel as if I've Been Able to Reinvent Myself' – A Biographical Interview with Doreen Massey". In D. Massey *Power Geometries and the Politics of Space-Time*. Heidelberg: Department of Geography, University of Heidelberg, 83–95.

Graham, J. 1998. "Spatial Divisions of Labour: Social Structures and the Geography of Production, by Doreen Massey". *Environment and Planning A* 30 (5): 942–3.

Hall, S. & D. Massey 2012. "Interpreting the Crisis". In S. Davison & K. Harris (eds) *The Neo-liberal Crisis*. London: Soundings, 55–69.

Hall, S., D. Massey & M. Rustin (eds) 2012. *After Neoliberalism? The Kilburn Manifesto.* London: Lawrence and Wishart.

Harcourt, W., A. B. Wilson, A. Escobar & D. Rocheleau 2013. "A Massey Muse". In D. Featherstone & J. Painter (eds) *Spatial Politics: Essays for Doreen Massey.* Oxford: Wiley-Blackwell, 158–77.

Harrison, S., D. Massey, K. Richards, *et al.* 2004. "Thinking across the Divide: Perspectives on the Conversations between Physical and Human Geography". *Area* 36 (4): 435–42.

Harrison, S., D. Massey & K. Richards 2006. "Complexity and Emergence (Another Conversation)". *Area* 38 (4): 465–71.

Harrison, S., D. Massey & K. Richards 2008. "Conversations across the Divide". *Geoforum* 39 (2): 549–51.

Hartmann, H. I. 1979. "The Unhappy Marriage of Marxism and Feminism: Towards a More Progressive Union". *Capital & Class* 12 (2): 1–33.

Hartsock, N. 1987. "Rethinking Modernism: Minority vs. Majority Theories". *Cultural Critique* 7: 187–206.

Harvey, D. 1987. "Three Myths in Search of a Reality in Urban Studies". *Environment and Planning D: Society and Space* 5 (4): 367–76.

Henry, N. & D. Massey 1995. "Competitive Time-Space in High Technology". *Geoforum* 26 (1): 49–64.

Hinchcliffe, S. 2013. "A Physical Sense of World". In D. Featherstone & J. Painter (eds) *Spatial Politics: Essays for Doreen Massey.* Oxford: Wiley-Blackwell, 178–88.

Ijams, B. W., J. Popke & K. Urch 1994. "Gender, Space and the Academy: An Interview with Doreen Massey, The Open University". *disClosure* 4: article 9. https://uknowledge.uky.edu/disclosure/vol4/iss1/ (accessed 3 January 2018).

John, P., S. Musson & A. Tickell 2002. "England's Problem Region: Regionalism in the South East". *Regional Studies* 36 (7): 733–41.

Lee, R. 2016. "Doreen Massey, 3 January 1944–11 March 2016". *Geographical Journal* 182 (3): 311–12.

Lovering, J. 1989. "The Restructuring Debate". In R. Peet & N. Thrift (eds) *New Models in Geography, Volume One.* London: Unwin Hyman, 198–223.

Martin, R., A. Markusen & Doreen Massey 1993. "Classics in Human Geography Revisited: Spatial Divisions of Labour". *Progress in Human Geography* 17 (1): 69–72.

Meegan, R. 2017. "Doreen Massey (1944–2016): A Geographer Who Really Mattered". *Regional Studies* 51 (9): 1285–96.

Mouffe, C. 2013. "Space, Hegemony and Radical Critique". In D. Featherstone & J. Painter (eds) *Spatial Politics: Essays for Doreen Massey.* Oxford: Wiley-Blackwell, 21–31.

Peck, J. 2016. "Macroeconomic Geographies". *Area Development and Policy* 1 (3): 305–22.

Peck, J., D. Massey, K. Gibson & V. Lawson 2014. "The Kilburn Manifesto: After Neoliberalism?" *Environment and Planning A* 46 (9): 2033–49.

Peck, J. & T. Barnes 2019. "Berkeley In-Between: Radicalizing Economic Geography". In T. Barnes & E. Sheppard (eds) *Spatial Histories of Radical Geography: North America and Beyond.* Oxford: Wiley-Blackwell.

Raper, J. & D. Livingstone 1995. "Development of a Geomorphological Spatial Model using Object-oriented Design". *International Journal of Geographical Information Systems* 9 (4): 359–83.

Routledge, P. & A. Cumbers 2013. *Global Justice Networks: Geographies of Transnational Solidarity*. Oxford: Oxford University Press.

Saldanha, A. 2013. "Power-Geometry as Philosophy of Space". In D. Featherstone & J. Painter (eds) *Spatial Politics: Essays for Doreen Massey*. Oxford: Wiley-Blackwell, 44–55.

Scott, A. 2000. "Economic Geography: The Great Half-Century". *Cambridge Journal of Economics* 24 (4): 483–504.

Scott, J. W. 1999 [1988]. *Gender and the Politics of History*. New York: Columbia University Press.

Smith, N. 1986. "Spatial Divisions of Labour: Social Structures and the Geography of Production, by Doreen Massey". *Geographical Review* 76 (3): 350–2.

Smith, N. 1987. "Dangers of the Empirical Turn: Some Comments on the CURS initiative". *Antipode* 19 (1): 59–68.

Soja, E. 1987. "The Postmodernization of Geography: A Review". *Annals of the Association of American Geographers* 77 (2): 289–94.

Sparke, M. 2007. "Acknowledging Responsibility *For Space*". *Progress in Human Geography* 31 (3): 395–403.

Storper, M. 1986. "Spatial Divisions of Labour: Social Structures and the Geography of Production, by Doreen Massey". *Progress in Human Geography* 10 (3): 455–7.

Telegraph 2016. "Doreen Massey; Geographer Who Examined How Places are 'Socially Constructed'". *The Telegraph*, 21 March: 33.

Vaiou, D. (ed.) 2017. "Special Section: Doreen Massey". *Geographies: A Biannual Review of Spatial Issues* 29 (Spring): 3–59 [In Greek].

Warde, A. 1985. "Spatial Change, Politics and the Division of Labour". In D. Gregory & J. Urry (eds) *Social Relations and Spatial Structures*. London: Macmillan, 190–212.

Werner, M. 2016. "Global Production Networks and Uneven Development: Exploring Geographies of Devaluation, Disinvestment, and Exclusion". *Geography Compass* 10 (11): 457–69.

Williams, K. & D. Thomas 1983. *Why Are the British Bad at Manufacturing?* London: Routledge and Kegan Paul.

PART I

CONTEXTS

CHAPTER 2

NORTH AND SOUTH: SPATIAL DIVISIONS IN A LIFE LIVED GEOGRAPHICALLY

Linda McDowell

TWO PLACES

The oft-repeated insistence by Doreen that "place matters" was as true of her life as her academic and political work. She lived the contradiction and inequalities of the UK's north–south divide in her everyday life: two diverse regions connected by the northwest train line between London and Manchester and onto the Lake District. Doreen was a frequent passenger, often seen on the train, learning Spanish verbs, checking birds that she might see on the way or reading papers, journals or novels. I travelled the same route – sometimes with Doreen, on the way to give a joint talk in Liverpool or Manchester, for example, as well as alone. I too was brought up in the north-west of England, just over the Mersey from where Doreen was born. She grew up in Wythenshawe, a local authority estate in Cheshire that was moved into Manchester in the 1930s, and I grew up in Stockport, an industrial town just in Cheshire, each place about seven miles from Manchester's city centre. Our joint origins were a source of common pleasure, as well as a basis for our joint interest in the changing UK economy, transformations in the landscape of labour and, in particular, the position of women in the workplace.

What further united us was our educational trajectory, in which social and spatial mobility combined to shift us across the class structure and into the southeast. We both moved from state schools into the elite spaces of Oxbridge, into charitable research organizations where for a time we both worked in London for the sociologist Peter Willmott, and then we both worked at the Open University, committed to increasing opportunities for people who had missed out on a university education when younger or who wanted a second chance at educational achievement. Here, however, we parted company in 1992, after 12 years as colleagues, when I left for Cambridge, seduced (in Doreen's view) by the fleshpots of a well-endowed institution. She never quite let go of her view that I had betrayed the cause of widening access, despite my frequent protests and insistence that Oxbridge too had a diversity agenda, and of course, we both owed a good deal of our own success as academics to Oxbridge accepting two northern state-schooled girls.

I often, although as it turns out not often enough, talked to Doreen about her early life. I assumed we would have many more years to meet and talk, sadly denied by her too-early death. In this chapter I want to explore some of the dissonances and challenges for young working-class women, like Doreen, who moved across social and spatial distance to enter a new world. I draw on our conversations, my own experience and some of the rich literature about the impact of higher education for the working class, especially, but not only, for young women.

PLACE MATTERS: GROWING UP IN THE NORTHWEST

An insistence on the specificity and particularity of the local is a key aspect of Doreen's theoretical legacy and so it is obvious that growing up in the north-west provided her with a particular lens on the world, especially on class and gender politics and their significance in debates about deindustrialization that were so central at the time that Doreen was writing *The Anatomy of Job Loss* (1982) and *Spatial Divisions of Labour* (1984). The early industrialization of the northwest was evident in its rural and urban landscape, where canals, eighteenth- and nineteenth-century mills, tall chimneys and acres of terraced housing defined a landscape that was also marked by poverty and inequality. This was the landscape captured by Engels in *The Condition* (1845), a city where the Chartist movement was important, where the massacre at Peterloo in 1819 (when a mounted force rode into crowds demanding better Parliamentary representation) is still commemorated and later where both middle- and working-class women were significant figures in the pre-First World War movement for female suffrage.

As part of the great housebuilding programme at the end of the First World War, new estates of housing built by the state for the industrial working class were established. In the 1920s a new estate in Wythenshawe provided homes for 10,000 households who were moved from the poorest areas of inner Manchester. It was largely composed of two-storey houses, and initially without community facilities such as pubs and social centres, or even shops. By 1944 when Doreen was born, the estate had expanded and was better provided with facilities. It was designed as garden city, and class differences were made concrete in the contrast with the old black and white Tudor manor house, Wythenshawe Hall, originally owned by the Tatton family, set in the parkland that bordered the new houses. These small houses were erected quickly, some of them prefabricated steel homes that lasted decades longer than at first planned, but in the main solid brick-built, slate-roofed two- and three-bedroomed houses: the type of house where Doreen lived as a child. The high-rise estates of flats that proved so unpopular were not built until the 1960s.

After the Second World War, Wythenshawe expanded, gradually acquiring all the amenities and facilities that the original planners had neglected to include. New schools, shops, pubs and churches were built and in 1948 the area also got its own hospital, Wythenshawe Hospital, which grew out of the earlier Baguley Hospital (and I now wonder if this was where Doreen was born). The area continued to change: a large shopping area was built in the 1960s, but it was not until 1971, after Doreen had left for Oxford and London, that a civic forum including a large library, a swimming pool, a restaurant, a bar and a theatre were added to the estate. Wythenshawe is now an area with a mixed reputation, marked by crime (gang shootings were not unknown in the early 2000s), it is also an area that is being gentrified, as houses once for rent were sold under the Thatcher Government's right to buy legislation, and new building occurs for owner occupation.

When Doreen was growing up in the forties and fifties, its reputation was perhaps best captured in the term "respectable". Her parents provided a loving home, although money was tight. Her father worked as a groundsman at the Northern Tennis Club and her mother was a secretary and later a personal assistant. Education mattered to them and they both took evening classes but it was through their daughters that their educational aspirations were realized. At the age of 11, Doreen passed the entrance examination for Manchester High School for Girls, a highly selective state-supported (though now independent) school, with a fierce reputation for academic excellence, a traditional curriculum, and a strict uniform policy. The school is the alma mater of, among others, three Pankhursts – Christabel, Sylvia and less well-known Adela, who was active in the Australian suffrage movement. Angela Brazil, author of stories set in a girls' boarding school, and the geographer Janet Momsen were also pupils.

This schooling, this landscape, provided the impetus for Doreen's influential work on divisions of labour and later her portrait of Kilburn (Massey 1991a), a neighbourhood in inner London, where she argued persuasively that place matters. She demonstrated both theoretically and empirically the ways in which the intersection of a locality's position in earlier divisions of labour and broader socioeconomic transformations over time and space both reflect and affect the range of employment opportunities available for women and for men, as well as the ways in which family life is constructed and maintained through the differential responsibilities of men and women in the workplace and the home. Drawing on her framework, in an early assessment of the varied geography of a woman's place, Doreen and I (McDowell & Massey

1984) wrote a chapter in an Open University text, *Geography Matters!* that I remain extremely proud of, and which still appears on reading lists. There we argued that women in the northwest played a significant part in the industrial history of the locality and so in the construction of that sense of place that distinguishes the northwest cotton towns in general and Manchester in particular from other places. We suggested that women's employment in the cotton industry was connected to the significance of suffrage politics in the city in the early twentieth century, when working-class, as well as middle-class, women were active in the movement (Liddington & Norris 1978). Their experience of going "out to work", as Kessler Harris (1982) also suggested in the mill towns of Massachusetts, gave "mill girls" a degree of independence and freedom from patriarchal control as single women, as well as some financial independence, making their participation in local politics easier than elsewhere at the same time. Manchester and the surrounding cotton towns were important in the suffrage movement. The Women's Social and Political Union (WSPU), the more militant wing of the suffrage movement, was founded by Emmeline Pankhurst in Manchester in 1903 and attracted working-class mill workers as well as middle-class members. Hannah Mitchell, for example, who was elected to Manchester City Council in 1919, had joined the WSPU in 1905, when she was working as a seamstress in Bolton. Annie Kerney, another working-class mill worker, joined at a meeting in Oldham, addressed by Christabel Pankhurst, Emmeline's daughter, who, between 1893 and 1897, was a pupil at the school Doreen later attended.

Growing up in the fifties

Growing up in the 1950s, however, was not a time when women's inequality and feminist politics were part of the social or political agenda. Many white, working-class women who had been active in the wartime labour market returned to their homes and contributed to the baby boom. Incomes were often low and the range of consumer goods now on offer would seem astonishing to families in the 1950s. Food and clothes rationing remained in place until 1953 and biscuits and sweets were often weekend treats. Houses were cold and draughty in the winter, often heated by a coal fire downstairs and unheated upstairs which made getting undressed for bed something of a torment in winter. Manchester, like most cities, also suffered from smogs in the winter, at least until the Clean Air Acts were introduced in 1956 and extended in 1968.

Little girls wore dresses to school, and at home, and often played indoors, with gender-specific toys, although the streets and green spaces of the Wythenshawe estate were also the grounds for outdoor adventures for both sexes, especially as traffic was not heavy during the 1950s. Doreen later argued though that she disliked the sylvan landscape surrounding the estate (see Pickles, this volume). Most people went to work and school by bus or bicycle, if it was too far to walk. In 1950 there were just under two million cars in Britain, owned by 14 per cent of households, not many of whom lived in Wythenshawe. Everyone who grew up then will have their own memories of the time – perhaps of welfare orange juice and cod liver oil – but this extract below from Paul Feeny's book about the 1950s (1999), despite its somewhat romanticized tone, captures the sorts of memories that women and men who grew up then will instantly recognize, or will if they came from the sort of modest background that typified Doreen's early years.

Despite the difficulties of day-to-day living people had great pride in and loyalty to their country and seemed to share a common purpose in life. Families stayed together through the hard times and everybody knew their neighbours and had a sense of belonging. They would routinely leave their street door on the latch and hang a key on a piece of string behind the letterbox when they were out for their children to come and go as they pleased.

Boys and girls played street games together, such as run outs, hopscotch and British bulldog. In the playground schoolgirls practised handstands and cartwheels with their skirts tucked up under the elastic of their navy blue knickers, while the boys played conkers.

We travelled in third-class compartments on train journeys to the seaside. In 1956 they were renamed second class. The change didn't move you any higher up the social ladder but it made you feel there was a bit less of a social gap. At the seaside you wore a knitted bathing costume on the beach.

For children the Saturday morning pictures provided the best fun. Every week, 200 to 300 unruly children would descend on a cinema for a couple of hours of film and live entertainment. The manager would regularly stop the film and threaten to send you all home if you didn't behave and the solitary usherette was often forced to run for cover. It was controlled mayhem with the stalls and circle filled with children cheering for the goodies and booing the baddies. It introduced us to The Lone Ranger and Zorro and the slapstick comedy of Mr Pastry and Buster Keaton.

Dusty, old-fashioned sweetshops had high wooden counters jam-packed with boxes of ha'penny chews and other sweet delights. Remember Lucky Bags and frozen Jubblys and getting a sore tongue from sucking on gobstoppers, aniseed balls and Spangles? Then there were those old Smith's potato crisps. The salt was in a twist of blue paper and you always had to rummage around for it at the bottom of the bag.

It is hard to identify the Britain of today with how it was back then. The whole appearance of the country has changed, particularly in inner cities where so much building and development work has been done over the years. The war-torn dilapidated houses, derelict land and bomb sites that were the forbidden playgrounds of postwar baby boomers are now long gone.

There was something cosy about growing up in the last decade in which most children retained their childish innocence to the age of 12 or 13 and enjoyed a carefree life full of fun and games. The stresses of adolescence and then adult life could wait. We were lucky.   (Feeney 1999: 4–5)

But it was a gender-segregated world as well, in which the message for young women from the pop songs of the time was a future defined by men. "I want to be Bobby's girl, I want to be Bobby's girl, that's the most important thing to me" was the relatively wholesome message from a track recorded by teenage singer Marcie Blane in 1962 (although the unnamed girlfriend was dumped by Bobby), as Doreen hit 18 and thought about leaving home, whereas the harder-hitting track "Leader of the Pack" by the Shangri-Las at least promised an escape on the back of a motorbike with a boy from the wrong side of town. But these were pipe dreams for young women at a school like Manchester High. Even these academic, selective girls' schools taught their pupils domestic science and needlework, although if you were clever enough you might escape the lessons aimed at a domestic future by earning a place in "the top set", where Latin or a second modern language rather than domestic science was taught. It was still unusual for young women, especially working-class women, to go to university. Most women still left school and went to work, leaving on marriage or the birth of their first child. In a large official survey of adolescent girls carried out in 1955 – Doreen was 11 then and not quite in the targeted age group – most of the 450 respondents expected to be married by their mid-twenties and saw making a home as their vocation (Wilkins 1955). In a similar survey at the end of the 1950s, 90 per cent of 600 young women questioned also expected to marry, but by that date, also suggested they would combine home-making with part-time employment (Joseph 1961). Doreen defied these predictions, never marrying nor having children, and unusually for a working-class girl of the time she went to university.

The life of "clever girls"

The school that Doreen attended from 1955 to 1962 was, however, not one that encouraged its pupils to consider early marriage and domesticity as a career. From the late 1940s onwards and into the 1950s, there was a long debate about how to school girls, initiated by Newsom's book *The Education of Girls* published in 1948, followed by the Crowther Report a decade later *Fifteen to Eighteen* (Central Advisory Council 1959) and Newsom's report in 1963 *Half Our Future* (Central Advisory Council 1963) on the education of average and less able children. In all three reports there was an assumption that, as Crowther noted, the curriculum should reflect girls' interests in dress, personal appearance and human relationships, including lessons in housewifery. Academic girls' schools, often staffed by educated women who had been early pioneers of girls' schooling and had themselves attended university, often had little patience with these arguments, run by women whom Newsom unpleasantly labelled as "normally deficient in the quality of womanliness and the particular physical and mental attributes of their sex" (1963: 1). It was these women who educated Doreen and encouraged her to aim for Oxford (see Photo 2.1). In these years when Doreen was at school, there were "clever" girls and "normal" girls and managing to negotiate the divide – at home, in the streets and through popular culture where clothes,

Photo 2.1 Doreen (second from right, front row) as a pupil at Manchester High School for Girls, 1959. Courtesy of the archivists, Manchester High School for Girls

make-up and music played a more important part of everyday social life than at school – was good training for a clever working-class girl who would soon find herself having to perform a different version of young womanhood in an elite university.

MOVING ACROSS BORDERS: OXFORD IN THE EARLY 1960s

In 1962 when Doreen left school there were only 118,000 students in total at university, 4 per cent of the age group. And women were the minority among these students – about 10 per cent of the students at university, although better represented at teacher training colleges. This was just before the Robbins Report (Committee on Higher Education 1963), that led to the growth of existing British universities and the foundation of several new universities, marking the beginning of the great expansion of tertiary education for young women. But escape from a domestic future and a life similar to that of earlier generations of women was possible and became increasingly common from the 1960s onwards. In 1962, Doreen won a scholarship to St Hugh's College, Oxford University where she once claimed she felt like a "space invader", although it is probable that her experiences at school had, at least to some extent, prepared her both for the class mix at this all-women's college and for the rigorous academic work demanded, despite her own relatively humble origins (and see also Barnes, this volume, for her opinion on what was then taught at Oxford). Nevertheless, moving from a local-authority housing estate into an Oxford College, and back and forth several times each year, was a huge dislocation and a leap across a social and cultural divide. Meals had different names, were eaten at different times, sherry was offered to 18 year olds, and a northern accent was a disadvantage. In the passage below, Keith Hart, an anthropologist who also grew up in Manchester, the same age as Doreen and a student in Cambridge rather than Oxford, captures the impact of belonging to two worlds.

> Nothing could have prepared me for the shock of going to Cambridge. The gap between Cambridge and my homeland was huge and I had to move between them three times a year, while desperately clinging to the notion that I had an identity, that I was not schizophrenic … I never got over that primal scene, juxtaposing that life in Manchester with a feudal Fenland society. They seemed to me then – and still do now – two completely different civilizations. In all my subsequent world travels, I have not come across a greater contrast. (Hart 2003: 419)

For Doreen, and for me, a student at Cambridge five or six years after Doreen went to Oxford, the shock was even greater, as we moved from all girls' schools, where we had been taught in the main by women, into a male-dominated world, where 90 per cent of our fellow students, as well as most of the lecturers, were men, even though at that time in the 1960s, women taught at the women-only colleges in these two elite institutions.

But the move across borders, however disorienting, proved to be the spring-board into a different life, new forms of gender relations and a different class position. This was the great leap forward in terms of social mobility: ignored in the classic studies such as that by Halsey *et al.* (1980) and others, who compared men who grew up in the decades after the Second World War with their fathers, excluding women altogether. A number of feminist scholars have produced memoirs of the extreme dislocation experienced by many women, especially those from working-class families, who were first-generation university students (Long *et al.* 2000; Reay 1997; Tokarczyk & Fay 1993). Despite moving within the same country, for women like Doreen social and geographical mobility meant that, as for all migrants, they were removed from familiar surroundings.

In *Girl Trouble*, an excellent survey of young women's lives in the twentieth century, Carol Dyhouse (2013) suggests that clever young women who made it to university between 1945 and the end of the 1960s often expressed frustration at restrictions they experienced. As she notes "clever girls cooped up in the women's college at Oxford and Cambridge often felt excluded, relegated to living on the margins of an extended, male-public-school kind of world" (Dyhouse 2013: 134). Dyhouse suggests that a novel by Margaret Forster, *Dames' Delight* (1964), captures the sort of life women university students of Doreen's age might have been living at the time. The novel depicts ageing, spinster, scholarly dons, focused on the intricacies of strip farming (a topic on my syllabus in 1968, and no doubt on the geography syllabus at Oxford too!) and failing to understand the lives of their students, worried by the shadow of the atomic bomb and for some of them, the fear of pregnancy, before the contraceptive pill was widely available.

Despite the feelings of dislocation and perhaps being on the margins of the masculinist university world, it was the three years at Oxford that both fired Doreen's academic imagination and solidified her class politics, as well as her nascent interest in feminist politics. As Ann Oakley has suggested, the women active in the second wave feminist movement in the UK typically were women educated in girls' selective schools, and often with sisters rather than brothers – Doreen exactly. Their entry into the world of higher education and later employment brought them face to face with forms of gender discrimination that they had not previously experienced. Neither at home, nor at school had "brains" been a disadvantage, even though for girls growing up in the 1950s, gender divisions were a central, and then largely accepted, untheorized, part of life, mirrored in the reports on girls' education outlined above. Looking back, I sometimes find it hard to remember how solid were the assumptions about how "respectable" girls were expected to behave and what a struggle it was to escape from them.

A LIFE IN LONDON

Apart from a year in Philadelphia studying for a master's degree in spatial science, which was to be the basis of her devastating critique of the assumptions that lay behind the type of quantitative analyses that had become important in geography from the late 1960s onwards (see Barnes, this volume), the rest of Doreen's life was to be spent in London. For many years she lived in Kilburn, a diverse, multicultural part of northwest London that was the stimulus for her influential article, "A Global Sense of Place" (Massey 1991a) and later the title of her left political manifesto – the Kilburn Manifesto – written with Mike Rustin and Stuart Hall (Hall *et al.* 2013). In London she worked variously for the Centre for Environmental Studies, seen as such a threat that its funding was withdrawn by Mrs Thatcher (and closed in the 1980s, after struggling as an independent think tank), in a research post at the LSE and for the Greater London Council. In 1982 she took up the chair of geography at the Open University, a post that she held until her retirement and which provided both the freedom and the stimulus for a series of academic books and papers and, increasingly for political activism beyond the UK: the subjects of later chapters in this collection. Her small flat in Ariel Road saw numerous discussions between many of the significant intellectuals of the years between the mid-1960s and the second decade of the twenty-first century. If only those walls could talk. They would be a wonderful record of Left politics in London and beyond and of a passionate proponent of a necessity of understanding how space and place matter. To this political legacy, another should be added – Doreen's huge impact on the discipline of geography and her continuing connections to past and future generations of students. This legacy lives on in her published writing. What those of us who had the privilege of working with her perhaps miss the most is her personal charisma, her direct and engaging style which made it feel as if she was speaking directly to you, however large the audience.

REFERENCES

For works authored and co-authored by Doreen Massey, please see Select Bibliography of Doreen Massey, beginning on p. 371.

Central Advisory Council for Education 1959. *Fifteen to Eighteen.* London: HMSO (the Crowther Report).

Central Advisory Council for Education 1963. *Half our Future.* London: HMSO (The Newsom Report).

Committee on Higher Education 1963. *Higher Education.* London: HMSO (The Robbins Report).

Dyhouse, C. 2013. *Girl Trouble: Panic and Progress in the History of Young Women.* London: Zed Books.

Engels, F. 1969 [1845]. *The Condition of the Working Class in England.* London: Panther Editions.

Feeney, P. 1999. *A 1950s Childhood.* London: The History Press.

Forster, M. 1964. *Dames' Delight.* London: Jonathan Cape.

Hall, S., D. Massey & M. Rustin (eds) 2013. *The Kilburn Manifesto.* London: Lawrence and Wishart.

Halsey, A., A. F. Heath & J. M. Ridge 1980. *Origins and Destinations: Family, Class and Education in Modern Britain.* Oxford: Oxford University Press.

Hart, K. 2003. "Manchester on My Mind: A Memoir". *Global Networks* 3 (3): 417–36.

Joseph, J. 1961. "A Research Note on Attitudes to Work and Marriage Among 600 Adolescent Girls". *British Journal of Sociology* 12 (2): 176–83.

Kessler Harris, A. 1982. *Out to Work: A History of Wage Earning Women in the United States.* Oxford: Oxford University Press.

Liddington, J. & J. Norris 1978. *One Hand Tied Behind Us: The Rise of the Women's Suffrage Movement.* London: Virago.

Long, M. L., G. R. Jenkins & S. Bracken 2000. "Imposters in the Sacred Grove: Working Class Women in the Academe". *The Qualitative Report* 5 (3). www.nova.edu/ssss/QR-3/long.html (accessed 19 January 2012).

Newsom, J. 1948. *The Education of Girls.* London: Faber and Faber.

Reay, D. 1997. "The Double Bind of the Working Class Feminist Academic". In P. Mahoney & C. Zmroczek (eds) *Class Matters: "Working Class" Women's Perspectives on Social Class.* London: Taylor and Francis, 19–30.

Tokarczyk, M. M. & E. A. Fay (eds) 1993. *Working Class Labourers in the Knowledge Factory: Women in the Academy.* Amherst, MA: University of Massachusetts Press.

Wilkins, L. 1955. *The Adolescent in Britain.* London: Central Office of Information.

CHAPTER 3

HER DARK PAST

Trevor Barnes

INTRODUCTION

In 1978, on a brisk, bright Friday morning in early spring, the dozen or so of us taking Jacquie Burgess's third-year philosophy of geography course at University College, London, trooped down Tottenham Court Road, took a right on Goodge Street, and walked the mile or so to the architecturally brutal modernist Polytechnic of Central London. We were there to participate in a day-long conference on ideology and geography. Sir Peter Hall (although then still a commoner), Ron Johnston (not yet a recipient of the Order of the British Empire, or OBE) and Doreen Massey (not yet having refused an OBE) were speakers. From my perspective, the event was perfectly timed. My course paper was on ideology and geography, pretentiously called, "Never mind the truth, here's the ideologues". The title was a riff on the Sex Pistols' album, *Never Mind the Bollocks, Here's the Sex Pistols*, released a few months before. Because of the record title's profanity, you had to buy the album in a plain brown paper bag. Ratcheting up the pretension, I handed in my paper to Jacquie Burgess in a plain brown paper bag. Like the Queen's reaction to the Pistols' version of "God Save the Queen", Jacquie Burgess was not amused. "Bollocks", she said. That is, Jacquie Burgess, not the Queen.

For my course paper, I read everything in Volume 5, issue 3, of *Antipode*, "Ideology in Geography", including Doreen Massey's (1973) contribution, "Towards a Critique of Industrial Location Theory". It was brilliant, the first piece I'd read in economic geography that took on the neoclassical economic underpinnings of spatial science. It showed how each of the presumed universal assumptions of neoclassicism were not so universal, their deployment producing only inconsistency and contradiction. Massey argued that the end of neoclassical theory, as well as industrial location theory, was not to explain the world but only to justify and legitimate ideologically what it studied, industrial capitalism.

On the podium to an audience of several hundred undergraduates in the capacious central hall at the Polytechnic of Central London, Massey repeated her arguments with vigour, clarity and in-your-face attitude. It was a punkish performance, loud and strident, and made more so by Doreen Massey's short-cropped blonde hair, black jeans and what looked like Doc Martens, although I couldn't swear to that.[1] She passionately believed in what she said. We the audience passionately believed in what she said. But ten years earlier in 1968 she believed and said something completely different.

In August 1968 at London's Centre for Environmental Studies, Doreen Massey published two single-authored working papers (Massey 1968a; 1968b). "Problems of Location: Linear Programming" was an exposition of industrial location theory of exactly the neoclassical kind (Massey 1968a). All its references were to the high clergy of neoclassical spatial science: Brian Berry, Duane Marble, William Garrison, Walter Isard, Ben Stevens and Martin Beckman. Moreover, this was not a one-off youthful indiscretion, never to be repeated. Later that very same month, Massey (1968b) also published "Problems of Location: Game Theory" that contained mathematics even more daunting than found in her linear programming paper (although she later claimed she could not understand mathematics; Massey 1993a: 11). This is Doreen Massey's dark past that I intend to explore in this chapter.

I do so by drawing on the explanatory framework with which Massey was most associated towards the end of her life, conjunctural analysis. It was worked out in part with her friend, political comrade and Open University (OU) colleague, Stuart Hall. "A conjuncture", says Hall, "is a period during which the different social, political, economic and ideological contradictions that are at work in society come together to give it a specific and distinctive shape" (Hall & Massey 2010: 57). The conjuncture that held during the immediate postwar period in the UK "was dominated by the welfare-state, public ownership and wealth redistribution" (Hall & Massey 2010: 57). It didn't last, though. Contradictions always present within that conjuncture condensed and fused, and from the early 1970s ruptured, producing disarray and upheaval, crisis. As Hall says, it was "Thatcherism, neoliberalism, globalisation, the era dominated by market forces, [that] brutally 'resolved' the contradictions [of the first postwar conjuncture] and opened a new conjuncture" (Hall & Massey 2010: 57). My aim is to understand Doreen Massey's intellectual life from the perspective of her own conjunctural analysis. I want

---

1. In an interview, Massey said, "if you are five-foot one, … fair haired, … female, and quite often you can barely see over the podium … you come in like a guerilla" (Ijams *et al.* 1995: 100).

to make sense both of her dark past as well as the later luminosity that shone so brightly that Friday spring day in 1978 at the Polytechnic of Central London.

EARLY LIFE (1944–68)

At around age one-and-a-half, Doreen Massey witnessed the beginning of the first postwar conjuncture, although likely she did not realize it. It arrived on the afternoon of 26 July 1945, when the landslide victory by the Clement Atlee-led Labour Party was announced. The election occurred on 5 July, but it was another three weeks before the vote was finally tallied because of the large numbers of the electorate who were still stationed abroad, involved in military occupation and even active fighting.

The conjuncture was assembled during the Atlee-government first term and involved: massive nationalization of heavy industry, utilities and transportation services (rail, road and canal); creation of the National Health Service; following the 1942 Beveridge Report, establishment of a cradle-to-grave social welfare system; diligent following of Keynesian economic policy prescriptions; initiation of the biggest ever public housing construction boom, adding around a million new houses within a six-year period; launch of the "New Towns" programme (1947) that would include Milton Keynes, where twenty years later the turf would be turned on the Open University, and where Doreen Massey would work from 1982 until her retirement in 2009;[2] and a 50 per cent increase in spending on public education used among other things to abolish fees for grammar schools (benefitting Doreen Massey who attended the Manchester High School for Girls), and to expand placements in higher education.

Massey grew up at the site of one of the beneficiaries of Labour's public-housing expansion, the Wythenshawe council estate in south Manchester (see McDowell, this volume). At one point, it was the largest public housing project in Europe (Massey 2001b: 460–1, says "in the world"). In an essay about growing up there, Massey suggests that her political radicalism was forged in part by her early exposure to Wythenshawe's "municipal socialism" (Massey 2001b: 460). What alienated her there was not the council estate but the "leafy country" immediately outside. It might have been an iconographic English landscape, "yet", as she said, "I felt utterly and totally – and both wilfully and not – excluded from it" (Massey, quoted in Telegraph 2016: 33).[3]

---

2. There was a second New Towns Act in 1965 that led to the development of Milton Keynes among other places (Clapson 2012).
3. An interesting contrast is with David Harvey (2002: 156), who grew up in Gillingham a decade earlier and for whom as a teenager the surrounding Kent countryside was a "central obsession." John Pickles (this volume) discusses Massey's later view of nature and environment and its indissoluble connection to the sociopolitical.

In 1963, Massey entered St Hugh's College, Oxford, where she read geography. Here there was not only the issue of class, but gender (St Hugh's was one of five all-female colleges at Oxford, with the other 33 all-male). Ninety per cent of Oxford students were men (McDowell, this volume). Massey won the Irene Shrigley Scholarship, supplemented financially by provisions within the 1945 Education Act. She felt like an outsider at Oxford, though, "agog at the beauty of Radcliffe Square … [but] angry and alienated at what it stood for" (Telegraph 2016). She reacted by becoming, in her words, a "raving socialist" (Telegraph 2016: 33). If that's true, it was not because she was imbibing Marx in her geography lectures. In fact, she was served up only the words of the enemy: "location theory, for example, people like Alfred Weber" (Massey, quoted in Freytag & Hoyler 1999: 83).

She was awarded a First at Oxford, winning a Special College Prize on her graduation in 1966. The best she could later say about her education, though, damning it with faint praise, was that it "was quite engaging" (quoted in Freytag & Hoyler 1999: 83). She recognized it gave her an analytical armature, but it was analysis without passion and political commitment. "What was most important to being at Oxford", Massey later said, "was not what conclusion you came to, but whether you had the intellectual agility to argue it" (quoted in Freytag and Hoyler 1999: 83). For Massey the conclusion was everything, though. Derogating it left her disaffected. As she later reflected, at Oxford, "I'd seen what I thought it meant to be an academic. And I didn't want to be that" (quoted in Freytag & Hoyler 1999: 83). What would she do, though? Go into a non-profit? Become a civil servant? Enter politics? She did none of these things. Instead, she became an "executive" in the computer department of a London-based market-research firm, AGB Ltd (Massey 1999a).

Her boss was Sir Bernard Audley, the A in AGB (the acronym was for Audley, Gapper, Brown, the last names of the three founders, although later the company name was changed, albeit still fitting the original initials, Audits for Great Britain). The firm had been created four years before Massey joined in 1966. Initially based in a "10ft by 10 ft" office in Aldwych, Central London, and presumably with no room for a computer, the company's goal was to track the consumer-durable purchasing patterns of UK households (Eiloart 1973: 277). The breakthrough came when AGB won the bid to monitor TV watching on ITN, then the sole commercial station in the UK. It developed a digital TV meter that recorded each household's television viewing habits, encoding it on a paper tape. Householders would mail in the tape to AGB HQ once a week, where presumably Doreen Massey and others fed it into the computer. It did not take long before Massey "hated" her work, however (quoted in Freytag & Hoyler 1999: 84). She left the firm in 1968, and went to the Centre for Environmental Studies, beginning her contribution to the first postwar conjuncture.

THE CENTRE FOR ENVIRONMENTAL STUDIES

Harold Wilson's Labour Party came to power in 1964 with a pledge to modernize Britain (O'Hara & Parr 2006: 295). The Tory governments that held power after Atlee's Labour administration had more or less kept the first conjuncture going, upholding the welfare state. But the UK economy was languishing compared to many of its counterparts in Europe and in North America. By the early 1960s, it needed shaking up, a sharp jolt. It needed the "white heat of the technological revolution", as Wilson called it as part of his pre-election platform.[4] That meant "bringing scientists and experts into the policy-making process" (O'Hara & Parr 2006: 295). British civil-service amateurism no longer sufficed, if it ever did. This was the moment for professional specialists, academic authorities, researchers and analysts to step up and shine. It was another element to add to the postwar conjuncture, scientific socialism.

One of Wilson's new ministries where scientific socialism was practised early on was Housing and Local Government, headed by Anthony Crosland, one of the intellectual architects of the new Labour government, who believed that to fulfil the politics of British socialism required an "army of outsiders, uninhibited by civil service procedures" (Crosland, quoted in Blick 2006: 354). The Centre for Environmental Studies (CES) was one of the sites where that "army" was gathered and mobilized. CES was jointly funded by the UK government, £600,000 over five years, and by the Ford Foundation, $750,000 over the same period. In 1966, Crosland had travelled to the New York headquarters of the Ford Foundation to negotiate the grant. By then, two years as Minister, he realized that not only British civil servants but also British planners were not up to the job: stodgy, traditional, insular and narrow (Clapson 2012: 43). They also needed to modernize, to think more broadly, and specifically, in Crosland's judgment, "to incorporate work of social scientists and economic geographers" (Crosland, quoted in Clapson 2012: 43). Ford specialized in funding exactly that kind of social-scientific research for practical ends in the United States. It was eager to do something similar in the UK, with results fed back to American researchers.

The resulting CES money was distributed across several sites and projects, including from January 1968, "an internal research programme" housed in a John Nash regency mansion overlooking London's Regent's Park (5 Cambridge Terrace, NW1; Area 1969: 11). The head of that programme overseeing a staff of 22, as well as taking on the role of Deputy Director of CES, was Alan

---

4. Harold Wilson first used the phrase "the white heat of the technological revolution" in his leadership speech at the Labour Party annual conference in Scarborough, Yorkshire, on 1 October 1963.

G. Wilson. Wilson was trained in theoretical physics at Cambridge in the early 1960s, but underwent a Pauline experience in the middle of that decade, converting to social science, and specifically to geography. After the scales fell from his eyes, his unlikely new vocation was probabilistic urban modelling based on the fiercely difficult calculus of entropy maximization. Wilson hired a variety of different experts, from "crystallographers" and systems analysts to "materials scientists;" he also recruited a geographer, by way of a market research company, Doreen Massey (Massey 1993a: 11).

1968 AND ALL THAT: DOREEN MASSEY AT CES

Massey immediately took up the mantle of scientific expertise at CES. Her working papers on subjects like linear programming and game theory (Massey 1968a, 1968b) were part of an effort to improve reality, to make it more efficient, to realize a set of pre-determined objectives. In an essay about 1968, Massey (1993a: 10) says, "it was a time of so much experiment and hope, … [of] the thrill of possibility". She continues:

> The nineteen-sixties was a glorification … of science and of technical expertise. Social democracy plus knowledge were in the United Kingdom at least to transform society; to modernise it but in a socially progressive way. Harold Wilson was the Prime Minister … [His government encouraged] not just "hard sciences" but also in the social sphere, new knowledge, new systems, new expertise were to transform society.   (Massey 1993a: 11)

That was her job description at CES: to provide new knowledge, new systems and new expertise to transform society, to make the postwar conjuncture better, to realize its promise. At least at first Massey believed she could do that by

> work[ing] on employment distributions, linear programming, and the intoxicating logical intricacies of how to split the basic from the service sector. People wandered about [CES] talking of modal split. Equations in coloured pens on white boards took the place of pictures in our offices.   (Massey 1993a: xx)

It was heady stuff. But as she also remembers, what she did during the day "in that classy terrace on Regent's Park" was irrevocably cleaved from what she did "in the evening [after she'd] walk[ed] home across the Park to another world which revolved around music and politics":

I was heavily involved in a small group which organised around Vietnam, and at night we'd wander to the Round House, come back, lie around and discuss the endless complexities, the politics and the delights of the world. Next morning across the park again, it was sometimes hard to be entirely convinced that all these complexities could really be coped with just by adding on a few more subscripts.

(Massey 1993a: 11–12)

That worry might have been an underlying concern, but it did not prevent her, also in 1968, from joining another venture that was all about "equations in coloured pens" and "adding on a few more subscripts", the founding of the journal *Environment and Planning* (later renamed, *Environment and Planning A*). Massey was the inaugural editorial assistant, and for a short period, the book review editor. *Environment and Planning* was in effect the British version of the American *Journal of Regional Science*, and the very embodiment of neoclassical location theory (see also Dear, this volume). Walter Isard, originator of Regional Science at the University of Pennsylvania, by working through Allen Scott and David Harvey in the UK had already established a British section of the Regional Science Association. It had no journal, though, until *Environment and Planning*.

As Allen Scott recalled of these heady days:

I was having lunch with David Harvey one day, and he said, "I met this guy called John Ashby." At the time, Pion [the publishing company that founded *Environment and Planning*] was an electronics firm ... But they had sold out for a pile of money, and wondering what to do, they thought maybe they should get into publishing ... So, David Harvey gave me John Ashby's card. I called him and went around to find these two guys, one on one side of the desk, the other on the other side.[5] And I persuaded them to publish [papers from a British section Regional Science conference] as an experiment.[6] They wanted to get into publishing and find some niches for themselves. I argued that Regional Science was a virgin niche, and their scientific ability would hold them in good stead. 						(Scott 1998)

---

5. The "other guy" was Adam Gelbtuch, Pion's joint owner with Ashby.
6. This became Pion's long-running monograph series, "London Papers in Regional Science".

The rest is history. John Ashby asked both Peter Haggett and David Harvey if they would take on the editorship of the new regional science journal to be called *Environment and Planning*.[7] Neither would. Instead they suggested approaching Alan Wilson. He said yes. But the journal also needed a "sub-editor with a suitable background" (Ashby 1993: 3). Wilson again came to the rescue. As Ashby (1993: 3) remembers, Wilson said "he had an assistant at CES, a very bright and promising young woman, recently from Oxford, whom he held in high regard, who could do the job very well … Her name was Doreen Massey. She agreed, and also with a smile".

Ashby doesn't say what kind of smile, however. He implies it was a result of the prospect of earning extra cash (Ashby 1993: 3). It might have been an ironic smile, though. Massey (1993a: 11) says she was "bemused" by the offer. While she recognized the "apparent precision of quantification, and the liberating potential of science" that *Environment and Planning's* regional-science approach promised, she had never been trained in its statistical and mathematical techniques (Massey 1993a: 11). Oxford had almost gone out of its way not to prepare her (Massey 1993a: 11). She might have smiled, then, because she was so underprepared for the job that she had just accepted. Massey (1993a: 11) later said, "I simply didn't understand a lot of the stuff that passed over my desk for the journal. In some of it there seemed barely a word of English between the equations, and what there was hardly of the flowing-prose variety". This was to change, however.

On "one very hot day" at the end of July 1971 she along with Michael Dear boarded a plane at Heathrow to Philadelphia via New York (see Dear 2000; this volume). They were going to Mecca, to the holiest site within regional science's world, to Walter Isard's Department of Regional Science at Penn within the Wharton School of Business. Massey had enrolled in the Master's programme to learn the statistical and mathematical techniques she hitherto had avoided, but crucial to applying science and technical expertise to the postwar conjuncture.

THE PHILADELPHIA STORY

It was even more blistering in Philadelphia once they got off the plane. Massey remembers it was so hot and humid that she would wrap her head in a cold wet towel. She might also have done that because her initial course regimen consisted of linear algebra, multivariate statistics and

---

7. The journal was to be called *Environment*. That title was already taken, however. Consequently, "*and Planning*" was tacked on.

general modelling techniques, some of which she took with Dear (see also Dear, this volume). They stuck with them in part because of their belief in the liberating potential of this kind of science. But at some point, remembers Dear,

> Doreen and I ... look[ed] at this stuff and said what the hell is going on here. [We're] learning statistics, ... linear algebra, ... microeconomic theory – all the things [we]wanted to be there for, but at the same time what ... [were we] doing with all this stuff? (Dear, this volume)

Massey's epiphany came not in Ron Miller's class on matrix algebra, or even Walter Isard's course on general theory, but from an elective course she squeezed into her regional science programme. It was in the French Department, on French philosophy, mainly on the work of Louis Althusser (Featherstone *et al.* 2013: 254). Even in the last interview she gave, in 2013, she continued to talk about that course and its influence. She said, "If there is one person that really influenced me early on, and this is a strange person to cite, it is Louis Althusser" (Featherstone *et al.* 2013: 253). He "utterly changed my view of life ... [It was] the moment when I woke up" (Featherstone *et al.* 2013: 254).

What most struck her reading of Althusser was his anti-essentialism, worked out through the notion of overdetermination, the idea that every element constituting a historical moment is determined by every other element. There is no final cause. As Althusser (1969: 113) put it, "the lonely hour of the 'last instance' never comes". For Massey, the implication would be that "nothing is given" (Featherstone *et al.* 2013: 256). To understand the economy, or for Massey, the geography of the economy, one must always see it as a historical conjuncture, the consequence of multiple determinations across many dimensions – gender, race, class, political institutions, even religious institutions. That was the brilliance of her later *Spatial Divisions of Labour* (Massey 1984b). Alan Warde (1985) argued at the time her approach in *Spatial Divisions* was based on a geological metaphor. It was Althusser who inspired her, not Charles Lyell, and the key was not sedimentary geology but, by way of Freud, the idea of overdetermination (Althusser 1969).

ALL CHANGE: MIND THE GAP

Most immediately, once she completed her MA degree in regional science in 1972 and returned to CES in London, Massey applied her Althusserian framework critically to precisely the kinds of modelling techniques that previously absorbed her. The first product was that paper that had been so influential on me, "Towards a Critique of Industrial Location". She took the Althusserian idea that "nothing is given" and used it as a scalpel to cut away the many "givens" assumed by neoclassical location theory. Neoclassical theory and its deployment in regional science was fundamentally based on accepting universals, like rational choice, from which conclusions were then logically derived. But Massey showed that nothing could be assumed. Maintaining universal givens, as neoclassicism and regional science did, produced only contradiction and inconsistency. Neoclassical and regional science theories were instead historical products, based not on universals but on the overdetermined result of conjunctural conditions. Their function was not to explain the world but to justify it ideologically. That was the prime insight of her paper. After she wrote it, she abandoned her interest in shoring up the postwar conjuncture through any kind of liberating universal science.

Instead, she initially drew on what she called "Alternative Frameworks for Analysis" based on forms of radical political economy (Massey & Batey 1977). At first that alternative was set within the institutional framework of regional science. The internal contradiction was too strong, however. In the introduction to the London Papers in Regional Science volume, *Alternative Frameworks for Analysis* with Peter Batey, Massey said her aim was to "break down ... the boundaries of regional science" (Massey & Batey 1977: 1). She said she was "not engaged in the unearthing of the eternal laws of spatial structure. The existence of such laws is ... rejected" (Massey & Batey 1977: 4). Consequently, she began to forge her own independent path, writing empirically rich and innovative critical analyses of land and capital (Massey & Catalano 1978), job loss (Massey & Meegan 1982), and regional economic transformation (Massey 1984b). Each of these works rejects spatial science as an analytical frame based on universal "givens". Instead, and following Althusser, they recognize various forms of openness, overdetermination and empirical contingency in the relationships of land to capital, in mechanisms of job loss and in processes of regional change.

The larger back story, of course, was the shift in the postwar conjuncture that began to break up just around the time Massey boarded that plane to Philadelphia in July 1971. This was the beginning of a transitionary crisis between conjunctures, the "ruptured unity" (Hall & Massey 2010: 57). The contradictions always present in the immediate postwar conjuncture, but more or less kept in check, became more acute through the 1970s, more destructive and debilitating: rising inflation, increasing unemployment, disinvestment, growing debt, labour strife, falling productivity, deindustrialization, corporate restructuring, power-cuts, three-day weeks. Massey was no longer concerned to contribute to the conjuncture, to patch it up. That possibility no longer existed. Rather, her aim – now post-Althusser, as well as after the postwar conjuncture – was to fashion new concepts to understand the multifarious forms of contradiction that were breaking out all over, and represented by the three books she published between 1978 and 1984.

The new conjuncture that began to replace the old postwar one started to solidify and to take shape from the very late 1970s. It was neoliberalism (Hall & Massey 2010), and signalled by the election of Margaret Thatcher's Tory Party in May 1979. For Massey, this also marked the beginning of the end for economic geography. After Thatcher came into power, and the new conjuncture began to congeal, Massey increasingly believed that the prospects for changing the world through economic geography had dimmed. She spoke of the "blind alley of economic geography" (quoted in Freytag & Hoyler 1999: 88). In part, it was personal. In 1979, CES, where she had worked for more than a decade, was closed down by the Tories in one of their first acts in power. CES could no longer be a site of progressive change. It was structural too. "Thatcherism, neoliberalism, globalisation, … [and] market forces" that constituted the new conjuncture as they bore on economic geography were just too powerful (Hall & Massey 2010: 57). Resistance needed to come from other sites one of which in 1995 became the journal *Soundings* that Massey co-founded with the cultural theorist Stuart Hall and the psychoanalyst Michael Rustin (see Rustin, this volume).

CONCLUSION

In an interview she gave in 2009, Massey said, "I am very wary of any stories of origin; biographies often search for the ultimate root" (Massey *et al.* 2009: 402). In this chapter, I tried to avoid identifying any ultimate root to explain Massey's own early biography. My approach was not to search for roots, but instead to track routes. Massey's travels from Wythenshawe to Oxford to London to Philadelphia back to London again left marks and traces, especially in combination with the larger historical, cultural, political and economic conjunctures in which those travels were undertaken. Massey's early life was overdetermined, multiply caused, produced and changed by the different paths on which she travelled. There was no single point of departure. Hers were multiple origins, shaped by roads not taken as well as those that were. Like any life, it was consequently shot through with contingency, and sometimes seeming contradiction. Her first job in a marketing firm! In their computer department! Editing highly mathematical papers at CES without knowing much maths! Taking a course on the French structural Marxist philosopher Louis Althusser at Penn's Wharton School of Business! As acts, they do not add up. Except of course that they do. They all contributed to making the Doreen Massey I saw that spring day almost 40 years ago, and who I found so compelling, and who I continued to find compelling. A dark past is maybe not so bad.[8]

REFERENCES

For works authored and co-authored by Doreen Massey, please see Select Bibliography of Doreen Massey, beginning on p. 371.

Althusser, L. 1969. "Contradiction and Overdetermination: Notes for an Investigation". In *For Marx.* (French translation, Bernard Brewster.) Harmondsworth: Allen Lane, Penguin, 87–128.

Ashby, J. H. 1993. "Environment and Planning A: A Silver Jubilee Vignette". *Environment and Planning A* Anniversary Issue: 1–6.

Area 1969. "Centre for Environmental Studies". *Area* 1 (1): 10–12.

Blick, A. 2006. "Harold Wilson, Labour and the Machinery of Government". *Contemporary British History* 20 (3): 343–62.

Clapson, M. 2012. *Anglo-American Crossroads: Urban Planning and Research in Britain 1940–2010.* London: Bloomsbury.

Dear, M. 2000. Interview with the author, Santa Monica, CA, January.

Eiloart, T. 1973. "Venture: Measure of Persuasion". *New Scientist*, 2 August: 277–8.

Featherstone, D., S. Bond & J. Painter 2013. "'Stories so Far': A Conversation with Doreen Massey". In D. Featherstone & J. Painter (eds) *Spatial Politics: Essays for Doreen Massey.* Oxford: Wiley-Blackwell, 253–66.

Freytag, T. & M. Hoyler 1999. "'I Feel as if I've Been Able to Reinvent Myself' – A Biographical Interview with Doreen Massey". In D. Massey *Power-Geometries and the Politics of Space-Time.* Heidelberg: Department of Geography, University of Heidelberg, 83–90.

Hall, S. & D. Massey 2010. "Interpreting the Crisis". *Soundings* 44: 57–70.

Harvey, D. 2002. "Memories and Desires". In P. Gould & F. R. Pitts (eds) *Geographical Voice: Fourteen Autobiographical Essays.* Syracuse, NY: Syracuse University Press, 149–88.

Ijams, B., J. Popke & K. Urch 1995. "Gender, Space and the Academy: An Interview with Doreen Massey, The Open University". *disClosure* 4: article 9. http://uknowledge.uky.edu/disclosure/vol4/iss1/9 (accessed 22 October 2017).

O'Hara, G. & H. Parr 2006. "Introduction: The Fall and Rise of a Reputation". *Contemporary British History* 20 (2): 295–302.

Scott, A. J. 1998. Interview with the author, Los Angeles, CA, March.

Telegraph 2016. "Doreen Massey; Geographer Who Examined How Places are 'Socially Constructed'". *The Telegraph*, 21 March: 33.

Warde, A. 1985. "Spatial Change, Politics and the Division of Labour". In D. Gregory & J. Urry (eds) *Social Relations and Spatial Structures.* London: Macmillan, 190–212.

---

8. I am immensely grateful to Michael Dear for his wonderful comments and editorial suggestions, including an amended title.

CHAPTER 4

TRAINSPOTTING IN BETHLEHEM

Michael Dear

On 29 July 1971, Doreen Massey and I boarded a flight from Stanstead airport near London bound for New York City's Kennedy airport, with an ongoing connection to Philadelphia where we had enrolled in the one-year master's degree in Regional Science at the University of Pennsylvania.

In those days, cross-Atlantic flights stopped to refuel in Maine or Newfoundland, but it was not the duration of the flight that lingered in my mind. What I most recall is the overwhelming humidity that struck us like a downpour as we exited the plane onto the tarmac at JFK. The sensation of drowning on dry land stayed with me for the next six weeks, as we laboured through what was called a Summer Mathematics Institute, but we called "remedial maths" (mostly calculus, linear algebra and microeconomics).

The culture shocks induced by adjusting to graduate education in the US were immense, but from the outset we were curious and omnivorous consumers facing the smorgasbord of opportunities offered by a challenging (and welcoming) faculty. But looking back, I see that as we learned together, we shared a growing discontent. Neither of us ended up disappointed in the education that Penn provided, and we never would dismiss or disown what we learned that year. Yet neither of us went on to pursue conventional regional science research.

This chapter seeks to explain these outcomes. It is a personal account of twelve transformative months in our lives and education, beginning from the summer of 1971. Needless to say, our experiences differed, and I make no pretense of accurately recording Doreen's story, nor of providing a comprehensive narrative of my own.

LONDON, 1966–69: INTRODUCTIONS

In the 1960s, London was described as "swinging", which I experienced as an energetic, experimental and optimistic spirit. Harold Wilson was prime minister, expounding what he called a "white-hot" technological revolution in Britain even though much of his time was spent trying to quell the old-time socialists scattered within the Labour Party. Typical weekends saw large and noisy demonstrations against American involvement in Vietnam. The Beatles became a sensation in the US, and the first episodes of *Star Trek* arrived on British television.

I came to London in 1966 to work on transportation policy and planning at the Greater London Council (GLC), a regional administrative government. One year later I was seconded to University College, London (UCL) to study for the MPhil degree in Town Planning. The course content largely followed the requirements of professional accreditation, which hindered innovation, but many visiting lecturers brightened the classroom, most notably Vida Nichols (later Godson), Alan Wilson and Allen Scott. Vida had recently returned from a year at Penn's regional science programme, and she gave an impromptu series of lectures on methods at UCL's Bartlett School. Alan Wilson taught a course there on transportation modelling. He was director of a newly established research unit called the Centre for Environmental Studies (CES), where Doreen Massey was on staff. I first encountered her through a 1968 publication in the CES Working Paper series, entitled "Problems in Location: Linear Programming". Allen Scott arrived at UCL en route to take up an appointment at the University of Toronto, but fortunately stayed long enough to teach a summer course on computer programming and statistical analysis, and to lodge the idea of study at Penn in my head.

This cast of characters came together during the summer of 1968 at the first London meeting of the Regional Science Association (RSA), which was a swell affair arranged by Scott. The roster of participants included many Penn personalities, including Walter Isard (in an unforgettable performance) and David Boyce (transportation modeller and big wheel in the RSA). I received lots of encouragement to apply for a master's degree at Penn, and Doreen did too.

LONDON, 1969–71: PREPARATIONS

After graduating from UCL, I returned to the GLC for two years to repay their investment in my education (for which I shall be eternally grateful). I ran into Doreen occasionally when visiting the CES, and kept track of her work through the working papers she produced between 1968 and 1971. These are important markers because they represented how she was exploring and evaluating the regional science "toolkit" which existed at that time.

Following her 1968 inquiry into linear programming, in quick succession Doreen authored or co-authored papers entitled:

- "Some Simple Models for Distributing Changes in Employment within Regions" (Massey 1969, with its now-quaint cover attribution: 'By Miss D. Massey');
- "Towards Operational Urban Models" (Cordey Hayes *et al.* 1970); and
- "The Basic: Service Categorization in Planning" (Massey 1971).

These careful explorations provide a sense of Doreen's evolving mission, which was to examine how the new techniques could be harnessed to the deeper purposes of public policy for land-use planning and community development. This was a time when everyone was avidly learning, and the learning curve was steep. I was independently writing reports at the GLC exploring the potential of linear programming, gravity and residential location models for regional planning. (I believe that my efforts were lost when Margaret Thatcher abolished the GLC in 1986.)

One other thing about those early, somewhat starchy CES working papers was that Doreen's sense of humour was already poking through. In WP-63, "The Basic: Service Categorization in Planning", her opening paragraph in its entirety reads:

> Much has been written on the basic:service categorization and especially the distinction founded in export base theory. The author is aware of this. (1971: 7)

Then the next paragraph shifts unequivocally: "The purpose of this paper, however, is rather different" (1971: 7). I do not know the exact story behind these caustic declarations, but evidently Doreen had run into criticism for her approach to the topic, and was taking this headline moment to underscore (a) that she knew the old stuff, and (b) that she was attempting something new. Her use of the academic third-person was a grinning fig-leaf.

The European meeting of the Regional Science Association took place in London during the summer of 1970. Not long after, Doreen and I prepared our separate applications for graduate studies at Penn. We kept in close touch, and were eventually accepted into the programme, Doreen with fellowship support and me with a research assistantship to work with Colin Gannon.

PHILADELPHIA, 1971–72: REGIONAL SCIENCE AT PENN

The summer heat and humidity did not diminish, and the first couple of months in Philadelphia were predictably chaotic. Doreen settled into International House, which seemed to be a pretty luxurious student dorm,

Photo 4.1 "Pennsylvania R.R.", cigarette card, no. 26 from a series of "50 Railway Engines" (*c.* 1924) [author's collection]

and I lived among herds of cockroaches at 43rd and Pine in West Philly. Confusion seemed to be our default emotion. We were impressed, for example, by the number of buildings emblazoned with Greek letters on campus streets – amazed that Classics studies should be so prominent at an American university.

We slowly came to terms with this strange new world by taking trips of discovery into Pennsylvania's industrial heartland. There we actually saw the famous "Bethlehem Steel" at work, at night, like some giant off-world cosmos. We closely observed trains with impossibly romantic names such as "Erie-Lackawanna Railroad", the "Atchison, Topeka and Santa Fe Railway", and the "Chicago, Rock Island and Pacific Railroad". Such titles had the force of magic because they had cast spells when we were kids in Britain, lifting our imaginations westward across the Atlantic. And now we were here.

Classes started on 9 September. A lot of courses were required for the master's degree, so we were basically on the same schedule, with classes in linear algebra, statistics, and something called "regional and social science theory", which was basically microeconomics. In addition, there was ongoing calculus instruction. This latter course just appeared on the schedule; I do not remember anyone telling us to do it, but we went anyhow. After all, this retooling was why we had come to Penn. On the other hand, the very notion of a *three-hour* seminar was alien to us, not only because three hours was an entire morning but also because the courses in that semester were almost uniformly didactic. There was not much of the critical learning or exchange that I expected of a "seminar", and sometimes it was hard going. Doreen was an inveterate doodler; I knew that when she stopped I should pay attention.

Time spent with friends outside the classroom helped to maintain our sanity and expand our knowledge of the US. A great friend and neighbour, Ken Ballard (a Californian in the economics programme), used to take us cruising in his Ford Mustang (yes!!) to exotic locales such as the Pine Barrens in New Jersey, the Pennsylvania Dutch country, and Valley Forge, the old battleground. Philadelphia is host to the birthplace of American independence, and we listened to a guide animatedly – and at great length – berate the colonial British. At the end of her presentation, the guide asked her international audience: "Is anyone here from Britain?" We kept our mouths shut, and sat on our hands.

The spring semester was a major upgrade. As usual we took the required classes, including location theory, spatial statistics, input–output models, etc. We were familiar with the routines by now, and classes were smaller, allowing for more give-and-take. Doreen took a class with Ben Stevens on residential location models, and looked at other courses outside regional science. Together with Nathan Edelson, another fine friend, we set up a small study group in statistics.

Part of the Penn programme that is too frequently overlooked is that many Master's and doctoral students at the Regional Science Department were actively involved with faculty research or consulting projects that had an applied focus. For example, as well as my assistantship with Colin Gannon to study the land-use and employment impacts of the Philadelphia-Lindenwold (NJ) high-speed transit line, Julian Wolpert set me up to work on an abandoned housing project for the Association of American Geographers' Metropolitan Analysis Project.

Exposure to such action-oriented research was an invaluable counter-weight to classroom theory and method, and not just because it helped pay the rent. We worked with real communities on real issues. Students at all levels were exposed to a research culture that included working as part of a larger team with direct responsibility for time-sensitive, project-related deadlines. Report-writing was mandatory, as was participation in the weekly rounds of project management. Preparing papers for formal publication was part of the job. More advanced students were often roped into preparing proposals for new funding.

Looking back, I think these project-related experiences provided insight into social science theory and practice that could only be gained beyond the classroom, providing important demonstrations of the limitations of regional science-style modelling. I have continued to draw on the experience of this theory/practice dialectic throughout my academic life, even (maybe especially) when I later moved into more humanities-oriented research and teaching. I will have more to say about Doreen's response to this dialectic in the next section.

There were other reasons why the second semester was memorable and consequential. Doreen still lived at International House among a large contingent of overseas students. Partly through them, we got involved in a variety of extracurricular seminars, readings and discussions. Louis Althusser made a big impact on Doreen at this time, plus there were myriad seminars on Gramsci, Kolakowski, Poulantzas and Miliband on the capitalist state, Bertell Ollman, the Frankfurt School and so on. It was non-stop, and all the more riveting because it was *not* regional science.

This last observation is not a complaint against regional science. Our extracurricular activities focused on an alternative Marxian epistemology that was ostensibly at least as comprehensive as that being offered by the Regional Science Department. More and more, we began juxtaposing a Marxian political economy (broadly understood) against regional science's powerful and appealing vision of a social science. We brought this *second* dialectic (the first was theory/practice, just noted) into our home department, where for the most part it was welcomed and engaged.

Certainly the most enjoyable instance of the ensuing epistemological exchanges occurred when Walter Isard was going through his *General Theory* (1969), which he and Tony Smith had spent years putting together. Isard always generously gave us the space to attack his beloved meta-theory. One day, as was our custom, Doreen and I were sitting next to each other, voicing our objections. The problem with today's sub-model, we opined, was that it did not incorporate emotions such as love. Walter pushed back, quite cleverly, by explaining how to develop the framework of a formal model to accommodate non-quantifiable intangibles. After some exchange, he meandered over to the equation-strewn blackboard and added a letter L, for love, at one corner of the scrawl (remembering to add expansionary sub- and superscripts!). Everyone laughed out loud at this solution, including Walter, understanding that it represented more of a truce than a treaty.

Isard's quest to build a single comprehensive mathematical model of society was misguided, we were certain. But we were simultaneously – and not coincidentally – reforming ideas about the limits of Marx, or more precisely, the way some people were those days using Marx. The intellectual audacity behind the two projects was inspiring, even though both were (there is no better way to say this) brutal in their oversimplifications and reductionism.

How *bratty* Doreen and I had become! At one point, we simply refused to do an exam because we judged that it was completely unconnected to any real-world concerns. (I thank Julian Wolpert for talking us down off this high ledge.) On another occasion we walked together out in the middle of an exam because she was not feeling well. I wince now when I recall these and similar events. How on earth did they put up with us at Penn? I suspect that some faculty members enjoyed our irregular behavior (after all, the beguiling seductions of a Cobb-Douglas production function are not infinite). In any event, I record here my appreciation of their forbearance, which helped us grow.

By summer's end in 1972, it was obvious that Doreen would be separating from mainstream regional science. She simply saw little of value or use in the more esoteric reaches of social-science modelling, and even though she still admired the spirit and purpose behind operational models and methods (such as input–output analysis), she now understood more about their limitations. I felt the same way. Both of us valued the regional science experience as an essential building-block in our education, one that would never be disowned or forgotten. Doreen returned to London, but I stayed at Penn to begin my doctoral studies in a different kind of regional science, with Julian Wolpert (Dear 1995).

My last Philadelphia dinner with Doreen was on 1 July 1972. After a summer of research in Massachusetts, I returned to Penn in late August, but by that time Doreen had already departed.

In November of that same year, Richard Nixon won a landslide re-election as President of the United States. Two years later after Watergate, he would resign the presidency in disgrace rather than face impeachment. In 1976, Margaret Thatcher became leader of the British Conservative Party.

AFTERMATH: 1972–2014

Back in London and her job at the CES, Doreen began publishing a spate of pathbreaking books which continued throughout her lifetime. One of her earliest major projects was an edited volume with Peter Batey, *Alternative Frameworks for Analysis* (1977), which contained selected papers from two RSA meetings held at London in 1975. Conference proceedings can be pretty blah, but the message conveyed by the editors of this volume was unequivocal, capturing the shifting intellectual climate of the mid-1970s, and Doreen's place in it.

The "Introduction" to *Frameworks* clearly distinguished the orientations of the two sets of papers in the collection: from the North West European Multilingual RSA; and from the annual meeting of the British Section of the RSA. The first set was framed as an explicit challenge to the established ideological and theoretical bases of regional science, originating from current Western European research and a praxis that was more policy-oriented, left-leaning and empirical. The second set of papers was characterized by the editors as representing more traditional preoccupations in regional science. The exceptions were two papers that questioned the mainstream, one being my article on "Spatial Externalities and Locational Conflict", a revision of public-facility location theory that examined how externalities and politics distorted the equitable distribution of public resources and infrastructure (Dear 1977).

Doreen and Peter closed their editorial introduction by synthesizing the "central characteristics" of an alternative regional science (1977: 4–5):

- that "attention must be paid to the analysis of present developments and changes [in the real world]";
- that there was no intention "to unearth eternal laws of spatial structure. The existence of such laws is indeed rejected";
- that any "'new' schools of thought must not be seen as constituting a single theoretical approach; nor one which can be provided 'ready made' [to different situations and places]"; and
- that "every distinct theoretical approach has its own concepts and terminology, and these must be understood and grasped before any evaluation can be said to be adequate".

And there you have it: a concise summary of the legacy of Doreen's thinking at Penn in the form of a prescient and forward-looking manifesto. It bore the fruits of our two Penn-based dialectics: theory/practice, and Marx/social science. In a nutshell, it prioritized application over theory, and a Marxian-informed empiricism over mathematics. The manifesto anticipated many of the intellectual upheavals that characterized the last quarter of the twentieth century, in which Doreen played a prominent role. Let me quickly sketch one instance where the Penn legacy and our friendship merged during these subsequent years.

In the early 1980s, plans were afoot to start a new journal that responded to the call for reconstructing social theory, method and practice. The first issue of *Environment and Planning D: Society and Space* appeared in 1983. I was its founding editor and Doreen straightaway accepted an invitation to become a member of the editorial board. She served in that capacity for almost a quarter-century until 2006, through two editors (my successor was Gerry Pratt).

Doreen's most memorable contribution to *Society and Space* during my term was her essay on "Flexible Sexism" (I still love that title). I had encouraged her and Rosalyn Deutsche (then at the Cooper Union's School of Art) to publish a feminist critique of the postmodern visions authored by David Harvey (resolutely anti-pomo) and Edward Soja (pro-, but with conditions). Doreen's essay is brilliant, ranging widely over several disciplines, postmodernism, feminism, art, film and politics. Deutsche's polished critique was felicitously entitled "Boys Town". The papers were published together as openers in the 1991 volume of *Society and Space*. Their impact was immediate, and both remain essential reading today.

Photo 4.2 Doreen, Topanga Canyon, Los Angeles, early 1990s [author's photograph]

In subsequent years, I saw Doreen intermittently, usually in conjunction with some conference or other. She enjoyed hanging out in the sunshine of Los Angeles, and early on understood LA as a world city warranting serious academic attention and political analysis. Later, after my move to Berkeley, we met in Tampa, Florida, where she was getting an award from the American Association of Geographers. Our lunchtime reunion pleasurably merged into an afternoon of gossip, reminiscence and politics – the kind that is possible only with long-time friends. Derek Gregory joined us, and stayed. The gossip quotient rose to critical levels of toxicity.

KILBURN: 18 MAY 2014

I last met Doreen at a cafe in north London on a bright morning almost warm enough to sit outside. We talked about the Kilburn Manifesto and the recent loss of Stuart Hall. She told me she had stopped watching Liverpool FC because the games made her too anxious. We compared the recent urban experiences of London and Los Angeles, her work in Central and South America and mine in Mexico, and regretted that neither of us had achieved a satisfactory fluency in Spanish. Talk drifted to how we had been drawn into more artistic, cultural and humanistic endeavours. Such career-long overlaps might strike others as verging on the uncanny, but neither of us seemed surprised by how much we had continued along shared paths.[1]

REFERENCES

For works authored and co-authored by Doreen Massey, please see Select Bibliography of Doreen Massey, beginning on p. 371.

Cordey Hayes, M., T. A. Broadbent & D. Massey 1970. "Towards Operational Urban Development Models". *Working Paper* 60, Centre for Environmental Studies, London.

Dear, M. 1977. "Spatial Externalities and Locational Conflict". In D. Massey & P. Batey (eds) *Alternative Frameworks for Analysis*. London: Pion, 152–67.

Dear, M. 1995. "Reinventing Regional Science". *International Regional Science Review* 17 (3): 355–60.

Deutsche, R. 1991. "Boys Town". *Environment and Planning D: Society and Space* 9 (1): 5–30.

Isard, W. & T. E. Smith 1969. *General Theory: Social, Political, Economic, and Regional with Particular Reference to Decision-Making Analysis*. Cambridge, MA: MIT Press.

---

1. Thanks to Trevor Barnes and Jennifer Wolch for helpful advice on an early draft of this essay.

CHAPTER 5

BECOMING A GEOGRAPHER: MASSEY MOMENTS IN A SPATIAL EDUCATION

Gillian Hart

Reading the "locality debates" in the late 1980s and early 1990s centred on Doreen Massey's work propelled my transformation from an economist to a geographer – along with my conception of the world more generally. I was utterly compelled by her feminist reformulation of space and place, which came to me at a crucial conjunctural moment: the end of the Cold War; the apartheid regime's unbanning of the African National Congress and other political parties; and returning to my native South Africa in 1990 after an absence of 19 years. It has profoundly shaped my research since the 1990s, and remains central to my teaching and political engagements.

Going back to re-read some of Doreen's work for purposes of this chapter has reaffirmed her powerful influence – but it has also made clear to me our different relations to Marxism, and how they have diverged more widely since the mid-late 1990s. Yet reflecting on these differences and divergences itself represents yet another moment in a spatial education – one that has pushed me to think more carefully about changing interconnections of political and analytical commitments in different spatiohistorical conjunctures.

MOMENTS OF CONVERGENCE

Let me start with a brief account of how Massey appeared on my radar through the locality debates. During the height of debate in the late 1980s and early 1990s, I was teaching at MIT where a different though related debate was raging around what the industrial future would look like after the implosion of the Fordist–Keynesian compromise and, what in retrospect, we can see as the neoliberal onslaught. On one side were Michael Piore and Charles Sabel, with their celebratory account of what they called flexible specialization in *The Second Industrial Divide: Possibilities for Prosperity* (1984), and their insistence that "industry should abandon its attachment to standardized mass production" and emulate small-scale, innovative forms of modern-day craft production such as those in central and northwestern Italy. Fiercely contesting notions of flexible specialization, Ben Harrison maintained that "contrary to prevailing wisdom, the big firm is not only alive and well but is becoming more flexible and efficient". His book *Lean and Mean: The Changing Landscape of Corporate Power in the Age of Flexibility* was only published in 1994, but by the late 1980s MIT had become a battleground on which the many students working on industrial restructuring felt compelled to line up behind one side or the other.

At the time I was teaching a graduate seminar on agrarian debates, going back to Lenin's and Chayanov's sharply opposed interpretations of the Russian *zemstvo* statistics in the late nineteenth and early twentieth centuries, and then circuiting through literature on agrarian transformations in Africa, Asia and Latin America. Students were quick to point out parallels with the Piore/Sabel vs. Harrison debate, and also to identify how these bodies of agrarian literature offered conceptual and methodological resources for engaging contemporary questions of industrial restructuring. I vividly recall wonderfully animated discussions in which we were all making what felt like new and exciting connections.

Students also led me to Massey and to geography. Late one afternoon a group of them showed up in my office carrying piles of blue-bound journals with bright pink Post-it notes indicating articles they thought I should read and the order in which I should read them. These were, it turned out, key interventions in the locality debates. On multiple occasions I have regaled my Berkeley colleagues with the story of how I was immediately captivated – and of how this was my moment of conversion to geography. While partly true, this is also something of an exaggeration. On going back to re-read at least part of the locality debates for purposes of this chapter, I realize that my conversion was a process that extended over the first half of the 1990s, in which Massey's contributions beyond the locality debates were also vitally important.

At the centre of the locality debates was a research programme entitled *The Changing Urban and Regional System in the UK* (CURS) in which Massey was closely involved, and which grew out of her book *Spatial Divisions of Labour* (1984b). In describing the political imperatives that drove CURS, Massey observed that, in the mid-1980s,

> Across the political spectrum, causal connections were being made between changes in employment and occupational structure and wider social, ideological and political changes. We were facing the end of the working class, the end of class politics, a new ideology of individualism, a politics of consumption, the

dominance of what were referred to as "new social movements". All this was being argued, most frequently, from national level statistics. Yet, quite apart from the difficulty of establishing such causal connections in the first place and the dubiousness of the economistic form in which they were usually proposed, the issues of spatial scale and spatial variation were usually ignored ... Something that might be called "restructuring" was clearly going on, but its implications both for everyday life and for the mode and potential of political organizing were clearly highly differentiated and we needed to know how. It was in this context that the localities projects in the United Kingdom were first imagined and proposed. It was research with an immediate, even urgent, relevance beyond academe.   (Massey 1991c: 269)

In my research in rural Java, it was precisely such sweeping claims about agrarian change on the basis of aggregate national level statistics that drove some of my battles with economists, leading me to show how the same sets of data could be interpreted in entirely different ways – and with important political stakes – depending on arbitrary assumptions about labour markets. Hence my argument about the importance of in-depth ethnographic and historical understandings of the intertwining in practice of labour relations with land, credit and other relations of power. Yet I still grappled with the question of how my intensive year-long study in a single village could be used to make more general claims. Along with many others, I thought in terms of relations between "micro" and "macro" levels, and recognized the need for historical and comparative work – but this framing remained deeply unsatisfactory because, as I came to see in retrospect, I lacked a critical understanding of spatiality. What was so immediately captivating about the locality debates was that they were grappling with very similar issues.

I was also intrigued by the intensity of debate. The CURS initiative quickly came under attack, with Neil Smith lobbing the opening salvo. He took aim at CURS for its "reluctance to generalize about the experience of restructuring" (1987: 63), pointing to the danger that CURS would do little more than repeat the empiricist studies of an earlier generation that refused to draw out theoretical or historical conclusions. Smith was also deeply critical of the selection of localities on grounds of scalar incomparability: "like the blind man with a python in one hand and an elephant's trunk in the other, the researchers are treating all seven localities as the same animal" (1987: 63). Lamenting the retreat from Marxist theory, Smith asserted that there is "nothing inherently or intellectually superior about the unique and the complex" (1987: 67). In the same issue of *Antipode*, Philip Cooke, the coordinator of CURS, quickly sprang to its defence with comments about the limits of Marxist theory and what he defined as the CURS strategy of "generalization within cases" (Cooke 1987) that served to fan the flames of dissent. David Harvey also entered the fray in 1987 in an issue of *Society and Space*.

Fierce and wide-ranging debates raged over the following four years. In their introduction to a special issue of *Environment and Planning A* in 1991, Duncan and Savage described the locality debates in terms of "the relation between theory and empirical research, the role of Marxism and postmodernism in social science, the difference that space makes, case studies and comparative research, economism versus culture, the contribution of realism, the definition of social objects, the boundary problem" (1991: 156). Their purpose was "to broaden the debate away from the narrow track to which it has recently been confined, and to indicate the wider conceptual and political issues which need to be introduced into the debate" (1991: 163).

In fact, it seems to me, in their contributions to this special issue Doreen Massey and Andrew Sayer effectively brought the locality debates to a close. In different though related ways, they both showed how a large chunk of the debates hinged on problematically aligning and conflating sets of dualisms:

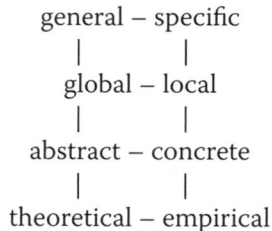

$$
\begin{array}{cc}
\text{general} - \text{specific} \\
| \qquad | \\
\text{global} - \text{local} \\
| \qquad | \\
\text{abstract} - \text{concrete} \\
| \qquad | \\
\text{theoretical} - \text{empirical}
\end{array}
$$

It is *not* the case, Massey pointed out, "that the study of locality is a necessary vehicle for, nor equivalent to, empirical research or the study of concrete phenomena" (1991c: 270). Thus, she goes on, the global economy is *general* in the sense of being a geographically large-scale phenomenon, to which can be counterposed internal variations. But it is no less *concrete* than a local one, in the sense of the product of many determinations (which, one might add, is a distinctly Marxist concept): "Those who conflate the local with the concrete, therefore, are confusing geographical scale with processes of abstraction in thought" (Massey 1991c: 270). Massey made two other moves in this essay, both of which she expanded and clarified in later publications that I discuss below: first that locality studies "are not *necessarily* part of the turn to the postmodern" (1991c: 272), and second a critique of Harvey's (1989) concept of place.

In his essay, Sayer usefully elaborated Massey's point about the problematic conflation of the local and the concrete (1991: 289–91), along with deconstructing a number of the other dualisms that underpinned the locality debates. He also engaged questions of generality in a way that I continue to find powerfully useful by distinguishing positivist from relational concepts of generality. Generality$_1$ refers to statistical representativeness, and is associated with positivist conceptions of discrete, bounded objects that stand in external relations to one another. Generality$_2$, in contrast, turns on a conception of internal relations in which differentiation and particularity arise from *interdependencies*. Hence, Sayer argued, participants in the localities and other debates have been talking past one another by using different concepts of generality:

> Thus, when we ask whether certain research findings from a particular case are "generalisable" we could answer in terms of generality$_1$, that is according to whether identical or similar findings are common elsewhere. In the case of research on localities the answer might often be negative. Yet, even if we thought that nothing was generalisable in this sense, it would not follow that the implications of the study were merely parochial and of no relevance for wider society, for they might be generalisable in terms of generality$_2$; that is, the particular or the unique might be internally related to some aspects of the whole or other parts of the system. In this second sense it is possible to argue that (some aspects of) the whole are "contained" in the part and even that the part imprints onto or structures the rest of the whole. For this reason, locality studies need not be solely of parochial interest.
>
> (Sayer 1991: 298)

Massey's and Sayer's interventions cut through much of the labyrinthine underbrush in which the locality debate had become entangled, and cleared the way for much sharper and clearer understandings of how intensive studies of specific localities can illuminate broader processes. In retrospect it's striking how contemporary Urban Studies debates over what Cindi Katz (2017) in her wickedly funny way calls "splanetary urbanization" (as in mansplaining) are traversing some of the same terrain as the old locality debates in a similarly gendered fashion.

On going back to re-read the locality debates, I can also see in retrospect how several further sets of interventions by Massey were crucial to my own efforts to come to grips with these questions: her *New Left Review* article on "Politics and Space/Time" (1992b); her critical engagement in "A Global Sense of Place" (1991a) and "Power-Geometry and a Progressive Sense of Place" (1993b) with Harvey's (1989) deployment of Heideggerian concepts of place and "the local"; and the superb General Introduction to *Space, Place, and Gender* (1994). Taken together, Massey's moves beyond the localities debates were central to my subsequent research, teaching and sense of myself as a geographer – along with a growing appreciation of the political stakes of critical conceptions of spatiality.

I vividly recall my excitement on reading "Politics and Space/Time", and the sense of being catapulted well beyond the locality debates. Most immediately I was struck by the powerfully elegant way Massey showed how both Ernesto Laclau and Fredric Jameson invoke seemingly opposite conceptions of space and time in relation to politics, while relying on similarly problematic dualistic (and Cartesian) counter-positions of space and time. She then went on to underscore the inherently gendered character of such dichotomies – and thence to sketch an alternative "view from physics", showing how Laclau's and Jameson's dualistic conceptions of space and time accord with those of Newton, in contrast to her own relational conception akin to that of Einstein: "It is not that the interrelations between objects occur *in* space and time; it is these relationships themselves which *create/define* space and time" (1992b: 79; emphasis in original).

It was this essay, along with Neil Smith and Cindi Katz's (1993) equally lucid exposition of the limits of spatial metaphors, that drew me to Lefebvre and *The Production of Space* (1991). In re-reading at least some of Massey's work for purposes of this essay, I'm struck that "Politics and Space/Time" is the most explicitly Lefebvrean piece of hers of which I am aware. In *For Space* (2005), for instance, Lefebvre makes a brief appearance in the introduction to a chapter on Bergson, Deleuze, Laclau and de Certeau – and then falls out of sight in the remainder of the book. What seems to be at stake here is what Arun Saldanha (2013: 44) calls Massey's "ambivalent relation ... with the Marxist legacy", on which I will reflect later in this chapter.

Also representing a leap beyond the locality debates was Massey's insistent refusal to separate space and place, and her elaboration in the early 1990s of an "extraverted sense of place" (1991a, 1993b, 1994) not as a bounded unit, but nodal points of interconnection in socially produced space:

> If ... the spatial is thought of in the context of space-time and as formed out of social interrelations at all scales, then one view of a place is as a particular articulation of those relations, a particular moment in those networks of social relations and understandings ... But the particular mix of social relations which are thus part of what defines the uniqueness of any place is by no means all included within that place itself. Important, it includes relations which stretch beyond – the global as part of what constitutes the local, the outside as part of the inside. (Massey 1994: 5)

In the first instance this formulation grew out of Massey's vigorous critique of Harvey's (1989) assertion of any focus on place and "the local" as necessarily reactionary. Subsequently of course it remained central to her work, even as she moved in the post-Marxist direction charted by Laclau and Mouffe – while, at the same time, insisting on "thinking radical democracy spatially".

Massey's explicitly feminist formulations of space and place constituted for me an incredibly powerful set of tools that were simultaneously analytical and political. Most immediately relevant was their close congruence with what Sayer (1991) called Generality$_2$, enabling what I have come to call critical ethnography and relational comparison (Hart 2002, 2006, 2016). Yet despite a close sense of affinity with her signal contributions, my efforts to further hone and elaborate these tools in relation to the political engagements and challenges I have confronted – mainly although by no means exclusively in post-apartheid South Africa – have taken me in a different direction from that in which Massey moved following the locality debates. In reflecting on how our trajectories have diverged in recent years, I've come to realize that it's necessary to reach further back in space-time to earlier conjunctures in each of our lives.

MOMENTS OF DIVERGENCE

In discussing Massey's political-intellectual formation, let me start with her own account in "'Stories So Far': A Conversation with Doreen Massey" in *Spatial Politics: Essays for Doreen Massey* edited by David Featherstone and Joe Painter (2013). In response to a question about key inspirations and influences, she began by emphasizing the political movements in which she had been engaged in the late 1960s and 1970s "with the emergence of Marxism, feminism, sexual liberation, being part of the GLC [Greater London Council] in the 1980s, or the kind of stuff that has happened more recently" and the urgent debates they provoked. But, she went on to say, "If there is one person that really influenced me early on, and this is a very strange person to cite, it is Louis Althusser" (Featherstone & Painter 2013: 253). Her affinity for Althusser, she explained, grew from her alienated response to readings of early Marx – more specifically, from "its intimations of a human nature, and as a feminist I couldn't buy it. So much of it was very essentialist about sexual divisions of labour and 'natural' divisions of labour. The heterosexual family was treated completely unproblematically. And so I found it difficult to buy into Marxism, even though I was strongly committed to issues of class" (Featherstone & Painter 2013: 254). She explained how Althusser "utterly changed my view of life and of Marxism". First, was his anti-essentialism embodied in the phrase "There is no point of departure": "as a young woman who was trying to escape the norms, who felt she didn't conform to any of the given descriptions of 'woman', and who wanted a way of challenging them – that first entry into anti-essentialism ... was utterly important" (Featherstone & Painter 2013: 254). She spoke as well of her affinity with Althusser's critique of the economic determinism of the French Communist Party.

Responding to a question about regional inequality, she called attention to *Spatial Divisions of Labour* as an expression of thinking relationally, which is "one of the things I have most taken from Marx" (Featherstone & Painter 2013: 257), and discussed how it grew out of her concern with regional inequality and uneven development – which, growing up in the northwest of England, she had "lived with, through and kind of in combat with" since childhood. The book was about trying to conceptualize a relational geography of power with the capitalist structure of class – and from the perspective of class. By the early 1990s with "A Global Sense of Place" the emphasis had shifted to other dimensions of difference, especially ethnicity: "It's an interesting shift and reflects a more general move within geography and the social sciences away from class and towards, especially, hybridity." She went on to say, though, that "Personally I think it is time for that balance to be redressed", pointing to "the shift from a social democratic Keynesian hegemony to a neoliberal one" (Featherstone & Painter 2013: 257).

Discussing the significance of concepts of "hegemony" and "social settlements" to her work, she reflected

> I guess another set of influences has come from Gramsci, or from a Gramscian school of thought, especially around Stuart Hall, Chantal Mouffe, Ernesto Laclau and others. Decades ago (I think it was in the early '80s) we were all members of a group called "The Hegemony Group" – another very challenging discussion forum ... It was related to that wider move to take culture and power more seriously – and the whole notion of the *construction* of a society and of its common sense; the way different instances both had a degree of autonomy and intersected; and of course the possibility of those moments of conjunctural rupture when the balance of social forces may be put in question and changed.
>
> (Featherstone & Painter 2013: 258)

She went on to note that

> through all of this I have been trying to weave a thread about the relation between space and power, about the nature of space and

the nature of place. The notion of hegemony, for instance, implies both place and a particular – contested – notion of place.

(Featherstone & Painter 2013: 258)

In the discussion of Gramsci and hegemony she specifically called attention to the work of Stuart Hall on Thatcherism, and noted that at *Soundings* they were trying to engage that kind of analysis again. This was, of course, the Kilburn Manifesto, launched in 2013.

Stuart Hall's work in relation to that of Gramsci has also been central to my own intellectual-political formation – although it has taken me in directions that diverge from Laclau and Mouffe, and from some of Massey's later work. These divergences derive from the conditions in which I came to embrace feminism in apartheid South Africa; from engagements with the US anti-apartheid movement from the late 1970s; and from my efforts to comprehend and participate in the twists and turns of post-apartheid South Africa since 1994 in which the past is far from dead.

Like Massey I was swept along by second-wave feminism in the late 1960s. Yet I also found it difficult to reconcile with my position of race/class privilege at the height of the most vicious period of apartheid, and lacked the conceptual resources to grapple with these tensions. It was only in the 1980s, when I became aware of the statement by the Combahee River Collective, that I was able to start confronting these challenges. Another important set of influences in my efforts to come to grips with feminism in relation to race and class was participating in discussions at the Bunting Institute at Radcliffe in 1983–4 that culminated in an innovative collection of essays on the intertwining of race, class and gender entitled *Women and the Politics of Empowerment* (1988), edited by Ann Bookman and Sandra Morgen.

From the late 1970s, by far the most important political commitment for me was the anti-apartheid movement. On arriving at Cornell at the beginning of 1972, I discovered staggering ignorance about apartheid South Africa. That changed in 1976 with the Soweto uprising, and the murder by the apartheid state of Steve Biko in September 1977 served to thrust South Africa further to the forefront of popular attention. It also provided the impetus for the movement focused on divestment and sanctions. At Cornell I participated in setting up an anti-apartheid movement in 1977–8, and then worked with the African National Congress when I moved to Boston in the fall of 1978. It was the divestment movement on campuses throughout the US – as well as churches and trade unions – that paved the way for sanctions with the passage of the Comprehensive Anti-Apartheid Act of 1986, when a number of Republicans joined Democrats in overriding Ronald Reagan's veto of the bill.

Through the 1980s, living in the Boston area, I was split between an Asian academic side and a South African political side that involved mobilizing students and teaching undergraduate courses on South Africa. As part of this activism I paid close attention to South African race v. class debates, and how they had evolved since I left the country at the end of 1971. On one side of this debate, in its crudest form, was a liberal argument claiming that racial oppression was an archaic holdover that would dissolve as black South Africans were drawn into the economy as producers and consumers – in other words, that apartheid was antithetical to capitalism. Essentially this was the argument invoked by many US university administrators and others opposing divestment of shares of companies operating in South Africa. Drawing on the so-called Sullivan Principles drawn up by a segment of corporate America, they asserted in good paternalistic fashion that divestment would hurt black South Africans more than it would help them – which was, of course, an entirely specious claim.

The counter-argument in the race v. class debate was that apartheid was functional to capitalism. Put forward most forcefully by Frederick Johnstone in a 1970 article entitled "White Prosperity and White Supremacy in South Africa Today", this argument was instrumental in my decision to leave South Africa in 1971. Yet I had also become painfully aware of the inadequacies of this position, which derived from its grounding in dependency theory, for understanding the massive uprisings against both capital and the apartheid state, starting with the resurgence of a militant labour movement in Durban in 1972 and gathering force over the 1970s and 1980s. It was this dynamic that US students urgently needed to grasp in order to situate their activism – and the race v. class debates, in their crude as well as more nuanced forms, were singularly unhelpful.

This is where Stuart Hall stepped into the picture. In 1980 Hall published an extraordinary essay entitled "Race, Articulation and Societies Structured in Dominance" that effectively transcended the South African race v. class debates, and for me has been profoundly formative. At the time, Harold Wolpe's "Capitalism and Cheap Labour Power in South Africa" (1972) represented by far the most sophisticated contribution to these debates, framed in neo-Althusserian terms of articulation of modes of production. By redefining the concept of articulation and shifting it to a Gramscian terrain, Hall's analysis of the relations between racism and class opened up vitally important political and analytical possibilities (Hart 2007). Hall's essay may well have been shaped by discussions in "The Hegemony Group" to which Massey makes reference, which included Laclau and Mouffe. Very importantly, though, it was also a product of Hall's deep association with the ANC in exile, with the British Anti-Apartheid Movement, and with Harold Wolpe, who escaped from an apartheid prison in 1963 and fled to Britain where he and Hall became close friends.

Hall's essay pointed me in directions quite different from the move that Massey made – along with Laclau, Mouffe, many other former Althusserians, and in part Hall himself in his later years – to poststructuralism(s) broadly conceived. First, reading Hall's 1980 essay in conjunction with his explication of Marx's notes on method in the 1857 Introduction to the *Grundrisse* (2003 [1973]) enabled an open understanding of dialectics sharply at odds with any sort of Hegelian teleology. Second, it was through Hall that I came to take Gramsci very seriously. In the process I came to a distinctively different reading of Gramsci from that of Althusser, Laclau and Mouffe – but closely compatible with the brilliant explication of Fanon's work by the Ghanaian scholar Ato Sekyi-Otu (1996). Speaking of Gramsci as a "precocious Fanonian", Sekyi-Otu foresaw with remarkable prescience the directions in which post-apartheid South Africa would move. Both theoretical orientations contributed to my growing recognition of the profound importance of a relational (dare I say dialectical?) understanding of the production of space/time indebted initially to Massey and subsequently to Lefebvre and other geographers.

Stefan Kipfer and I have tried to suggest the mutually synergistic relations among these approaches as well as some strands of feminist theory (Kipfer & Hart 2013), along with their political stakes. Propelled by the horrors of Trumpism, I am now focusing on how resurgent nationalisms and populist politics in South Africa and India since the end of the Cold Ward can illuminate contemporary forces in the US – and that can also speak to problematic tendencies to analyse Trumpism primarily in terms of either race or class/neoliberalism. This work builds on my effort to extend the method of relational comparison in a more explicitly conjunctural direction (Hart 2016), and remains grounded in critical understandings of space/time and place.

Finally and perhaps most importantly, I want to emphasize that the point of outlining how my intellectual trajectory has both converged with and diverged from that of Doreen Massey has emphatically *not* been to assert the "correctness" of one theory (or reading of theory) over and above another. Rather, it has been to underscore that our theoretical predilections are always partial, situated and politically driven, and that the key criterion is always that of *usefulness* – in other words, what is the *work* that different conceptions of the world can do? Where Doreen and I are in complete agreement is that critical understandings of space and place along the lines she charted are not just part of academic debates, but constitute key analytical and political resources in the increasingly dangerous conditions in which we find ourselves.

REFERENCES

For works authored and co-authored by Doreen Massey, please see Select Bibliography of Doreen Massey, beginning on p. 371.

Bookman, A. & S. Morgen 1988. *Women and the Politics of Empowerment.* Philadelphia, PA: Temple University Press.

Combahee River Collective 1977. *The Combahee River Collective Statement.* http:// americanstudies.yale.edu/sites/default/files/files/Keyword%20Coalition_Readings. pdf (accessed 7 January 2018).

Cooke, P. 1987. "Clinical Inference and Geographic Theory". *Antipode* 19 (1): 407–16.

Duncan, S. & M. Savage 1991. "New Perspectives on the Locality Debate". *Environment and Planning A* 23 (2): 155–64.

Featherstone, D. & J. Painter 2013. *Spatial Politics: Essays for Doreen Massey.* Oxford: Wiley-Blackwell.

Hall, S. 1980. "Race, Articulation, and Societies Structured in Dominance". In *Sociological Theories: Race and Colonialism.* Paris: UNESCO.

Hall, S. 2003 [1973]. "Marx's Notes on Method: A 'Reading' of the 1857 Introduction". *Cultural Studies* 17 (2): 113–49.

Harrison, B. 1994. *Lean and Mean: The Changing Landscape of Corporate Power.* New York: Basic Books.

Hart, G. 2002. *Disabling Globalization: Places of Power in Post-Apartheid South Africa.* Berkeley, CA: University of California Press.

Hart, G. 2006. "Denaturalizing Dispossession: Critical Ethnography in the Age of Resurgent Imperialism". *Antipode* 38 (5): 977–1004.

Hart, G. 2007. "Changing Concepts of Articulation: Political Stakes in South Africa Today". *Review of African Political Economy* 34 (111): 85–101.

Hart, G. 2016. "Relational Comparison Revisited: Marxist Postcolonial Geographies in Practice". *Progress in Human Geography* 42 (3): 371–94.

Harvey, D. 1987. "Three Myths in Search of a Reality in Urban Studies". *Environment and Planning D* 5 (4): 367–76.

Harvey, D. 1989. *The Condition of Postmodernity.* Oxford: Basil Blackwell.

Johnstone, F. 1970. "White Prosperity and White Supremacy in South Africa Today". *African Affairs* 69 (275): 124–40.

Katz, C. 2017. "Splanetary Urbanization". Paper presented at the American Association of Geographers annual meeting, Boston, April.

Kipfer, S. & G. Hart 2013. "Translating Gramsci in the Current Conjuncture". In M. Ekers, G. Hart, S. Kipfer & A. Loftus (eds) *Gramsci: Space, Nature, Politics.* Oxford: Wiley-Blackwell, 321–43.

Lefebvre, H. 1991 [1974]. *The Production of Space.* Oxford: Blackwell.

Piore, M. & C. Sabel 1984. *The Second Industrial Divide: Possibilities for Prosperity.* New York: Basic Books.

Saldanha, A. 2013. "Power-Geometry as Philosophy of Space". In D. Featherstone & J. Painter (eds) *Spatial Politics: Essays for Doreen Massey.* Oxford: Wiley-Blackwell, 44–55.

Sayer, A. 1991. "Behind the Locality Debate: Deconstructing Geography's Dualisms". *Environment and Planning A* (23): 283–308.

Sekyi-Otu, A. 1996. *Fanon's Dialectics of Experience*. Cambridge, MA: Harvard University Press.

Smith, N. 1987. "Dangers of the Empirical Turn: Some Comments on the CURS Initiative". *Antipode* 19 (1): 59–68.

Smith, N. & C. Katz 1993. "Grounding Metaphor: Towards a Spatialized Politics". In M. Keith & S. Pile (eds) *Place and the Politics of Identity*. London: Routledge, 66–80.

Wolpe, H. 1972. "Capitalism and Cheap Labour Power in South Africa: From Segregation to Apartheid". *Economy and Society* 1 (4): 425–56.

CHAPTER 6

WHY DID SPACE MATTER TO DOREEN MASSEY?

Michael Rustin

The great significance of space to Doreen Massey – it is one of the main themes of her lifetime's work – reflects its relevance to the entire society in which she lived and worked. She was one of the most influential and most respected geographers of her generation, and is one of the most-cited scholars in contemporary geography. The themes she explored in her work had resonance for a large number of people in her academic field, in political life and more widely. This chapter will ask how one can explain the significance of the idea of space for her and those whom she influenced.

EXPERIENCES OF MOBILITY

One should first recognize that Doreen Massey's life itself involved some abrupt and challenging relocations in space. First the transition from the Wythenshawe housing estate where she grew up, to one of Manchester's leading state girls' schools, and then to the upper-middle-class environment of Oxford University, at the "women's college" St Hugh's, which admitted its first male students in 1986 many years after Massey's time there (see McDowell, this volume). Many writers have reflected on the transition from a working-class upbringing to the milieu of a university, and Oxford and Cambridge universities in particular, as a massively disorienting and alienating one, although for many it was also an enabling one. Raymond Williams, Richard Hoggart and Richard Sennett each wrote about the effects and meanings of cultural relocations like this.

Writers, including several with values similar to Doreen Massey's, have deployed different intellectual resources to reflect on such transitions. For Raymond Williams, born in 1921 near Abergavenny in Wales, the initial resource was his training in literature and literary criticism. His own working-class origin was a key point of reference for him in his major work on the English novel and on drama (Williams 1968, 1970). He reinterpreted the "Great Tradition" of the novel which he encountered in the Cambridge English curriculum, through the perspective of class relations. He described the emergence of a serious working-class presence in the novel, redefining F. R. Leavis's (1948) influential view as essentially liberal in its assumptions. His work on drama similarly noted the significant theme, within the European tradition, of ideological and moral impasses, brought about by the inherent limitations of a bourgeois world-view, for example most starkly in the plays of Ibsen. Massey appreciated Williams's work, although she was critical of its gendered assumptions (Massey 1994).

Williams's experiences of mobility in a class society were equally important in the development of his theory of culture, which he redefined no longer in the high cultural terms of the tradition of Matthew Arnold ("The best that has been thought and said in the world") and T. S. Eliot, but instead as the meanings embodied in a "whole way of life", in which he included the democratic practices of the institutions of the working class. In his work Williams made a seminal contribution to the democratic redefinition of culture and to the emergence of the field of "cultural studies" which embodied this perspective. Williams (1960, 1964, 1979) also explored the meaning of experiences of transitions like his own in three novels, with the opportunities and losses which went with them. For Richard Hoggart (born in Leeds in 1918), literature was also a primary resource. He focused the particular sensibilities he had learned from literary study to describe the lives of his own working-class community of upbringing. He contrasted its solidities and certainties with what he wrote about as the risks of alienation and rootlessness of "the scholarship boy" (Hoggart 1957).

Stuart Hall (born in Kingston, Jamaica, in 1932), from the background of a light-skinned, middle-class Jamaican family of colour, also chose English literature as his initial idiom for understanding himself and his social world, as well as the political discourses about colonialism and race, which, as he describes in his memoir, *Familiar Stranger* (2017), had an early significance for him. Hall was a friend and colleague of Massey's for many years. The American sociologist and writer, Richard Sennett wrote in biographical rather than autobiographical terms to describe such a transition. The central character of *The Corrosion of Character* (1998) was the upwardly mobile, yet dissatisfied and rootless son of the working-class janitor, who was a central figure in his earlier book (with Jonathan Cobb), *The Hidden Injuries of Class* (1972). The son had in effect succeeded in making the social journey which his parents had worked hard to make possible, yet the contrast in generational experiences does not suggest that upward mobility brought unqualified benefits.

For many other people experiencing social transitions like these, the discipline of sociology provided a serviceable language to think with. In the

1960s and 1970s, one of the attractions of sociology for many of its students and teachers lay in the way it made it possible for them to reflect on and understand the changing and conflicted place of class relations in British society, and later of relations of gender and ethnicity too, as these were being experienced in their own lives.

Doreen Massey, a "scholarship girl", to extend Hoggart's term, studied geography, and this subject, from her time at university onwards, became her principal frame for understanding the social spaces she encountered. Geography was the principal "intellectual space" which she chose to occupy and remain loyal to throughout her life.

Oxford University was no doubt in many ways an alienating experience for her – she never wanted to have much to do with Oxford later in her career, where she could probably (if she had wished) taken up a senior academic post. One value to her of the field of geography was that it provided her with the intellectual resources to understand the nature of and relations between the different social worlds she was encountering. One can see that her work on *Spatial Divisions of Labour* (1984), emerged from her lifelong identification with her Northern working-class origins, and her wish to understand and contest the social and economic disadvantages of her community of origin compared with the more privileged locations of London and the southeast, with which her exceptional abilities had brought her into close contact.

Massey's biographical journey through many different spatial locations brought her into contact with many social worlds. After Oxford, she moved to the Centre for Environmental Studies (CES) in London, and while there she studied for a year in the United States, in order to better understand neo-classical economics, her lifelong adversary (see Barnes, this volume; Dear, this volume). She worked for decades at the Open University – she was deeply committed to its distance-learning systems of adult education, its connection with a wide student public, and its collaborative methods of work. At CES and the Open University, she found intellectual homes in institutions whose ethos was essentially social democratic, although, as her work at the Greater London Council when Ken Livingstone was its leader showed, she was more radical than most social democrats. She had a great affection for the multi-ethnic and multiclass district neighbourhood of Kilburn in London, where she lived. Her feminist intellectual networks, and participation in radical politics in Nicaragua, Venezuela, and Mexico (she learned Spanish to make the most of this) were contexts in which she found she could do creative work with different people who shared her values and commitments. Each of these became "locations" – they were intellectual and political as well as physical in nature – whose wider relations and connections became subjects for her writing and research.

91

THE NEW GEOGRAPHY

Massey (1985) has written about the development of geography from the time when she first encountered it as an academic field in the 1960s. She described human geography as then pervaded by a somewhat inert traditionalism, dominated by a focus on the specificities and differences between places and regions. The field was divided between physical and human geography, although the momentum of development at this time lay with the latter. Massey retained her commitment to the entire discipline, and later sought to reconnect and reintegrate what had become its divided branches (see Magilligan *et al.*, this volume). She perhaps found at university that the somewhat scientific orientation of the geography discipline allowed her to keep a little distance between herself and the dominant Oxford culture. Later she contributed greatly to the broadening of the agendas of her field in ways which enabled her to challenge establishment assumptions of many kinds. She retained the enthusiasms of a physical geographer, for example, writing about the geological history of Skiddaw, a famous peak in the Lake District, describing her early love of maps, and enjoying intense experiences in travel to remote places, such as the Namibian desert. Once she explained to me (it was unfortunately necessary to do so) that aircraft did not fly over the earth in straight lines, as some maps might suggest, but in curves, because of the earth's shape.

Massey seems to have first recognized the importance of ideological constructions of her field when she encountered, in the 1960s, spatial location theory, which occupied a dominant position in urban and regional geography at that time (see Barnes, this volume; Dear, this volume). This approach represented spatial inequalities and deprivations as "natural" phenomena, the outcome of impersonal forces. Its implication was that regional inequalities were the consequence of deficits of capacity among the inhabitants of disadvantaged areas. It seems that the neoclassical takeover of the field of economics which took place in the 1960s had an analogous development in geography. The implicit assumption of the dominant style of location theory was that spatial patterns could be explained by the forces of supply and demand, in effect by the exercise of market forces. Massey's critique of standard location theories in geography – that it was relations of power, and especially their influence on production, which explained differences between places – was a response to this assertive ideology. Her involvement in the critique of neoliberalism thus extended over 50 years. Her rejection of this "common-sense of the age" – the ideology of the free market – had come early, before she had read Gramsci.

How did it happen that during Doreen Massey's career, geography evolved to become one of the most influential and powerful of the social sciences, when few people would previously have seen its scope in this way? Massey's biographical journey coincides remarkably with the development of her entire discipline, she being one of the most important actors in its development. In this period, the idea of space was advanced as a key and hitherto under-used explanatory concept in the social sciences. The previously dominant assumption, that the social world is best understood in terms of grand temporal periodizations, was put in question by this new geographical, space-oriented perspective. These periodizations had taken different forms – the stages of feudalism, capitalism and socialism of classical Marxism, each corresponding to the domination of a specific social class; liberal modernization theory which demarcated pre-modern from modern social forms; sociological models of pre-industrial and industrial societies; and theories related to the "Third World" of colonial and postcolonial societies. The new geographers' assertion of the neglected dimension of space allowed attention to be focused on differences which were no longer classified by reference to positions in an onward march towards progress, within whatever ideological idiom this idea was expressed.

Massey has explained that the idea of differences, and thus of spaces, literal or metaphorical, was initially explored by her in the context of the complexification of standard Marxist models of economic and class determination, and the historicist tramlines that were inferred from them. Althusser, with his idea of partially autonomous and multiply interdependent "levels and instances" of the social formation, was a key theoretical resource for Massey for thinking through these issues (see Barnes, this volume). Differences in regard to gender and gender relations were an equally important focus. Indeed these two issues were closely connected, since it was the gender blindness – the unquestioned masculinist assumptions – of much class theory at the time that led to Massey's partial distance from Marxist approaches, such as those of David Harvey, the most influential radical geographer among her contemporaries (see Dear, this volume).

What has come to be widely recognized as the "postmodern" rethinking or outright rejection of Marxist theories of transformation was the context for some of this development, with geography as one of its intellectual locations. However for many of its key figures, including Doreen Massey and certainly David Harvey, the aim was not to reject or displace Marxism, but to make its analysis and model of change more adequate to the complex realities of modern capitalism through a deeper understanding of differences located in the dimensions of both space and time. The journal *Soundings*, which she co-founded in 1985, was opposed to one-dimensional definitions of markets, the state, and politics. Its first editorial stated:

> Our object in *Soundings* will be to register changes taking place
> in many domains of life … We shall argue that social change can

be achieved in many social spaces besides that which is normally designated as political.   (Hall *et al.* 1995: 15)

Massey maintained an energetic commitment to *Soundings* from its foundation to the end of her life, even when she became less well. She introduced many contributors to the journal, and was active in its editorial debates. In later years, she was crucial to its critique of neoliberalism.

The influence of the spatial conceptions of the new geography was felt far beyond academic geography itself. Sociologists also became keenly interested in the relations of space. In fact in the 1970s and 1980s in some fields of debate it was sometimes difficult to see a clear boundary between geography and sociology. Manuel Castells, for example, is described as a sociologist, but his theory of networks is deeply shaped by the awareness of transformed spatial relations. Anthony Giddens is one of the leading sociologists of the time, yet one of his most influential ideas concerns the effects of globalization and changes in what he called "time-space distanciation" (1991). John Urry's sociological writing gave sustained attention to changing spatial relations within modern societies, as a key transformative aspect of their development. His book *The End of Organised Capitalism* (with Scott Lash, 1987), his work on the meanings and effects of tourism, on car cultures, on the "offshoring" of financial assets and financial powers, and his redefinition of sociology for modern times as the study of "mobilities" as their normal condition, all attached crucial importance to spatial and thus geographical issues.

Massey has described how, traditionally, geographers were content to study "places" and "regions" and their particularities, differences and similarities. Sociologists, in the first decades after the Second World War, had for the most part confined their studies to the societies defined as nation-states, and circumscribed by national boundaries. A subsidiary topic was the comparative study of the societies of different nations, though in truth few sociologists in any nation then seemed to know much about the society of any other.

But this situation changed, as the factors which influenced structures and processes in any one location seemed increasingly to have their origins in causes external to them. Reducing costs and increasing speeds of physical communication, and the emergence of the almost instantaneous communication of the electronic sphere, began to transform the hitherto localized and slow-moving relations of space. Globalization gave rise to new kinds of mobility, of financialized capital but also of labour, and of populations fleeing from damaged places to seek opportunities to labour. This was surely the historical moment when spatial perspectives became indispensable to other disciplines, and when geography's new centrality as an integrative social science began.

This new geographical emphasis in the social sciences did not come about merely because of the inventive powers of major geographers and geographical sociologists and those whom they influenced. This body of work, not least Massey's, has reflected and explored changes taking place in human societies themselves and in their changing relations to space. Individuals' experiences of, for example, social mobility and opportunities for travel, of the advantages and disadvantages of place and region, of unacceptable assumptions about gender, of the untenability of received beliefs about class and political agency, gave rise to insights into the wider changes which lay behind these. What C. Wright Mills famously described as "the sociological imagination" has its close parallel in a "geographical imagination", such as Doreen Massey's, which is able to recognize the "public" nature of private troubles, and undertake intellectual labour which then illuminates these situations. Writing does not have to be personal (like that of autobiographers) to bring out these implicit connections (although Massey's writing sometimes was). The fact is, as feminists proclaimed, the personal *is* political, in the broadest sense of that term.

Marx made a much-used distinction, although not in these precise words, between "a class in itself" and a "class for itself".[1] This axiom describes a more universal relationship between objective social facts, and their recognition as subjectively and collectively meaningful facts. Class differences and antagonisms explored in the work of Raymond Williams, or ethnic and diasporic differences in the writing of Stuart Hall, become recognized to be of wide significance because the situations of transition and mobility which expose them are experienced by many people, even though it requires courageous and exceptional writers to articulate them clearly. It was probably not incidental to John Urry's understanding of the modern significance of "the tourist gaze" in transforming many places that the University of Lancaster, where he worked, is on the edges of the Lake District whose touristic consecration and transformation began with the writing of Wordsworth and Coleridge more than two centuries ago. The development of the successive stages of feminism can also be understood in these terms. "Women" became a "class in themselves" (through entry into the workplace and through enhanced educational opportunities) before they could become a "class for themselves", through the writing and advocacy of feminists. This happened

---

1. Marx put it thus: "Economic conditions had first transformed the mass of the people of the country into workers. The combination of capital has created for this mass a common situation, common interests. This mass is thus already a class as against capital, but not for itself. In the struggle, of which we have noted only a few phases, this mass becomes united and constitutes a class for itself. The interests it defends become class interests. But the struggle of class against class is a political struggle" (1847: 173).

in several stages, over 100 years or more, as women from different classes and class fractions gained a measure of economic and social autonomy, and became able to assert their claims.

We can in this way trace many links between the personal and professional trajectory of Doreen Massey's life and work, and the significance she came to see in space and the relationships of many kinds – economic, political, cultural and ecological – which the idea of space enables us to understand.

SPACE AND TIME

How was the dimension of time to be accommodated with this new geographical, or sociogeographical perspective, which was giving so much emphasis to space and spatial relations? There was a traditional disciplinary division of labour between geography (focused on places and spaces and the relations and differences) and history (focused on periods and differences and transitions between them). To some degree, the "new geography" came about as a displacement of an early dominant focus on temporalities. These took different ideological forms – attachment to traditions and to the values of early social formations among conservatives; the idea of an advancing present and normative state of modernity, among liberals; and the idea of an evolutionary or revolutionary transformation to be hoped for, among socialists. Explicitly "modernist" social theories tended to prioritize the dimension of time and temporal succession, explicitly "postmodernist" theories by contrast prioritized the idea of differences, located across different physical and social spaces. Massey was convinced by neither of these opposed positions, since she was committed both to the recognition and positive value assigned to differences, and to the progressive transformation of society, and especially to overcoming the hegemony of the neoliberal form of capitalism whose manifestations and expressions, in Britain and across the world, she analysed and critiqued in nearly all of her work. Theoretically and intellectually, the problem was how to integrate and achieve a synthesis between these two axes of explanation, as she saw them.

One argument that was formative for her was with her friends Ernesto Laclau and Chantal Mouffe. Laclau argued that space, by definition, meant fixity and stasis. It was in time that choices and decisions were made, political actions devised, equivalences and antagonisms constructed. Massey (1992b) argued that this was a false distinction. How could "equivalences" be discursively constructed, through political thought and action, if there were no distinct and different entities to be brought into symbolic equivalence or identity? Laclau and Mouffe had rejected, in *Hegemony and Socialist Strategy* (1985), an essentialist, one-dimensional theory of class and class agency, correctly in Massey's view. That view had involved the greatest possible sim-plification of "difference", into a dichotomous model of antagonism between social classes. The value of the contributions of Althusser and Gramsci lay in their complexification of this model, and in their recognition of the many different components and interconnections of social systems which had to be taken into account if their forms of domination, and potential resistances to this, could be understood. Massey thought that Laclau and Mouffe had gone too far in their dismissal of the salience of class relations, but that the recognition of differences, which needed to be interpreted in their spatial embodiments, and connected up through political definition and action, had been one of the most valuable contributions of their work.

Massey's continuing emphasis was on the necessity to hold on to the dimensions of both space and time in geographical and social analysis. It was their separation and dichotomization that she thought was fatal to understanding and to the construction of a viable politics. One way in which she sought to avoid a potentially destructive choice between "time-oriented" and "space-oriented" explanatory models was by turning to physics, and its idea of "space-time" as an indivisible entity (Massey 2005). This approach has the value of recognizing that no geographical or sociological analysis makes any sense unless the dimensions of time and space are both included. Any historical description which is sensitive to particulars (Massey was insistent on the relevance of these) is implicitly referring to entities located in space, whether this be physical or metaphorical. Likewise, all descriptions of entities located in space are necessarily referring to entities with a past, present and future, which have been and are subject to change. It seems that only a misleading kind of disciplinary or ideological over-specialism, or fetishism of time or space, can have overlooked these obvious facts about these two necessary dimensions of any social fact or event.

However, a massive disciplinary separation and splitting of these dimensions there has been, and one of Doreen Massey's main projects throughout her career was to put an end to this. In many respects she succeeded, not least in establishing the need to restore the domain of space and spatial relations to social science and politics. And also in showing how analysis can be conducted, from the example of the spatial division of labour in Britain and across the world, in ways which take proper account of the dimensions of both space and time. I am a little less confident that Massey found a satisfactory conceptual solution to this problem of a failed integra-tion, however, despite the breadth of scholarly reading she brought to this task, especially in *For Space* (2005). In the concluding section of this essay I will set out some thoughts on this.

It seems to me that space and time are unlike societies, cultures and material things, in so far as they are not in themselves entities with causal powers. Massey in effect acknowledges this, by virtually always referring to "social relations in space", and indeed "social relations in time", and not time and space in themselves as possessing causal agency. She is emphatic that it is relations between entities, not isolated entities in themselves which determine events, qualities and outcomes. What then is the ontology, or essential meaning, of time and space, which appear to exist as both essential yet indefinable categories in Massey's most theoretical writing (1992b, 2005)?

One of the solutions she sought to this problem of integration lay in a recourse to science, physics in particular and its idea of "space-time", which she suggested might overcome the destructive space-time dichotomy. The physicists' idea of "space-time" as the true context of existence and experience surely arose from the theory of relativity, the idea that the perception of objects in space and their relation to one another is mediated by the transmission of light and its velocity. The ordinary common-sense understanding of the separate dimensions of space and time "works" for practical purposes, within a universe which assumes that the means of perception do not exceed the speed of light. Is it not only when the transmission of light is regarded not as a constant, but instead as, hypothetically, a variable, that complications ensue, and paradoxes arise such as the disruption of the idea of linear and irreversible temporal succession?

This way of thinking derives an undoubted metaphorical power from the fact that human relations of space and time have indeed been transformed over the last century. Communications that once took weeks, now take micro-seconds. Places that once seemed firmly located in one "temporal zone" can now experience an accelerated development that modifies all previous understandings of succession. But while it is essential to understand these substantive transformations in what is happening in modern locations of space and time, it does not follow that the space-time conceptions of theoretical physics have much explanatory relevance to them. It may be that for practical purposes, it is sufficient to remain within the Newtonian space-time universe, and that little is gained, in regard to combinations of geographical and historical explanations, from calling this into question. In a similar way, understanding of the physical world in the terms of quantum physics – for example Richard Feynman's (1998) descriptions of objects as masses of jiggling electrons – does not help one much in one's encounters with things in everyday life.

PHILOSOPHICAL REALISM, OBJECTS AND CATEGORIES

There may be ways of formulating these questions which provide more adequate resolutions of them. The underlying issue is the nature of space and time themselves. Are they entities and objects like others in the universe, with causal powers? An ontology or philosophy of science of a "realist" kind (Bhaskar 1975; Keat & Urry 1975; Harré 1986) suggests that may be a category mistake to suggest that they are. Realist ontology and epistemology proposes that the world consists of objects and structures with causal powers, some of which may be directly observable while some are hidden or deep structures known only through their effects.

The point is that neither space nor time are objects or structures possessing causal powers. They are, by contrast, the "dimensions" or "containers" within which such causal powers are exercised by objects. Everything that happens takes place "within" (as we say) both time and space. Certainly, the mode of distribution and potency of objects within time and space is highly significant for the respective outcomes of their causal powers. There are, it is obvious, different "spaces". These can be both literal geographical spaces (e.g. New Zealand), but also entities that are tied to geographical spaces only in complicated and indirect ways (e.g. the system of intellectual labour which gives rise to neoclassical economic theory, or Handel's *Messiah*). There are also, although this is less obvious, and it was one of Althusser's contributions to recognize this, different "times". Thus the temporality, and the mode of organizing life through time, of aboriginal peoples on the continent we call Australia, is or was different from that of Australia's "modern" citizens, although it is framed by some of the same material limits (length of days, seasons, human lifespans) to which human temporality has to be accommodated in some way. But what we refer to when we describe different times and spaces are differences that are effected by the relations between the objects with causal powers (these "objects" include human belief systems as well as, for example, the possession of firearms), not the entities of time and space as such, which have none.

In the early days of the new synthesis between geography and sociology, a realist position like this was clearly set out by John Urry (1985), who refers positively to Massey's work in developing his argument. However philosophical realism is absent as a topic from Doreen Massey's most theoretical book, *For Space* (2005). Even so, it seems to me that she fully recognized in terms of practice and application the significance of these distinctions, and thus avoided assigning causal powers to space and time as such. But this does not necessarily mean that she found an adequate conceptual frame with which to clarify these rather abstract and metatheoretical issues.

The science of physics was one location through which illumination could be sought, in exploring the ontological and epistemological issues raised by the ideas of time and space. It seems to me that philosophy, and especially that of Kant, offers a better route than post-Newtonian physics to clarifying these issues. Kant, in *The Critique of Pure Reason*, identified the categories of space, time and identity as those essential to the capacity for rational thought. (These concepts were given a very clear reformulation in Philip Strawson's *The Bounds of Sense*, 1966.) Reference to Kant seems to be infrequent in the geographical literature: perhaps this was because the idea of the fundamental categories of human thought was deemed to be relevant more to psychology than to geography. But to recognize the distinction between categories (framing or containing principles) and objects or entities (which possess causal powers) might help us to resolve crucial conceptual problems. Geography could then be understood as the study of objects primarily located within the categorical dimension of space, and history as the study of objects located within the categorical dimension of time. Of course, it is then obvious that all objects of whatever kind have their existence in both dimensions. One of the virtues of Marxism as a theoretical system was that at its best it sought to integrate rather than separate these.

The third primary Kantian category of identity may also be relevant. From what point of view or perspective do human beings experience, or reflect on, objects, including themselves, in space and time? Clearly, from the point of view of their practical interests and purposes, individual or collective, the human life-cycle seems a crucial organizer of these experiences. (In a previous volume dedicated to Doreen, I suggested parallels between her commitment to relational explanations and those central to some varieties of psychoanalysis (Rustin 2013).) At the beginning of life, proximity (between infants and carers) is everything, or life does not survive, and a similar condition occurs at the end of life. In the middle stages of the life cycle, both spatial and temporal horizons grow larger. This is a psychosocial analogue to the variable relations of collective human subjects to time and space which were Doreen Massey's central and wonderfully productive field of study.

REFERENCES

For works authored and co-authored by Doreen Massey, please see Select Bibliography of Doreen Massey, beginning on p. 371.

Bhaskar, R. 1975. *A Realist Theory of Science*. Leeds: Leeds Books.
Featherstone, D. & J. Painter 2013. *Spatial Politics: Essays for Doreen Massey*. Oxford: Wiley-Blackwell.
Feynman, R. 1998. *Six Easy Pieces*. London: Penguin.
Giddens, A. 1991. *The Consequences of Modernity*. Cambridge: Polity Press.
Hall, S. 2017. *Familiar Stranger: A Life Between Two Islands*. London: Penguin.
Hall, S., D. Massey & M. Rustin 1995. "Editorial: Uncomfortable Times". *Soundings* 1 (Autumn): 5–18.
Harré, H. R. 1986. *Varieties of Realism*. Oxford: Blackwell.

Hoggart, R. 1957. *The Uses of Literacy*. London: Chatto and Windus.

Keat, R. & J. Urry 1975. *Social Theory as Science*. London: Routledge.

Laclau, E. & C. Mouffe 1985. *Hegemony and Socialist Strategy*. London: Verso.

Lash, S. & J. Urry 1987. *The End of Organised Capitalism*. Cambridge: Polity Press.

Leavis, F. R. 1948. *The Great Tradition*. London: Chatto and Windus.

Marx, K. 1847. *The Poverty of Philosophy*. Moscow: Foreign Languages Publishing House.

Rustin, M. 2013. "Spatial Relations and Human Relations". In D. Featherstone & J. Painter (eds) *Spatial Politics: Essays for Doreen Massey*. Oxford: Wiley-Blackwell, 56–69.

Sennett, R. 1998. *The Corrosion of Character*. London: Norton.

Sennett, R. & J. Cobb 1972. *The Hidden Injuries of Class*. London: Faber.

Strawson, P. 1966. *The Bounds of Sense: An Essay on Kant's Critique of Pure Reason*. London: Methuen.

Urry, J. 1985. "Social Relations, Space and Time". In D. Gregory & J. Urry (eds) *Social Relations and Spatial Structures*. London: Macmillan, 20–48.

Williams, R. 1960. *Border Country*. London: Chatto and Windus.

Williams, R. 1964. *Second Generation*. London: Chatto and Windus.

Williams, R. 1968. *Drama from Ibsen to Brecht*. London: Chatto and Windus.

Williams, R. 1970. *The English Novel from Dickens to Lawrence*. London: Chatto and Windus.

Williams, R. 1979. *The Fight for Manod*. London: Chatto and Windus.

CHAPTER 7

ONTOLOGY AND THE POLITICS OF SPACE

Andrew Sayer

When you've known someone for a long time it's hard to separate the person from their work. And in Doreen's case it would be unfair too, because her influence extended beyond her writing to her way of being – as an academic, an activist and a person. From a working-class background, she grew up during the exceptional decades of growth and increasing equality that followed the Second World War. Like many others, she celebrated the end of deference; she showed that one could be excellent without being elitist, and profound without being pompous or obscure. There was nothing incongruous in laughing at a silly joke one moment (often a Pythonesque send-up of academia) and being completely serious the next; in fact this was the way to work and live. Round about 1989, Doreen and I and Open University colleague John Allen formed our own discussion group – "The Brighton Pier Social Theory Group" – that met several times in Brighton, and once in Copenhagen. The ingredients of the meetings were social and spatial theory, ontology, coffee and pastries, and irreverent laughter, combined in any order. While being with her could be fun, she could also be scarily intense, with a formidable ability to encapsulate the complex briefly and to zero in on the weakness of others' ideas, so you often felt you needed to be on your toes when you were with her. She had an insatiable appetite for learning, for writing, for engaging in politics, and getting the most out of life. While her championing of the importance of space linked her firmly to geography, she was also a post-disciplinary thinker whose work defied identification with any single -ism, and she was exceptionally widely read. Whether the subject was space, politics, social theory, Marx, feminism, Bergson, geology, or bird watching – all were approached with her inimitable *relish*.

Her range of interests was wide, but she always strove to find connections, relations among things. Hence her overtures to physical geography (Massey 1999g). She used to carry a geology map of the UK with her in her travels so she could check out the underlying rocks wherever she was. In her short piece on the sense of space, sociability and "lagom" (Swedish for sufficiency, "enough"), she used the example of a simple public birdwatching hide to discuss what architects and society in general might learn from it regarding the balance between simplicity and excess (Massey 2003a). But her interest in the world was never simply academic, for she also strove to change it. In this piece I want to discuss some of the ways in which the thinking that informed her academic and political work was relational.

One form that this took was methodological. In the early 1970s, Doreen, like many others at that time, rejected the idea that the only path to rigorous knowledge was to seek out regularities or correlations among empirical data. Instead she focused on analysing how things were *substantially* related or interdependent through connections and flows of people, objects, information and money. This was consistent with both Marxism and critical realism. Thereafter in British human geography and also in some other social sciences, extensive methods of statistical analysis of covariation among discrete data began to take second place to intensive methods that traced connections wherever they led (Sayer 1992).

For years it had widely been assumed in economic geography and allied subjects that changes in output and employment would roughly correlate so that job losses could be taken to reflect industrial decline. In *The Anatomy of Job Loss*, written at a time when Margaret Thatcher's attempt to weaken labour by undermining British-based manufacturing was getting underway, Doreen Massey and Richard Meegan presented a simple three-way distinction between job losses due to falling demand, those due to automation and those due to intensification of labour processes through speed-up and increased efficiency (Massey & Meegan 1982). (Doreen had a knack of coming up with simple but powerful distinctions.) What seemed like a straightforwardly uniform process was actually the outcome of quite different tendencies. Job losses could even occur while demand and output increased. These were classic realist arguments. Contrary to positivism, causation did not entail correlation, for different processes can lead to the same result; job losses could arise from quite different combinations of causes. Conversely, the same process could lead to different outcomes. The effects of the pressure of the law of value or competition on firms – a general tendency intrinsic to capitalism – would always depend on what the specific characteristics of the firm and its line of work were, and much else; the impact on clothing firms would be very different from that on chemical firms. Both the general tendencies and the specific outcomes could be explained. One didn't need to deny one to affirm the other. Theory could explain differences as well as empirical regularities. Explanation was not about finding regularities, but identifying mechanisms – the way that causal powers, when activated, interacted with whatever circumstances they encountered. Although, to my knowledge, Doreen did not endorse critical realism unambiguously in print, I know from conversations with her that she was familiar with it, and didn't reject my critical realist glosses on this methodology.

This approach implied a relational, rather than an atomistic, ontology. However, it was different from the philosophy of internal relations advocated by Bertell Ollman and promoted by David Harvey, because it did not imply that *all* relations are internal, or (absurdly) that *every*thing is linked to everything else (Ollman 1971), just that everything is linked to some other things. There is always also contingency – accidental juxtapositions of externally related phenomena. Geography deals with complex open systems in which today's activities take place in contexts not of current agents' making, but are nevertheless constrained and enabled by them. In such systems there are at best only temporary and local regularities. Yet one could still explain what happens in them by tracing connections and looking for causal mechanisms. This is effectively what Doreen did in her work on industrial and regional geography.

A second aspect of relationality that she emphasized particularly in her work in the late 1970s and 1980s was the connection of relationality and *difference*. There cannot be many books that have more emphasis on difference than *Spatial Divisions of Labour*. Over and over again, it emphasized how outcomes of capitalist processes were different in different cases, or how superficially similar outcomes could result from different processes. Difference had been a nuisance for positivism, given its focus on regularities. Yet although Marxism broke with positivism, in the 1970s and 1980s it often took a reductionist form in which anything that couldn't be reduced to the abstract terms of Marxist theory was treated as unimportant, and not worthy of "theory". Much early radical geography had often simply "read-off" empirical instances from theory: there are the means of production; there is the reproduction of labour-power (households, housing, schools etc.); there are the general conditions of production (infrastructure, public utilities) etc. In so doing it was reductionist or "pseudo-concrete": the specificities of the concrete as the *combination* of multiple elements or determinations were treated as reducible to just those determinations dealt with by Marxist abstractions (Sayer 1982). As Sartre said of a different version of this tendency: "Valéry is a petit-bourgeois intellectual, no doubt about it. But not every petit bourgeois intellectual is Valéry. The heuristic inadequacy of contemporary Marxism is contained in these two sentences" (Sartre 1963: 56). Reductionism can only find in the world what it puts in it.

Doreen's work on industry and regional development with Richard Meegan (Massey & Meegan 1982) and in *Spatial Divisions of Labour* broke out of this stifling tendency. At first some radical geographers misunderstood this as a retreat from theory, or even a return to the empiricist, atheoretical, idiographic approaches from which geography had escaped in the 1960s. In part this was a difference about Marxism; while Doreen was well versed in Marx and Marxism, she was never theological about it or closed to other theory. But it was also a reflection of the reductionist character of much Marxist work at that time: as E. P. Thompson once said the fundamentalists were like Antarctic explorers fearful of leaving their theory tent for fear of getting lost in a blizzard of facts.

In *Spatial Divisions of Labour* differences were always shown to have been produced through relations between processes or things. So instead of abandoning theory, it put theory to work in interpreting actual instances in all their specificity, rather than stripped of everything not anticipated by Marxist theory. Some relations, such as those between capital and wage-labour, are internal in the sense that the nature of the relata depends on their relation. Others may be external, accidental or contingent, like the relation between a firm and the prior economic history of its local context, yet still be consequential. To understand a concrete situation, such as changes in employment in a particular region, one had to combine a range of abstractions to understand the concrete as a spatially structured combination of the forces identified by many different concepts. Hence difference was not to be understood as discrete and unique or as something to be passed over in the search for regularity, but as produced through sociospatial relations. Nor, as Gillian Hart discusses (this volume), was the concrete to be limited to the local, or the abstract to the non-local, as many assumed. The concern with difference was also evident in the many locality studies that were inspired partly by Doreen's work (e.g. Cooke 1989). Again, it worried those who seemed to want every piece of research to come back to affirming Marxist theory and who saw such empirical research – or "empiricism" as it was sometimes mislabelled – not only as a desertion of theory but of radical politics, though radical geography was always a broad church including many social democrats (e.g. Harvey 1987).

A third relationalist theme in Doreen's work concerned the relational nature of space and place. This was argued through discussions of a range of concrete cases, but also at a more abstract level in her book, *For Space* (Massey 2005). In my view, the former were more persuasive than the latter.

The ideas that space is constituted by the relations and interactions among things and that society and nature are spatially constituted had become widely supported in geography by the end of the twentieth century, but Doreen took this further. Space is also the dimension of multiplicity in the sense of "contemporaneous plurality" and "coexisting heterogeneity". However, this is not a multiplicity in which everything is connected to or dependent on everything else: there are always many loose ends and missing links. Further, space is always being re-made. It is not the realm of stasis, as some assumed, but the realm of a multiplicity of processes of change, of the "simultaneity of stories so far" (Massey 2005), stories that in various ways were different yet often connected.

I agree but I would add some mundane but often overlooked points. First, material things offer varying degrees of resistance so that two things cannot be in exactly the same place at the same time. They also have spatial extension and certain characteristic forms or shapes. As we discussed in the Brighton Pier Social Theory Group there is a limited number of spatial forms that certain objects, such as a house, can take and still serve the functions in virtue of which we define them. For a boat to float it mustn't be shaped like a sieve. Others, like air, are capable of maintaining their characteristic properties in a wide range of spatial forms. Secondly, there are spatial and topological arrangements such as betweenness, separation, encirclement, connection, closure and openness, that make a (causal) difference to what happens, though the difference that these make always depends on the nature of the objects constituting these forms: a wall between two people has different effects from air (Sayer 2000). This is why it's hard to say much about space at an abstract level: the difference it makes always depends on the concrete things constituting it. But space wouldn't be space without spatial extension, and without arrangements or configurations. To ignore these points is to risk allowing space to become merely metaphorical rather than a universal aspect of physical being or becoming. There isn't just multiplicity and heterogeneity: there are spatial, and indeed topological configurations, as John Allen has recently argued (Allen 2016). And although we may be accustomed to thinking of these spatial forms as static, we do not have to do so. Nothing stays exactly the same forever. Even continents move and deform, as Doreen described in an essay on (among other things!) the geology of the mountains of the English Lake District (see Rustin, this volume).

In the last two decades of her career, Doreen gave growing consideration to how we think about time as well as space, and to the idea that everything is always changing. This fitted with a wider shift in social science away from an excessive emphasis on social structures as givens rather than as contingent products whose reproduction was never assured, and an emphasis on *becoming* rather than just being. Hence the growth of interest in ontologies of assemblages and flows.

However, there's a danger here, albeit one avoided by Doreen, of flipping from an overly structured view of the world to one in which it is represented as wholly unstructured and uniformly fluid, or no more than a mess. (Exaggeration is the besetting sin of avant-garde social theory.) Rates of change vary enormously. While some things constantly flow and mutate, others are decidedly constant (the elements – despite the movements of particles within them), or at least sticky. In geology, James Hutton's principle of uniformitarianism – that the same processes that can be observed operating now also operated in the past, so that the present is the key to the past – has been invaluable. Certainly the anthropocene (capitalocene?) has brought novel changes but these do not change physical laws. And as Doreen acknowledged in *For Space*, non-Bergsonian science has been remarkably successful. Structures can be characterized as slow processes, but their relative durability compared to other processes cannot be ignored. To be sure, the future is open and structures depend at every moment on actions for their reproduction, but structures of interlocking elements make it difficult to act in ways that do not reproduce them. We can acknowledge this and still accept that there are *also* loose, constantly shifting assemblages of externally related objects. Cities, for example, combine both. Rather than opt for ontologies which declare everything internally related or everything externally related, so that we are asked to opt for *either* structures or assemblages and flows, we should use ontologies that are plural enough to allow them to make sense of variety in the fabric of the world.

Doreen devoted considerable efforts to countering any tendency to associate space with the realm of stasis, in contrast to time as the realm of change. This can happen where we associate space with "a slice through time", a snapshot representation. For some social theorists – typically ones who were not familiar with geography – only time was associated with change. But after relativity theory, it's no longer tenable to think of space and time as mutually exclusive opposites. Actually, we do not need to get into relativity theory to see this. If we think of space concretely, for example in terms of the relation of British consumers to Chinese producers, or of someone to their neighbour, then it's obvious that social space is constituted by people doing things, by multiplicities of simultaneous change.

Certainly, many social scientists and other writers have typically attached more importance to history than geography – an emphasis reflected in and reflective of those disciplines' relative social standing in academia and society. But it is not so much that they should be thought of as equal in importance and status, but that we shouldn't even think of them as separable. All histories are geographical: all geographies are historical. Yet I have to say I am puzzled why Doreen paid so much attention to theorists who thought that space was the realm of stasis. Take Laclau's typically gnomic pronouncement that "Politics and space are antinomic terms. Politics only exist insofar as the spatial eludes us" (quoted in Massey 2005). It's so obviously absurd that I'm baffled why it should have been given the time of day.

I remember some discussions with her about her view of space in which I was surprised by how strongly she felt about it. Ontological discussions do not usually generate much heat, but she was angry that while I agreed with her account, I seemed underwhelmed by it. She saw views of space as static as not merely mistaken but as dangerously reactionary. Later I came to realize what she was getting at. Although again I think the point is made better through concrete cases than abstractly, there are indeed significant dangers in static treatments of space – some of them political and not merely academic. For example, the idea, central to notions of the pastoral and nostalgic patriotism, that there are special places where time stands still, where ways of life go back centuries, where people are straightforward, honest and dignified, and deeply rooted to the land. (In the UK, *The Dalesman*, a magazine celebrating Yorkshire rural life, is a good example of this reactionary tendency.) The stories appeal to a desire for attachment, which as Doreen noted is understandable, though not necessarily only capable of being met locally (Massey 1994). Such stories are not only half-truths but can be reactionary, for they can easily gloss over oppression and inequalities and represent narrow-mindedness as commonsense, while implying that change, difference and otherness are threats. In particular, they can support the idea that migration – a common phenomenon throughout human history – is a threatening abnormality. Doreen herself had an obvious love of many particular places – their appearance and atmosphere, their local accents, geology, flora and fauna and history, but she was always acutely aware of their relationality, so her attachment to them was never parochial or escapist.

Postcolonial theory has a parallel message here. It shows how colonialist, Orientalist discourse tends to assume that the space of the other, the Oriental, is fixed in space and time, and the behaviour of its peoples no more than an expression of an unchanging and unchangeable essence, whereas the colonizers, by contrast, are progressive, open-minded, rational and self-determining. At the same time, the respective spaces of both the colonized and the colonizer are misrepresented as basically homogeneous. Difference, variety are flattened and forced into a simple binary opposition that ignores the relatedness, indeed the interdependence, of colonizer and colonized. Thus we have histories and geographies of the industrialized world that ignore their dependence on the global South (Said 1978; Connell 2007; Bhambra 2014; Radcliffe 2017). Doreen's birthplace, Manchester, was of course the product of relations with (among others) cotton and slavery in the Americas and Africa. Again, this is why the relational character of space matters.

Doreen made the political implications of how we think about place vivid in one of her most famous essays, "A Global Sense of Place". Based on her own home, Kilburn, in London, it showed how places are constituted and continually remade through their relations to other places, near or distant. As she put it:

> This is not a question of making the ritualistic connections to "the wider system" – the people in the local meeting who bring up

> international capitalism every time you try to have a discussion about rubbish-collection – the point is that there are real relations with real content – economic, political, cultural – between any local place and the wider world in which it is set.
>
> (Massey 1991a)

Moreover, many of the political concerns of residents were not parochial; concern for others was not simply localized, but strongly influenced by connections, many of which were to distant others, particularly postcolonial ones.

Politicization is often said to depend on relating the specific experiences of individuals' lives to wider social forces so that people realize that their problems and those of others are not just personal misfortunes, but common products of these forces. Doreen took this further, reminding us that these forces are spatial and take particular forms that we may need to address if we are to understand and change them. Here, ideas as much as material connections and flows are important. For example, ideas of the City of London as the golden goose that has saved the UK, and of the provinces as backward have been central to political ideology in Britain. The implicit spatial imaginaries are part of what has to be challenged. At the same time, I'd argue that there are times when it's useful to abstract from space, and talk about common social relations that are unequal, oppressive or exploitative, whatever their spatial form. Thus, the City of London's relationship to the rest of the UK and many other places includes a form of economic social relation – creditor–debtor – that systematically and parasitically centralizes wealth to the top 0.1 per cent in the income and wealth distributions. (Less than 15 per cent of bank lending in the UK funds productive investment: most lending is against existing property.) Even if instances of this relation were equally dispersed spatially, rather than focused on the City, they would still be problematic. The most general point to be made about this situation is that it's parasitic; the credit system functions as an engine of wealth extraction and concentration, and it's predicated on future economic growth despite contributing little to that growth. Actually, I remember from talking to Doreen that she didn't object to the idea that sometimes it's harmless and indeed useful to abstract from space, at least temporarily, and she did this in her own writing at times. Too often in economic geography and related subjects, the exploitative or extractive nature of many common forms of economic social relations is overlooked because of a preoccupation with their particular concrete, technical, institutional and spatial forms. The concrete is the unity – in particular configurations – of many processes and relations that can be identified individually at a more abstract level. Both abstract and concrete analyses are needed.

Sometimes thinking through political issues in terms of spatial relations can be unsettling. She reminds us that immigration is always also emigration, and the consequences of the latter are no less important than those of the former (see also Sayad 2004). It's not just a matter of what happens when immigrants settle somewhere. For example, while we might celebrate London's ethnic mix, it comes at a cost. When the UK draws upon migrant workers from poorer countries – whether they are skilled, as in the case of doctors and nurses, or unskilled, as in the case of fruit pickers – the UK is free-riding on the education, training and care provided by those countries; it is being subsidized by poorer countries. I would suggest that particularly in the case of skilled labour, rich countries like the UK should be net exporters rather than net importers. In terms of culture, such immigration may contribute to London's vitality, but in economic terms – remembering that London is defined by its relations to other places – it is problematic. As Doreen noted, this creates problems for the common view that the Left or progressives ought to be unqualified supporters of immigration. Much migration derives from processes of uneven development and unequal exchange that are unjust and dysfunctional in themselves, and they encourage emigration that often worsens the situation (Massey 2007). To focus only on immigration and its effects is to miss not only other effects but also their causes. For these reasons, Doreen and certain others argued that countries that benefit from the perverse subsidy have a responsibility to pay restitution to the immigrants' countries of origin. Of course, that's a lot to ask in today's political climate, but the argument itself is a strong one and needs to be put. Sociospatial relations frequently raise questions of justice and responsibility for others, distant or near (see, for example, Young 2013). What was also welcome about these discussions of responsibility is that they begin to break down the division between political theory, which deals with normative questions of justice, but usually abstracts from actually existing economies because of its preoccupation with "ideal theory", à la Rawls, and political economy and economic geography, which deal with those economies but usually leave the normative arguments to others.

All this leads to the idea that instead of assuming that political movements must be tied to particular places, regions, of countries, they can and should embrace key relations between different places: "a politics of place beyond place" (Massey 2007). Particularly through her activism in forging links with Nicaragua, Mexico and Venezuela, Doreen set an example of this. Divide and rule was a principle of British imperialism, and is still applied to both domestic and foreign policy. The UK's network of tax havens, masquerading as distant, independent microstates, are an integral part of the City of London's financial network and were constructed precisely to enhance the power of finance and the rich; the ambiguity about their status is not accidental. Of course, divisions that support domination can arise inadvertently too, but whether they come about by design or accident, so often, "it is space, not time, that hides consequences from us" as John Berger famously said (Berger 1974: 40).

Doreen Massey took a leading role in countering this myopia and revealing what is normally hidden. I feel privileged to have known her and to have worked in overlapping areas. I miss those conversations about these matters, and others – including the birds and the geology. And like so many other people who knew her, I miss the stimulus of her extraordinary intellectual energy.

REFERENCES

For works authored and co-authored by Doreen Massey, please see Select Bibliography of Doreen Massey, beginning on p. 371.

Allen, J. 2016. *Topologies of Power*. London: Routledge.

Berger, J. 1974. *The Look of Things*. New York: Viking.

Bhambra, G. 2014. *Connected Sociologies*. London: Bloomsbury.

Connell, R. 2007. *Southern Social Theory: The Global Dynamics of Knowledge in Social Science*. Cambridge: Polity.

Cooke, P. 1989. *Localities*. London: Unwin Hyman.

Harvey, D. 1987. "Three Myths in Search of a Reality". *Environment and Planning D: Society and Space* 5 (4): 367–76.

Morgan, K. J. & A. Sayer 1988. *Micro-Circuits of Capital: "Sunrise" Industries and Uneven Development*. Cambridge: Polity.

Ollman, B. 1971. *Alienation*. Cambridge: Cambridge University Press.

Radcliffe, S. (ed.) 2017. "Special Issue on Decolonising Geographical Knowledges". *Transactions of the Institute of British Geographers* 42 (3): 329–484.

Said, E. 1978. *Orientalism*. New York: Pantheon.

Sartre, J.-P. 1963. *Search for a Method*. New York: Knopf.

Sayad, A. 2004. *The Suffering of the Immigrant*. Cambridge: Polity.

Sayer, A. 1982. "Explanation in Economic Geography". *Progress in Human Geography* 6 (1): 68–88.

Sayer, A. 1992. *Method in Social Science*. 2nd edition. London: Routledge.

Sayer, A. 2000. *Realism and Social Science*. London: Sage.

Young, I. M. 2013. *Responsibility for Justice*. Oxford: Oxford University Press.

CHAPTER 8

DOREEN MATTERS: WAYS OF UNDERSTANDING AND BEING IN THE WORLD

Núria Benach and Abel Albet

If there were one slogan that set our pulses racing in our university days, then it was "Geography Matters!" The title of an Open University publication (Massey & Allen 1984), it captured in just two words the potential, not only of the discipline of geography, weighed down by its centuries of service to imperialism, but also of a way of understanding and being in the world. "Geography matters" was a call to arms, to defend the relevance of the questions that Geography has always raised, regardless of the responses provided over the ages, and to highlight the pressing geographical issues currently on the table awaiting a solution. Over the years, and having gotten to know Doreen and the depth of her spatial thinking, we could not resist doctoring that title and reconverting it into *Doreen matters*, and so merge the author with her work. Little did we imagine that the happiness with which we took up this task would quickly be converted into such deep sadness as we dedicate *Doreen matters* in the wake of her death.[1]

In 2012, we published a book on the life and intellectual trajectory of Doreen Massey in Spanish (Albet & Benach 2012). In the process of preparing that book, we undertook an exhaustive reading of her work and we shared many hours of conversation with Doreen that allowed us to explore her ideas further and to learn how to use them in our own work. But, as important as her intellectual legacy is, we liked Doreen for the way she was, for the constant example she gave us of how to belong to an academic world without submitting to its unjust and unjustifiable rules, and how to intervene in political debates and public life with a frank commitment and without personal benefit. As we hope to have captured in the title of this article, Doreen Massey showed us a way of understanding the world that was inseparable from her way of being in it.

---

1. "Doreen matters! En la desaparición de Doreen Massey." 2016. Espais Crítics March 14. http://espaiscritics.org/doreen-matters-en-la-desaparicion-de-doreen-massey/.

In 2005, Massey published *For Space*, her detailed, personal and theoretical reflection, which was not at all a work on space per se, but rather a work that argued for the "possibilities and potentialities enabled by space(s)" (Anderson 2008: 229). Massey began her book by stating, "I've been thinking about 'space' for a long time" (2005: 1). A suspiciously straightforward beginning, yet a statement, in fact, which reveals much about her personality and her work. Her directness was not only manifest in her manner; it was also the way in which she positioned herself publicly in her writing.

With one foot in academia and another in politics, Massey wrote and published prolifically although, as she herself claimed, she spent less time writing and publishing than she did thinking. It is impossible to imagine her sitting at her typewriter in front of a blank page (let alone in front of an empty screen: she hated computers). Doreen worked with ideas, moulded them, compared and contrasted them, confronted them one with another, presented them, and went back to them after they had been discussed. Perhaps, for this reason, her work remains full of life, pulsating to the rhythm of an almost musical beat: one ... two ... three ... the rhythm that was the result of the ordering of her ideas scattered in her numerous notebooks that sooner or later would find their way into print.

At the same time, by placing space between inverted commas in that deceptively simple opening, Doreen wished to make it clear from the outset that she intended take up a combative stance *against* a single, natural definition of the term. Her reflections on space, seen through different prisms, but always with the intention of breaking with that dominant view, closed, static and devoid of any political significance, are the ones that guided Doreen's thinking for more than thirty years and have placed her at the forefront of geographical ideas where she displayed one of the most attractive, innovative and influential ways of thinking in geography, the social sciences and the humanities.

In this chapter, we wish to explore the life and intellectual trajectory of Doreen Massey by examining her contributions to three major questions: what is the purpose of geography, why is space important, and what is the role of the intellectual?

GEOGRAPHY MATTERS

Massey was not known for the great lengths she was prepared to go to defend her discipline; rather, she strove to provide concepts that might serve as bridges with other branches of knowledge and which could be useful to them. She did not advocate the dissolution of the discipline's boundaries, something she considered naive and even dangerous, but instead preferred to breach these boundaries in as constructive a manner as possible (Massey 1999c). Thus, in defending geography, she was ultimately defending a certain way of understanding the world and not an academic discipline weighed down as it was with its history and its power relations.

The current relevance of geography as knowledge, Massey claimed, should be beyond any doubt in an age faced by the crisis of globalization, by the dramatic inequalities both between and within countries, and by environmental crisis and climate change, all of them issues that call for the crucial insights of geographical analysis. Moreover, the need to integrate knowledge of the "physical" and "social" world is now undeniable. Massey was an early proponent of this necessity, and claimed that geography was one of the few disciplines that could occupy this space between the human and the natural sciences in order to address the complex environmental problems that the world faces. Indeed, she noted that the established divide between "human" and "physical" sciences was leading to a growing distance between physical and human geography, which left her decidedly perplexed (Massey 1999g: 261).

Massey's perspective on geography was shaped by her dynamic conception of space (of which, more below). Breaking with the traditional Kantian opposition between space and time, between geography and history, Massey deployed an entire conceptual apparatus and applied it to concrete situations concerning globalization, places and regions. The divisions between the physical and human sciences, between space and time, were not the only ones that Massey set out to question and overcome during her career, however. Her work constantly sought to avoid the use of established, but rarely questioned, dichotomies: economic-cultural, regional-international, gender-class, general-specific, differentiation-homogenization. Likewise, the oppositions between theory and practice and between abstraction and reality do not appear in her work, which is characterized precisely by breaching these false boundaries that make stagnant categories out of what are in fact constructions of a single reality (Smith 1986: 350). Massey's academic ideas could become very sophisticated, but they never ceased to be practical: they always had their feet on the ground (Wills in Allen *et al.* 2009). In fact, Massey's work was always based on strong empirical foundations, always preferring the intensive analysis of phenomena as the basis for the building of hypotheses and new theories.

Her position was not that far removed from the postulates of the realist approach to geography proposed by Sayer (1992 [1984]), who defended the explanatory capacity of the discipline to explore specific situations. In examining the idea that to generalize is not the same as to explain, Sayer advocated identifying explanatory elements by exploring specific cases, which should be studied in all their complexity (see Sayer, this volume). Likewise, Massey's methodological approach was not that different from the rejection of the grand narratives typical of poststructuralism and postmodernism. Edward Soja, a staunch advocate of a radical postmodernism with a spatialized vision of reality, even went so far as to present Massey's vision as being fully in tune with this postmodernization of geography. With his typical enthusiasm, Soja referred to various studies that seemed to share this perspective: "They are telling us that 'geography matters', that space and place make a critical difference, that *nous sommes tous géographes!*" (Soja 1987: 289).

Massey pointed out, however, that her rejection of grand narratives and her unwillingness to adopt grand explanatory frameworks (such as progress, modernity, modes of production etc.), positions that characterized postmodern thought (and also, of course, poststructuralist geographies that reject space as something static, associating it, instead, with dynamism and heterogeneity), should not prevent our seeing the broader picture and the general implications of what we analyse. Because "a big story" is not the same, Massey claimed, as "a grand narrative" in which the end is always known (be it progress, globalization, communism etc.); in a big story, all we know is that there are "big things" taking place in the world, allowing us to reach a higher level than that of micro interrelations (Massey in Allen *et al.* 2009). To find this general framework, there is no other path than in-depth empirical work, because proximity to the concrete is precisely what differentiates the big story from the grand narrative. The analytic framework that Massey used to interpret all these pieces of information was a very profound and well-orchestrated notion of space that helped to make sense of even the apparently more insignificant details.

A RELATIONAL VIEW OF SPACE

Massey's conceptual proposals about space have contributed to transforming what forty years ago was unquestioningly assumed as true. As Ash Amin points out, an idea as widely accepted today as the notion that space occupies a central position in the constitution of the social was by no means that obvious when Massey was beginning to formulate her conceptualizations of space (Amin in Allen *et al.* 2009). In the sixties and seventies, places were seen either as closed historical entities, products solely of their own past or, at times of neopositivist effervescence, as abstract points on a geometric surface. Massey, taking a completely different approach, was one of those that contributed most to politicizing space by means of three basic ideas: space as the product of relations, as the sphere of multiplicity and as always under construction (see Massey 1999e, 2006a, and also, more extensively, Massey 2005).

First, space is defined as the product of the interrelations and interactions at all scales: "space is a product of practices, relations, connections and disconnections. We make space in the conduct of our lives" (Massey 2006a: 94). Space is thus seen not in an essentialized way, as something already constituted, but rather as being constructed through existing relations. Thus, identities, subjects, and spaces do not pre-exist but are built through the relations with other subjects and with other spaces; this is the basis of Massey's relational conception of space. This relational vision has been criticized on occasion – especially from within economic geography – for being excessively concerned, it is argued, with networks, flows and change and for being more interested in the personal and in social relations than in mechanisms and underlying macrostructures.

Curiously, on other occasions, critics have pointed out that it was the insistence on the specificities of place which did not adequately accommodate this "more global" space of flows (Taylor *et al.* 2008). Yet, neither criticism takes into account that relational thinking pursues precisely the convergence of everything (the micro and the macro, the contingent and the institutionalized), so that any approach that would lead to their separation would necessarily be inadequate. For Massey, there should not be any incompatibility in the dichotomy that Manuel Castells (1989) identified between space of places and space of flows. Her reconceptualization of space consisted precisely in thinking relationally without denying the existence of territorial entities such as nations, regions, places etc. It was, ultimately, about not bringing territories and relations into opposition, but thinking about territories relationally.

Second, space is also the sphere of the possibility of the existence of multiplicity, the sphere therefore of coexisting heterogeneity. And not only that, but the same space is also constituted by multiple dynamics and sources through which cities and regions are formed: historical legacies, external influences, institutional forces, connections between places, local social relations etc. Space, then, is the dimension of multiplicity, the dimension of the social. So, the fundamental sociopolitical question is that raised by the very existence of space: "How are we going to live together, to coexist? It is space that presents us with the challenge (and the pleasure and the responsibility) of the existence of 'others'" (Massey 2009b: 18). In such a conception, the analysis must necessarily be conducted from below, to be able to trace the existence of networks and relationships and to detect the plurality of actors that, in short, *make* places. Space is therefore a political arena in which power relations cannot be reduced solely to those of the powerful; relations are significant among those who do not wield dominant power, in the aspirations for collective and truly democratic power (and therein lies the political power of a concept that typifies Massey – power-geometries).

Third, and finally, this conception of space as the product of relations is, as a result, something that is never closed nor finished, but is always under construction, in a constant process of being made. We have already noted how Massey vehemently defended a vision of geography in which space and time were intimately united. Its antithesis, consolidated over centuries during which the temporal was identified with the dynamic and the spatial with the static, was called into question and reappraised by Massey. Using the concept of space-time, Massey sought to show that without space there is no time, there is no production of history, there is no possibility of political change (Massey 1992b; 2006a). One of the ideas that Massey most wanted to combat was precisely the predominant modernist notion of space as an atemporal dimension (Massey 2003b). Massey affirmed that space is a cut through the myriad stories that occur at any one given moment: "a simultaneity of stories-so-far" (Massey 2005: 12).

BEYOND THE PURELY ACADEMIC

Despite the remarkable quality of her theoretical studies, Massey always sought to avoid theorizing for the sake of theorizing. As she herself claimed, her sources of inspiration could not be closer than those of her immediate world: often these sources were small, everyday occurrences. The reality that inspired her was, after all, the eventual focus of her work, so it would not have made much sense if the theory had come from anywhere else.

One aspect that the speakers invited to *Spatial Delights*, the 2009 workshop organized to engage with the work of Doreen Massey, all stressed was the inseparability of Massey the intellectual from Massey the activist. Massey explicitly and intentionally took up a position beyond the confines of academia, but not without first questioning how to do so, what role to adopt, what responsibilities to take on (Massey 2008). As she stressed, it is not that we have to study what is going on in the street, it is that the very issues emerge in the street! Here, we shall examine some instances in which she showed how her role as an intellectual also had a very obvious political dimension: in her feminist thinking, in her "local" activism in her home city of London and in her joint undertakings with people in other parts of the world.

Massey was a feminist from the outset, but it was something that became more obvious after entering elitist Oxford, where she experienced first hand the double oppression of class and gender. Her anti-essentialist thinking (which she attributed in part to Althusser) nourished her vision of social groups, whose construction owed more to the spatial organization of social relations than to any intrinsic characteristic. Her militant feminism was always more closely related to this vision than to any dedication to gender studies per se. For Massey, her feminist geography represented much more than a simple dedication to gender studies: the feminist perspective should be present in all occupations; it was a way of doing things. "I want to have feminists everywhere, in nuclear physics, in geomorphology, in human geography, etc. Studying everything as a feminist, not just studying women or gender" (Massey with HGRG 2009: 405). As was said in a review of a collection of her essays, *Space, Place and Gender*: "It could be said that her essays are academic-personal, that the academic and the personal are inseparable, that one is integral to the other. This would be a gendered geography" (Slater 1995: 224).

It was in London that Massey was most explicitly engaged in political activity. In the seventies, she was not only actively involved in the feminist movement, but also sat as an expert on the Labour Party's National Executive Committee. Massey spoke of her difficulties of reconciling herself politically with the committee (she found herself much further to the left than the official positions supported by the party) and of having to make her own appearance ("a short blonde girl who didn't even wear a suit..." as she described herself) a statement of class and gender (Massey 2008).

In the 1980s, one of the episodes that Doreen Massey and her comrades-in-arms recall with greatest satisfaction was their role on the Greater London Enterprise Board, the economic arm of the Greater London Council, which Margaret Thatcher would abolish in the middle of that decade to counter the enormous political weight of the London metropolis. Massey spoke of just how important and difficult it was to be obliged to take up constructive and responsible positions.

But Massey would never accept that constructing politics from below – that is, having to negotiate on multiple fronts at the same time, while trying to steer a more progressive line – should be labelled as "applied work" (which would suppose the existence of a prior study or theorization that could simply be applied without further ado). A constant feature of Massey's work was that it was not something that had been conceived and undertaken in academia in order to then be applied in practice. The non-unidirectionality of her work was evident even in her central, foundational role in *Soundings: A Journal of Politics and Culture*, a non-academic journal (albeit with the conspicuous presence of many university professors among its contributors). It was also evident in her explicit will to intervene in the political debates of the day, at the same time as she organized seminars, discussion forums, and more aimed at promoting genuine communication.

London was the trigger of her theoretical thinking over the decades, as it would be once again in her ideas about the responsibility of the financial metropolis in a global world. In *World City* (2007), apparently a book centred on London (although not strictly about London, as Massey warned, but about subjects that arise from London), she brought together many of the issues that had concerned her throughout her long career. The book has received a lot of praise, much of which is related to Massey's ability to produce a text that is theoretically shrewd, politically astute and accessible to the lay reader, amply demonstrating the sort of contribution that can be socially useful. In 2006, she summed up her arguments in the article "London Inside-Out" published in *Soundings* and based on her interventions in the European Social Forum and other forums of public debate (Massey 2006c). Massey wanted to show that the financial and economic success of the city was related to growing world poverty and inequality. She exhorted her readers not to forget what it means to be a metropolis in these times of neoliberalism and globalization (Peck 2008: 375). To understand London, we need to understand what goes on beyond London by showing the "other side of the picture" of the global sense of place: "This is the other geography, the external geography if you like, of a global sense of place" (Massey 2007: 7).

In the negative reviews that this book received (for example, Paul 2008; Short 2008), the academic world demonstrated a certain misunderstanding of the type of work that Massey sometimes undertook. Massey deliberately produced studies that broke out of the academic straitjacket, spurning the obligation to demonstrate her knowledge of this or that contribution, and instead concentrating on the idea that she really wanted to convey, often without distinguishing between her academic and non-academic audiences. In this regard, Phil Hubbard (2008) identified her as one of the few genuine "public intellectuals" in British geography, which alluded to a social role that transcends the narrow framework of the university. Indeed, it was a characterization with which she herself did not feel entirely comfortable. In this case, her exploration of the role played by London in the global system led Massey to reflect on the changes that had occurred in the world over the preceding 30 years. This is the perspective she took in *World City*:

> We are told that the current crisis is global and was made in the USA (in other words, don't blame us). It is certainly global in its repercussions and certainly the initial trigger was subprime markets in the USA – though it could easily have been something else. But the preconditions for it, the immense cultural changes it wrought, were, in part, invented and established in the UK, in the City of London, whose unique selling point has been its lack of regulation.
> (Massey 2009a: 141)

In later years, and publishing now mainly in non-academic journals like *Soundings*, Massey sought to present an in-depth analysis of the global financial crisis (Massey 2010, 2011a). One of the aspects she was to focus her attention on was the examination of a conjuncture characterized by neo-liberal ideological hegemony and on the possibilities for bringing about a break that might shift the balance in social power. For Massey, the ideological terrain that dominated analyses of the crisis had been tilled for decades with the invaluable help of the mass media, so that all economic questions had acquired a nature of inevitability, to the point that they no longer formed part of the political debate. For this reason, the only possibility of initiating change (of creating a "conjunctural rupture", as Massey would say) was to change this conceptualization of all things rendered economic, and with the economic serving as a kind of external natural force about which there could be no dis-cussion. And what Massey sought was to clear the way for such discussions. First, Massey insisted on questioning the ideas, which were blindly accepted, about economic growth (and especially, about the accepted sequence: first we grow, then we redistribute). Second, she called into question the false comparisons between equality and freedom that often led to the denoun-cing of the egalitarian experiences of other countries as non-democratic, forgetting that equality is one of the basic principles of the democratic trad-ition. Massey went so far as to say that "neoliberalism represents a threat to democratic institutions" (Massey 2011a: 37). And, third, she advocated for the recovery of the lost sense of collectivity, both culturally and politically, undermined by decades of individualism, based on a sensitivity to the new forms of collective organization that were emerging. And, as always, reflec-tion should serve for debate and for action. Massey herself shared these ideas during the occupation of the steps of Saint Paul's Cathedral in London in October 2011, when she referred to the dramatic consequences of the dom-inance achieved by the financial sector over the last thirty years and, in par-ticular, to the "invasion of the imagination" that blighted the possibility for alternative ideas and readings.

Her involvement in political work and her commitment outside her own country were, according to Massey, a great challenge for determining how she should position herself in relation to them. Massey's diverse experiences in Nicaragua, Mexico, South Africa and Venezuela forced her to consider extremely carefully the role that she, as an intellectual based in the global North and writing in English, could play. Here, we should stress, that she was at pains to undertake these experiences from the perspective of cooperation and mutual learning, rather than from that of an expert who provided advice on the actions to adopt in these different contexts affected by a great diver-sity of problems. Perhaps for this reason, her travels through these countries left few publications beyond a number of reports and other texts – published in English and Spanish – and turned into interesting teaching materials, as illustrated by some examples gathered in her Open University materials.

Her Venezuelan experience was, perhaps, an unexpected exception even by Massey's standards, because of the leading role that her concept of power-geometries was to acquire as one of the main axes of the government's political action (see also Perla Zusman, this volume). Following an initial call from the mayor of Caracas, Massey was directly faced with the use of her ideas by other people and in contexts quite different from those in which they originated. To her great surprise, power-geometries had been adopted as one of the axes of the Chávez government! The sentiments of the geographer were justifiable: "there's nothing like trying to use one of your own concepts in a real political actuality" (Massey in Allen *et al.* 2009). A deep, yet abstract, concept found a form of expression in practice through a far-reaching restructuring of territorial organization, the aim of which was to change the nature of power, bolstering its popular and communal dimension, which, not unexpectedly, gave rise to many fronts of political resistance. There can be no doubt that geographical concepts and the connection between space and power so often defended by Massey had truly penetrated the Venezuelan government structure.

According to its many critics, however, this was simply an ideological proposal that sought to concentrate, centralize and monopolize economic power (Estaba 2007; Banko 2008). Rojas López and Pulido argued that power-geometry was nothing but:

> a confused adaptation of the discourse of Radical Geography, which while it sees territory as a strategic category to understand and overcome the contradictions that emerge among the social actors, also rejects a[ny?] pyramidal conception of power, since it defends the principle of territorial subsidiarity and the specificities of local governments in global contexts. [...] In essence, it was an attempt at imposing a new political-administrative division without any precedent in the history of the country, created from above, a parallel structure that reduced the powers, functions and resources of the federal state governors and the mayors of the municipalities.   (2009: 93; our translation)

Massey (2011c) was not entirely averse to these criticisms. Indeed, she had shown herself to be critically sympathetic of Chávez's Bolivarian project and, in particular, of the application of the power-geometry concept to Venezuela's internal geography. For example, one objective of the Bolivarian Revolution with which Massey naturally sympathized was to give voice to those who had never before been heard by establishing political entities at the community level (the so-called *consejos comunales* or communal councils). However, Massey was keen to point out that it was not enough to define territorial entities to achieve this goal, and that "popular power" required establishing the rules of relation between and within different entities as well as clarifying the terms by which each was to exercise power. Another example: Massey stressed that in the Constitutional Reform, the new power-geometry concentrated a lot of power in the hands of the president, in what was a rather centralized geometry of power relations (as the anti-Chavista opposition also indignantly pointed out); and while this had the counter-weight of growing popular power, the latter occupied a very different time-scale: "although the formal structure (the president's powers) is established in very little time, the development of a proper political culture requires a lot of time and demands political work" (Massey 2009b: 23). And what is most interesting about the case is that Massey made all these observations while in Venezuela in published material that was neither governmental nor academic, but instead pamphlets that were circulated among the people and debated by them. With all her doubts and reservations, Massey's commitment to Venezuela represented the way in which a public intellectual felt; she ran the risks of such involvement (of being misinterpreted, of being criticized, and even of falling into self-contradiction). Massey showed that if geography is going to be relevant and shape the public sphere on par with its theoretical arguments, its theoretical concepts must be tested, enriched and modified in practice: they must, necessarily, be concepts that can go the distance.

REFERENCES

For works authored and co-authored by Doreen Massey, please see Select Bibliography of Doreen Massey, beginning on p. 371.

Albet, A. & N. Benach 2012. *Doreen Massey: Un Sentido Global del Lugar*. Barcelona: Icaria.

Allen, J., A. Amin, J. Peck, *et al*. 2009. *Spatial Delights / An Engagement with the Work of Doreen Massey*. 11 March. London: Royal Geographical Society. www8.open.ac.uk/researchcentres/osrc/events/spatial-delights (accessed 11 January 2018).

Allen, J., D. Massey & A. Cochrane 1998. *Rethinking the Region*. London: Routledge.

Anderson, B. 2008. *"For Space* (2005): Doreen Massey". In P. Hubbard, R. Kitchin & G. Valentine (eds) *Key Texts in Human Geography*. London: Sage Publications, 227–35.

Banko, C. 2008. "De la Descentralización a la 'Nueva Geometría del Poder.'" *Revista Venezolana de Economía y Ciencias Sociales* 14 (2): 165–81.

Castells, M. 1989. *The Informational City: Economic Restructuring and Urban Development*. Oxford: Blackwell.

Estaba, R. 2007. "El Sacudón Territorial en Venezuela". *Scripta Nova* XII (716). www. ub.edu/geocrit/b3w-716.htm (accessed 10 January 2018).

Hubbard, P. 2008. "Introduction. Book Review Forum: *World City*". *Area* 40 (3): 411–12.

Paul, D. A. 2008. "Imagine There's No Countries". *International Studies Review* 10: 635–8.

Peck, J. 2008. "Book Review on *World City*". *Economic Geography* 84 (3): 375.

Rojas López, J. & N. Pulido 2009. "Estrategias Territoriales Recientes en Venezuela: ¿Reordenación Viable de los Sistemas Territoriales o Ensayos de Laboratorio?" *Eure* XXXV (104): 77–101.

Sayer, A. 1992 [1984]. *Method in Social Science*. 2nd edition. London: Hutchinson.

Short, J. R. 2008. "Book Review on *World City*". *Annals of the Association of American Geographers* 98 (4): 949–50.

Slater, F. 1995. "Book Review on *Space, Place and Gender*". *Gender and Education* 7 (2): 222–4.

Smith, N. 1986. "Book Review on *Spatial Divisions of Labour*". *Geographical Review* 76 (3): 350–2.

Soja, E. 1987. "The Postmodernization of Geography: A Review". *Annals of the Association of American Geographers* 77 (2): 289–94.

Taylor, P., M. Hoyler, K. Pain & J. Harrison 2008. "A Global Sense of Flow? Book Review Forum: *World City*". *Area* 40 (3): 413–15.

CHAPTER 9

JUST CARRY ON BEING DIFFERENT

Susan M. Roberts

Former US Speaker of the House of Representatives, Tip O'Neill is credited with the observation that "all politics is local". Geographer Doreen Massey, more than anyone, has brought clarity to the complicated conceptual and political questions that this well-worn phrase brings in its train. Her work addressed questions such as: What is the local? How do we conceptualize places? Can politics be simultaneously rooted and progressive? What kinds of complications do feelings of belonging, home, and place bring in a globalizing capitalist world? Are such attachments bound to be reactionary and/or exclusionary? How do we theorize in ways that take into account social inequality and differences of all sorts? In addition to developing carefully worked-out, politically progressive, arguments on these issues, Doreen Massey articulated her arguments in distinctive ways. Doreen Massey's voice and her presence were different from those of other leading scholars, and this difference mattered.

PLACE, DIFFERENCE AND DEBATE

Questions about the local, and about place, globalization and difference remain important questions, but they were especially pointed in the 1980s and early 1990s. I will start by discussing the significance of Doreen Massey's thinking through personal recollections of the impact of her thinking on me as a young geographer during those very stimulating times. In the mid-1980s, I was teaching at a technical college in Cambridge (1986–7 academic year) and I used to sneak into Open University lectures and other free talks at Cambridge University given by the likes of Derek Gregory, David Harvey and Edward Soja. Their lectures were all about retheorizing capitalism, about structure and agency, and about the role of space in social life. These scholars (among others) were bringing ideas about postmodernism and postmodernity into geography. Their readings of French and German theorists and their arguments about space and how the experiences of space and of time were changing seemed fresh, exciting and important – and they certainly changed geography.

The end of the decade, 1989 to be exact, saw the publication of Edward Soja's *Postmodern Geographies* and David Harvey's *The Condition of Postmodernity*. These books remain landmark contributions to social thought and still stand as key texts, and they more-or-less defined and prompted much of the most exciting debate in human geography in the late eighties and early nineties. Identifying the time as one of unprecedent-edly rapid capitalist globalization, with associated immense sociocultural changes and reconfigurations of urban space, Soja and Harvey (and other contemporaries such as Michael Dear) variously drew upon theories of postmodernity developed by Fredric Jameson and others to make sense of the on-going changes they observed, captured in the phrase "time-space compression", and the ways these changes were experienced – especially by inhabitants of cities in the global North.

1989 was also, I realized later, when Catherine MacKinnon's *Toward a Feminist Theory of the State* was published, and it was the year Nancy Hartsock's "Postmodernism and Political Change: Issues for Feminist Theory" paper came out. Linda Nicholson's *Feminism/Postmodernism* (an edited collection) appeared in 1990, the same year as Judith Butler's *Gender Trouble,* and Patricia Hill Collins' *Black Feminist Thought.*

But though some in geography were engaging the theoretical advances in Marxist feminism and Black feminism in these and other texts, these ideas were not even acknowledged, much less engaged, in the ballroom sessions at the annual Association of American Geographers' meetings at that time. That encounter was happening only in a few of the smaller rooms in the far-thest reaches of the conference space. At the end of the 1980s, I was a student in a US graduate programme and while feminist ideas were being discussed in fields such as anthropology and sociology on my campus, they were not on the agenda in geography. The graduate seminar reading lists I encountered on geographic thought, and those that introduced contemporary debates, may have occasionally included works by the likes of Linda McDowell and Doreen Massey, but not as centrally featured key texts.

Nonetheless, I remember closely reading and studying Doreen Massey's 1991 paper "Flexible Sexism" – along with Rosalyn Deutsche's "Boys Town" which appeared in the same issue of *Society and Space.* Both papers were wholesale critiques of the dominant geographical treatments of postmod-ernity, and both drew deeply on feminist thought. I also remember finding Doreen Massey's 1992 paper "A Place Called Home?". It was clear from these and other papers that Doreen's understanding of what was happening at that time, especially – but not only – in cities of the global North such as London, was very different from Soja's and Harvey's. And her approach was quite different. She wrote "A Place Called Home?" in counterpoint to what she characterized as: "The most commonly argued position" regarding "the vast current reorganizations of capital". She reported that she saw that a "special style of hype and hyperbole has been developed to write of these matters. The same words and phrases recur; the author gets carried away in a reeling vision of hyperspace." Massey said she didn't write that way because it allowed a kind of blindness to how the times were experienced by those not in charge of, or benefitting from, globalization. She reminded her readers that

> amid the Ridley Scott images of world cities, the writing about skyscraper fortresses, the Baudrillard visions of hyperspace ... most people actually still live in places like Harlesden or West Brom. Much of life for many people, even in the heart of the first world, still consists of waiting in a bus-shelter with your shopping for a bus that never comes. Hardly a graphic illustration of time-space compression. (Massey 1992a: 8)

I never forgot Doreen's person waiting at the bus stop. It is such a poignant and powerful image, and it has been stuck in my mind since I first read it. I even imagined a kind of Edward Hopper visual to go along with the scene, in which it is (of course) dreary and raining, and the person in the bus-shelter is an older woman, probably a pensioner in a rather worn coat, waiting with her heavy shopping bags.

But, it is not just a question of writing style, or of privileging one type of experience. Rather, as Doreen was at pains to point out, the main problem with dominant geographical works on globalization was that they did not recognize difference and the significance of difference.

As she put it: "Different social groups and different individuals are located in many different ways in the new organization of relations over time-space" (1992a: 9).

> From jet-setters, to pensioners holed up in lonely bed-sits, to Pacific Islanders whose air and sea links have been cut, to inter-national migrants risking life and livelihood for the chance of a better life ... all in some way or another are likely to be affected by the shifting relations of time-space, but in each case the effect is different; each is placed in a different way in relation to the shifting scene. (*ibid.*)

127

The way Doreen Massey approached difference was deeply empathetic, based as it was on her recognition that, for many people, time-space compression was not bringing exciting new opportunities. Her recognition of the struggles of others was also deeply critical. Drawing on feminist thinking – in this paper Massey wove together insights from bell hooks, Toni Morrison, Nancy Hartsock, Elizabeth Wilson and others – she showed how difference matters. Moreover, difference for Doreen Massey was not a flat mosaic; rather it was constructed through asymmetrical and unequal "power relations" – and she mentioned in particular the importance of social relations built upon sexism and colonialism (11). So, in the 1992 paper, Doreen Massey argued that how time-space compression is experienced depends not just on the actual changes in the spatial organization of society and where one is positioned in relative space, but also on one's position in social relations.

To return to the person in the bus-shelter for a moment: as I mentioned, I in fact had imagined the person at the bus stop to be a woman, and this might be in part because I muddled up this image with another that Doreen Massey conjured when she was depicting what she so brilliantly called the "power-geometry of time-space compression" to illustrate this point about difference and its centrality to understanding sociospace.

"A Global Sense of Place" is a beautiful 1991 paper that many of us know and have no doubt used to help students discover their own geographical imaginations. Doreen's writing here is quite poetic, and her imagery arresting. She issued an invitation to her readers, thus:

> Imagine for a moment that you are on a satellite, further out and beyond all actual satellites; you can see "planet earth" from a distance and, unusually for someone with only peaceful intentions, you are equipped with the kind of technology which allows you to see the colours of people's eyes and the numbers on their number plates. You can see all the movement and tune in to all the communication that is going on. Furthest out are the satellites, then aeroplanes, the long haul between London and Tokyo and the hop from San Salvador to Guatemala City. Some of this is people moving, some of it is physical trade, some is media broadcasting. There are faxes, email, film-distribution networks, financial flows and transactions. Look in closer and there are ships and trains, steam trains slogging laboriously up hills somewhere in Asia. Look in closer still and there are lorries and cars and buses, and on down further, somewhere in sub-Saharan Africa, there's a woman – amongst many women – on foot, who still spends hours a day collecting water. (1991a: 25)

The imagery her words conjure, and the groundedness of her observations are so powerful. She went on:

> Now I want to make one simple point here, and that is about what one might call the power geometry of it all; the *power geometry* of time-space compression [...] Different social groups have distinct relationships to this anyway differentiated mobility: some people are more in charge of it than others; some initiate flows and movement, others don't; some are more on the receiving-end of it than others; some are effectively imprisoned by it.
>
> (1991a: 25–6)

At its most basic, Doreen was right: this is a simple point, but for her to articulate it so clearly and to make it at a time when those whose work she was criticizing were setting the agenda and were the stars of geography, took courage and persistence.

This point, about the significance of the different ways people find themselves positioned vis-à-vis sociospatial relations, built logically on Doreen's consistent emphasis on the importance of geographical variation. In the Introduction to 1984's *Geography Matters!* (edited with John Allen), Doreen wrote, in her characteristically crystal-clear prose, that: "Spatial distributions and geographical differentiation may be the result of social processes, but they also affect how those processes work. 'The spatial' is not just an outcome; it is part of the explanation" (1984c: 4). This, of course, is an argument elaborated in Doreen's *Spatial Divisions of Labour* (1984b) and also in her chapter in John Urry and Derek Gregory's edited collection of the same year, *Social Relations and Spatial Structures* (Massey 1984a). In her work in the 1980s, Doreen was particularly interested in how to understand and explain uneven development, and the drastic divisions between the fortunes of regions and their inhabitants in Mrs Thatcher's Britain (1979–90). Thus, questions about how best to conceptualize the region and place were not only intellectual but also purposely deeply political questions. Because, "the issue of conceptualization is bound up with social form and social order, and that in both maintaining and challenging social orders different forms of conceptualization of both space and nature are frequently at stake" (1984c: 10). Reflecting on this period, in the 1991 paper "The Political Place of Locality Studies", Doreen wrote "Something that might be called 'restructuring' was clearly going on, but its implications both for everyday life and the mode and potential of political organizing were clearly highly differentiated and we needed to know how" (Massey 1991c: 269).

Doreen saw the connections between how space (and time) are conceptualized and how political possibilities (or impossibilities) are understood and realized. Likewise, she identified how thinking about space and place is deeply imbricated with thinking about gender and gender relations, race and colonialism (see, for example, Massey 1992b).

STRETCHED-OUT SOCIAL RELATIONS

An important aspect of Doreen Massey's way of conceptualizing place and space-time, was its expansiveness. By this I mean that Doreen's approach to places as bundles of connections, as intersections of movements of all sorts, past and present, prevented her analysis from being constrained by the boundaries that characterized the subdisciplines of human geography at the time. During the 1980s and 1990s, economic geography was largely focused on issues impacting countries of the global North. Moreover, dominant Anglophone economic geography focused on Anglophone countries, and within them, on cities. And, as Doreen (and other feminist geographers) showed, even within this setting, the lives and work of many ordinary people (notably women) were not considered central. So, the actual empirical grounding of economic geography was quite a highly circumscribed swathe of human experience. Development geography, in contrast, focused on the so-called Third World, and the problems and challenges of economies and societies in the global South.

Doreen, though, did not restrict her analyses to just what was going on in the places in the global North she studied, such as London, Cambridge or the Midlands of Britain. She also studied and was deeply involved in policy-making in Nicaragua, for instance. More particularly, even when her subject matter was the urban experience in the global North, a crucial part of Doreen's distinctive conceptualization of place was its formative situatedness in webs of "stretched out" social relations intertangling people, households, neighbourhoods and cities. Famously, in some of her writings, including a "A Global Sense of Place", Doreen used her own neighbourhood, Kilburn in north London, as a way to explain how rethinking place in this way is possible.

> It is (or it ought to be) impossible even to begin thinking about Kilburn High Road without bringing into play half the world and a considerable amount of British imperialist history (and this certainly goes for mining villages too). Imagining it this way provokes in you (or at least in me) a really global sense of place.
>
> (1991a: 28)

Later on in "A Global Sense of Place", Doreen urged readers to get back in your mind's eye on a satellite; go right out again and look back at the globe. This time, however, imagine not just all the physical movement, nor even all the often invisible communications, but also and especially all the social relations, all the links between people. Fill it in with all those different experiences of time-space compression. For what is happening is that the geography of social relations is changing. In many cases such relations are increasingly stretched out over space. Economic, political and cultural social relations, each full of power and with internal structures of domination and subordination, stretched out over the planet at every different level, from the household to the local area to the international.   (1991a: 28)

JUST CARRY ON BEING DIFFERENT

It is difficult to separate the impact of Doreen's thinking from the impact of her as a person. So, to conclude this chapter, I will turn again to this theme, and especially to the considerable and formative force of her presence as a scholar.

All these debates about space, place and politics were going on as I was finishing my PhD, and in 1991 I got a job at the University of Kentucky. I had a chance to invite speakers and the first two people I invited were Susan Christopherson and Doreen Massey. Susan gave a talk on globalization in our Department of Geography colloquia series and Doreen came as part of a multidisciplinary seminar run by the Committee on Social Theory in Spring 1994. As part of her time in Kentucky, Doreen gave a brilliant lecture to a packed room – a lecture that I can remember vividly. She also joined in intense classroom discussions with students and colleagues. Doreen enjoyed going out and about in Lexington with colleague Rich Schein, curious about how the tobacco industry worked, tramping through pungent warehouses full of tobacco bales and chatting easily with the men who worked tobacco auctions that were still going on in Kentucky at that time. While she was in Kentucky, Doreen was interviewed by three graduate students in the social theory programme, including geographer Jeff Popke. In that interview, she spoke of how she had "tried to do something different".

I think in a way I've tried to do something different. It's not to speak exactly as I would have liked to have spoken, had the context been different. I can't speak like some of these big guys do. If you are five foot one, and you are fair-haired, and you are female, and quite often you can barely see over the podium, then just physically and materially you cannot be imposing in the same way that you can when you are six foot five and have a big male deep voice. The very physicality and materiality of it, as well as the fact that they just take those people more seriously than they take us, starts you off in a different situation. So what I've tried to do is just carry on being different.   (Ijams *et al.* 1995: 101)

As someone who never heard a lecture from a woman professor as an undergraduate, who had to go outside geography in graduate school to encounter female professors, and who was the only woman faculty member in her department at that time, I found Doreen's frankly critical but resolutely practical and unpretentious approach to "carry on being different" something I could immediately and profoundly relate to.

In that same interview, Doreen talked with the students about how she saw important differences between being an intellectual and being an academic. She spoke about how her time at the Greater London Council was exciting but also challenging:

as an intellectual and an academic, some of the toughest questions were posed to me by that very practical situation of having to decide where we were going to put investment money, having to decide which groups we were going to support, having to decide what to do when a firm went bankrupt. Those decisions ask much tougher questions about your theoretical work than just working as an isolated intellect; I really do believe both that you shouldn't separate theoretical work from empirical work, and that the engagement in the rest of life is part of being an intellectual.
(Ijams *et al.* 1995: 100)

She went on:

It might help us then, to separate out different kinds of inter-relations between being academic and being intellectual. I do see them as very different. I see myself as being an intellectual in the sense that I am very interested in the kinds of intellectual ideas I am working in and working around, and my particular take on things is one which is often biased in that direction. That is where I would see I have developed particular skills, enjoy oper-ating, and can contribute. But I think that is utterly different from being an academic which is to be bound up in the institutions, competitions, and the career structures …

(Ijams *et al.* 1995: 100)

These observations distinguishing academic and intellectual identities were very refreshing: idealistic but pragmatic. They remind us of the importance of commitments to ideas and to dialogue, and warn against the pressures of competitive individualism that riddle work in higher education.

Explaining her goals in her writing and engagements, Doreen said: "What you want to do is set up a dialogue, have some influence, get people thinking" (in Ijams *et al.* 1995:101). Doreen certainly did that. Geography, and each one of us, is so much the better for her.

REFERENCES

For works authored and co-authored by Doreen Massey, please see Select Bibliography of Doreen Massey, beginning on p. 371.

Butler, J. 1990. *Gender Trouble: Feminism and the Subversion of Identity.* London: Routledge.

Collins, P. H. 1990. *Black Feminist Thought: Knowledge, Consciousness, and the Politics of Empowerment.* New York: Routledge.

Deutsche, R. 1991. "Boys Town". *Environment and Planning D: Society and Space* 9 (1): 5–30.

Hartsock, N. 1989. "Postmodernism and Political Change: Issues for Feminist Theory". *Cultural Critique* 14 (Winter): 15–33.

Harvey, D. 1989. *The Condition of Postmodernity: An Enquiry into the Origins of Cultural Change.* Cambridge, MA: Blackwell.

Ijams, B. W., J. Popke & K. Urch 1995. "Gender, Space and the Academy: An Interview with Doreen Massey, The Open University". *disClosure* 4: article 9, 97–115.

MacKinnon, C. 1989. *Toward a Feminist Theory of the State.* Cambridge, MA: Harvard University Press.

Nicholson, L. J. (ed.) 1989. *Feminism/Postmodernism.* New York: Routledge.

Soja, E. 1989. *Postmodern Geographies: The Reassertion of Space in Critical Social Theory.* New York: Verso.

PART II

CONJUNCTURES

CHAPTER 10

FROM "THE" NORTH TO "THE" SOUTH: SPATIALIZING THE CONJUNCTURE IN BRITISH CULTURAL STUDIES

John Pickles

INTRODUCTION

Doreen Massey had a lifelong fascination with the social and environmental lives of places, lived space, the dimension of multiplicity and radical simultaneity, and the political possibilities of thinking space-time relationally and dynamically. This brought her into close collaboration with British Cultural Studies and its theories of the conjuncture. Drawing particularly on economic geography, the New Left, Althusserian Marxism, British Cultural Studies, the feminism of Luce Irigaray (among others) and later post-Heideggerian anti-essentialist theories, her work was important in inflecting these perspectives with a deep and abiding commitment to the "mattering" of space.

In this chapter, I focus on Doreen's contributions to these crucial progressive movements in contemporary British Cultural Studies. In conjunction with her colleagues at the Open University, her writings on critical feminist geographies of place, space, environment, gender and geography have come to define the practices and goals of a conjunctural analysis for the Left, and have challenged taken-for-granted explanations of regional decline, peripheralization and inequality, the privileging of high-tech fantasies, the reworking of development logics around neoliberal economics and the domination of financialization logics and interests in framing London (and wider British) futures.

PLACE, SPACE AND POLITICS

From her position at the heart of economic geography, and throughout her academic, professional and wider social life, Doreen Massey was a key contributor and interlocutor in British Marxism, the New Left, feminism,

British Cultural Studies and a series of major institutions they influenced. Like others influenced by critical readings of Marxism and postwar continental philosophy, Doreen was committed to challenging the hegemony of a temporal understanding of radical politics in favour of what she variously referred to as a relational understanding of place, dynamic concepts of space-time, non-binary concept of nature, and non-functionalist epistemologies of causality (Massey 1973, 1978, 2000b). If time was the dimension of sequences of actions and experience, space was the dimension of radical simultaneity and multiplicity. As she argued repeatedly with Ernesto Laclau, understanding change purely in temporal terms elided the complex interweaving of space and time that drove the actual content of concrete events (Massey 1995c; Mouffe 2013). Such ways of understanding space-time and its related concepts of place, region and nature offered new opportunities for radical democratic politics to question the ways in which power worked. Since space is never flat, static or inert, but was always dynamic, active and unfinished, it is constantly produced and reworked (Massey 2005; Massey & Warburton 2013; Featherstone *et al.* 2013). In this view, social and spatial relations were not separate, one determining the other, but always overdetermined in relationship to each other (Rustin 2013). Relations of power operated between people shaping the historical material forms of social life, producing complex, contested, mutable and open places (Massey 1992b, 1993b, 1995c). The result was a relational understanding of place and space that demanded an ethics focused on spatial responsibility (Massey 1994).

In order to "build" this relational understanding of the economy Doreen sought to rework how we understand and write about place, region, space and identity (Massey 1994, 2011b). Her questions were always about how places and localities can be understood in terms of the importance of rootedness and affiliation without being defensive and reactionary, and how we can understand identity if we accept that concepts such as lagging region are thoroughly social and relational constructs. As she noted in a 2013 interview: "A lot of what I've been trying to do over the all too many years when I've been writing about space is to bring space alive, to dynamize it and to make it relevant, to emphasize how important space is in the lives in which we live, and in the organization of the societies in which we live" (Massey & Warburton 2013).

This was why "Place Matters" to people living their everyday lives, making sense of the world around them, struggling to defend or change the common spaces and resources that are fundamental to their existence. It is the purely quotidian and often unacknowledged condition of possibility for movement, interaction, meaning, community and survival that is the material and symbolic context within which action and politics unfold.

Place, in this sense, is the locus of the lifeworld; the pre-given context of entanglements and commitments that make up the taken-for-granted of our lives in particular localities. But it is also a nexus of practices and meanings shaped by the broader regional space-economy and political ideology of the state.

CONTEXT AND CONJUNCTURE

Especially in conjunction with her colleagues at the Open University, Doreen insisted on the importance of understanding change in places in relation to these broader regional and state apparatuses and ideologies. But, as Peck (2005) has warned, the replacement of "place" with a broader concept of "context" would be insufficient unless the meaning of context was also more fully elaborated. As I have suggested elsewhere (Pickles 2012, forthcoming) this required a conjunctural theory of place that took seriously questions of context and causality (see Clarke, this volume).

In developing this dynamic understanding of regional economic geography and a richly spatialized historical materialism, her work eschewed the formulaic abstractions of both regional science and structural Marxism, in favour of the New Left's engagement with social history, cultural theory, the production of space, hegemony and democratic politics, anti-essentialism, deconstruction and feminism. If her writings were inflected by, and often produced in collaboration with her Open University colleagues, and her intellectual models included E. P Thompson, Henri Lefebvre, Antonio Gramsci, Louis Althusser, and later Heidegger, Foucault and Derrida, her interlocutors were found among parts of the British Labour Party and the Greater London Council, her colleagues in British Cultural Studies such as Stuart Hall, John Clarke, Michael Rustin, Lawrence Grossberg, Ernesto Laclau, Chantal Mouffe and Hilary Wainwright among others, and feminist theorists, particularly the political psychoanalytics of Luce Irigaray.

Here Doreen began to work more fully with a concept of the conjuncture that understood context as the concrete, temporary, contingent and over-determined specificities of a problematic, and she contributed substantially to its spatial understanding of context among these various groups (Clarke, this volume). For Grossberg (2006: 4–5):

> a conjuncture is always a social formation understood as more than a mere context – but as an articulation, accumulation, or condensation of contradictions ... If reality is relational and articulated, such relations are both contingent (i.e., not necessary) and real, and thus, never finished or closed for all times.

As such the concept of a conjuncture offers an understanding of space and politics as always open, without guarantees. In this sense, as Grossberg (2013) has pointed out, despite its name cultural studies is never about culture but about given socio-geo-historical contexts; in its attempt to articulate and clarify the contingent relations and contradictory causalities that constitute the conditions of life in particular places. It begins with the assumption that the world did not have to become what it is.

In conjunctural analysis, place and locality matter in at least four ways (Grossberg 2010, 2013). First, place is always relational; places are produced through the constant struggle to make, unmake and remake complex, mobile and often contradictory configurations. The regional challenge of the North in Britain is, in Massey's hands, transformed into a complex assemblage of relations and forces driven as much by the political power of London as by the internal conditions of industrial decline. Second, such concepts of place reject reductionist or deterministic analyses that attempt to impose direct and simple causality in attempts to explain how places are produced; place is always produced by the articulation of multiple and often contradictory forces operating at a variety of scales (see Sayer, this volume). In this sense, the global city and the declining region are crystallizations of complex interrelated and differently scaled projects. Third, conjunctural studies of places are always about concrete socio-geo-historical formations that are strategic, not universal, concerned with producing new interpretations and concepts that allow for other forms of political engagement and action (Mezzadra & Neilsen 2013). Fourth, places could always have been and still can be otherwise, not as something we can just choose, but as something to be produced through hard work. Only through such work are the relations and structures that constitute a particular "real" and the social forces mobilized to sustain it changed. In this sense, life is the continuous actualization – construction and reconstruction – of effective structures that constitute specific geo-historical realities of power and real places. By rearticulating a context – whether in reading the North against developmental logics of lack, understanding underdevelopment through complex power-geometries, or reading space not as theorists of gender or race, but as always gendered and raced – conjunctural analyses is always about asking the question; what alternatives were and are possible? (Massey 1991b, 1995a).

In asserting that place matters and that she was *"for space"* Doreen was focusing attention on the material and spatial forms of such a conjuncture, reading it as an articulation, accumulation, and condensation of different currents or circumstances occurring at different scales. Much like the "immigrant rocks" of the Lake District, place and nature are both highly materialized, often fixed, but always moving on, never to be revisited in the same way.

Hers is a thoroughly non-essentialist project sensitive to the contextual and embedded nature of cultural practices, but focused strategically on processes and dynamics of socioeconomic change. It asks how we can be attentive to the rhythms and contexts of everyday life, and to the diversity of forms, practices and class processes that constitute this matrix of diverse and alternative ways of living (the deepening of contemporary capitalisms, the proliferation or demise of non-capitalisms, or their articulations in concrete regional economic geographies) (Massey 2005) (see Sayer's realist interpretation and caveat in this volume). Such lines of force, determination, resistance and movement, with their own temporalities and spatialities, are what she refers to as power-geometries (Massey 1999b, 1999e, 1999g, 2005). It has come to define precisely what we mean by "place" when we say that places are always in the process of being made, always contested.

The Anatomy of Job Loss, Rethinking the Region and *Spatial Divisions of Labour* (Massey & Meegan 1982; Allen *et al.* 1998; Massey 1984b) variously repositioned the question of work and employment in the context of broader regional economic dynamics. Eshewing the reductive accounts of employment change that had come to characterize regional science, she elected instead to read regional uneven and combined development in terms of interregional dynamics, the politics of circuits of capital and the cultures and transformation of particular places (Massey 1987b; Peck 2013). *Space, Place and Gender* (1994) showed how sociospatial relations were always gendered as well as classed and raced, and hence were more open than traditional regional analysis had reflected. In "Global Sense of Place" (1991a) places were the crystallization of flows and interactions occurring at various scales; important sites of struggle, always being reworked, not annihilated, by processes of globalization. In *For Space* (2005) she developed an even more elaborated engagement with feminist and poststructural theory to elaborate her ideas on the dynamic open nature of place and space, in the process attracting the attention of feminist critiques of capitalocentrism, while crucially informing economic geographies of industrial change (see Anderson 2008; Peck *et al.* 2014). And in *World City* (Massey 2007) she focused on the ways in which the increasing ideological role of financial markets and the associated regional power of London has affected the city, its inhabitants and the wider regions of the UK. Throughout, her deployment of the concept of "geometries of power" drew attention to the complex and differential networks (or what Escobar (2001) called "meshworks") that shape economic change (Massey 1993b, 1999e).

Viewing regional uneven and combined developed in such relational terms provided important tools for political practice. In her regular writings in *Marxism Today* (Massey 1983a, 1983b, 1988a, 1991a) and later in *Soundings* (Massey 1997a, 1999f, 2010, 2011a; Hall *et al.* 1995), she stressed the need for the Left to question the deeply embedded spatial logics and stories that made neoliberalism appear to be "normal" and, instead, to focus on the structuring role of London and its nexus of economic and political power. At the Open University, she and her colleagues were crucial to reframing geographies of region, place, economy and power (e.g. Massey & Meegan (1982); Allen *et al.* (1998)). In her interactions with Ken Livingstone and the Greater London Council, particularly as a member of the Greater London Enterprise Board (GLEB), Massey was able to push for more socialist democratic policies in municipal government with particularly emphases on the needs of labour, fair wage campaigns and access to housing in the face of the increasing dominance of financial elites (Massey & Livingstone 2007; Wills 2013). In her engagement with post-development theories, she encouraged this kind of analysis, but cautioned against the all too easy acceptance of romanticized notions of place as sites of cultural defence (Escobar 2001; Massey 2002a). And in her collaborations in Nicaragua, Brazil and South Africa, among other countries, her insistence on relational ontologies of space and place opened new avenues for thinking and acting in regard to uneven and combined geographical development that placed concepts of power-geometry at the very heart of the politics of autonomous development (Massey 1986, 1987a; Menendez 2013; Saldanha 2013; but see her warning against counterposing concepts of dynamic space to place as fixed; Massey 2002a).

When Doreen asserts that "place matters" it is not a claim about the essential or unique character of a locality. Instead, place is a nodal point of a relational network that has effects. It is – as she so eloquently told us – the "throwntogetherness"; the articulation and crystallization of relations and flows occurring at many scales. Place functions as the material and symbolic horizon of locale (home, neighbourhood, village, locale, region) and of the institutional practices that give it meaning (the nation, the people, us-not-them).

We move from place to place, settle in place, materialize those movements and settlements, trade among places, dream of close and distance locales, work daily in proximity with others and travel annually to distant places, we carve out regional economies and regional systems, and we defend territory often on the basis of landscape markers and visions, which we come to identify as ourselves and as our society that must be defended. In this sense, place and community are not purified concepts bounded spatially by the locality, a point she explained in terms of the intensities of flows and diversity of actors that shaped particularly locales. It was this work on locality studies and her commitment to complex analyses of over-determination that initially attracted many (see Sayer, this volume; Hart, this volume), while also becoming the focus of strong criticism, notably from David Harvey

(1987: 369), who saw in this focus on localities a turn to empiricism and particularisms instead of a focus on the underlying structural dynamics of capital accumulation (see Featherstone & Painter 2013: 8–9).

The throwntogetherness of postwar Manchester and the North where Massey grew up was – like all localities – a complex sociospatial and potentially open formation; one she constantly returned to in thinking the conjuncture of Thatcherism' and its geographically and socially uneven efforts (see McDowell, this volume). Wartime and immediate postwar shortages gave way to gradually expanded social consumption as industries were rebuilt with state investments, new forms of social investment expanded education, health care and other public services, and consumer goods became ever more readily available and central aspects of popular culture. The combination of social welfare gains with the emergence of a Fordist virtuous cycle of technology, production practices and expanded consumption ushered in a period of collective hope in the possibilities of a national economy that might deliver economic growth and a more just society.

In the process, Northern labour markets were transformed. By the 1950s, the return of male workers from war duties stripped many women of jobs and opportunities as they were dislodged from more technical areas of manufacture. But skills learned in and around Manchester in munitions, aircraft and textiles, skills themselves based on prior decades of industrial work, also strengthened the hold of women on some areas of work. For a while, beaming, weaving and other manufacturing jobs were sustained, while service and commercial opportunities expanded in the growing towns of the North. The rapid expansion of free schooling and health care, along with subsidized council (or social) housing, also transformed the old cotton factory towns of Lancashire, opening educational and other social opportunities for working-class kids, particularly as mill, mine and shop futures were under increasing threat from international competition and changing attitudes to British trade policies, at the same time creating new outlets for working women (see McDowell, this volume).

The "North" in this context was a site and symbol of intense structures of feeling; working-class histories shaped in part by longstanding and well-rehearsed Manchester legacies of Marx and Engels, the deep effective histories of labour and feminist organizing, and a mix of complexly gendered social and labour organizations such as independent socialist halls, working men's clubs, British Legion halls, rambling, birding and cycling associations etc. (and lifelong if misguided affections for a particular football club), which together generated deep attachments to new forms of urban industrial life. Regional attachments to place were deeply inflected with dialect, close familial associations, masculinist speech and strong matriarchal households, and with them a varied agglomeration of norms and practices that marked the identities of places (payday Thursday, dancehalls on Friday night, football on Saturdays, baths on Sundays, market-days on Mondays, among many others).

NATURE

The environment also played a complicated role in this analysis of the conjuncture. From her early writing on the political economy of capital, land and landownership in Britain (Massey & Catalano 1978), class dynamics were integrally interwoven into questions of nature, access to resources and ownership of land, and each was to be read both as historical and geographical. Thus, when she asks whether there is another way of thinking about physical geography she is not evoking a conception of countryside as bucolic release from the grime of factory life (Massey 1999g, 2005, 2006b). While for many workers, the countryside around the industrial towns offered late afternoon or weekend opportunities for cycling, rambling, and birding groups to explore the countryside, for Doreen the countryside was also always an integral part of their political and social landscape.

This should not be surprising. One hour directly east of Wythenshawe lies Kinder Scout in the High Peak District, one of the main sites of land seizures and enclosures by a feudal landed gentry in the face of rapidly expanding city life. Kinder Scout was also the specific site of the mass trespass by largely working-class Mancunians in 1932 claiming their rights to ramble on public pathways. Described in 2003 by Roy (Lord) Hattersley as the most successful act of direct action in British history, the mass trespass erupted over conflict between the feudal landed gentry who had gradually been appropriating public lands and a militant working-class intent on defending their traditional rights of access. Led by members of the Communist Party British Workers' Sports Federation in Manchester, the mass trespass consolidated the rights of public access, restricted the refeudalization of land, and eventually led to the Peak District centred on Kinder Scout becoming the first national park in Britain (Toft 2012).

Here the valley towns of Doreen's industrial North were integrally embedded in constant conflicts over rights of access, use and ownership. Many industrial workers had come from surrounding sheep and dairy farms and family stories continued to relate their centrality to their experiences in their transition to industrial life. In these families, town and country were not binary juxtapositions, but integrally interrelated experiences. The view from the valley was across the moors or the fields to the hills behind, the weekday was bookmarked by weekend walks in the countryside, and wartime training and schooling meant that every worker handled compass and map easily in regular excursions and explorations into the countryside. This countryside was a site for recuperation but also a resource for political mobilization, as generations of worker education programmes had long realized.

This interwoven social and geographical sensibility meant that the environment itself could never be separated off from the sociospatial processes of life, and it was this rich spatiotemporal understanding of change that came to frame Doreen's later efforts to think physical geography anew; an intervention perhaps no better exemplified than in her interesting, and at first strange, reading of the erratic, immigrant rocks and the mutability of the postglacial landscapes of the Lake District (Massey 2005). Even the rocks and strata were movable. Surely neoliberalism was!

FROM "THE" NORTH TO "THE" SOUTH

As with her account of landscape in which geological time is invoked to show the mutability and relationality of glacial landscapes, the "story of the world cannot be told (nor its geography elaborated) through the eyes of 'The West' alone" (Massey 2005: 10). Instead, Eurocentric histories of modernity must be understood for what they were; culturally specific ideas whose claim to universality erased the many histories of others without pen or voice.

Reversing Tobler's first law of geography (that things closer together interact more), Massey insisted that there was a historico-geographical responsibility to such others at a distance. Since place was the crystallization of flows operating across space and linking people and places in a diverse range of social relations, and the production and circulation of value was always geographical, the question of spatial responsibility and the ethics of place were crucial ones.

If Manchester and other older mining and textile areas grew around a specific set of class, gender and environmental politics, they were also shaped from their beginnings by various forms of internationalism, particularly in their inception with competition with Indian cottons, the supply of fibres from the American South, later competition from American producers, and the constant movement of peoples from farm to town, from other regions to the North, and from country to country (Massey & Wainwight 1985).

Postwar Lancashire was quickly drawn into a postcolonial world. Competitive pressures drove the search for innovation and efficiency, leading to a deepening of manufacture with emerging European technology centres and machine suppliers who –like the Swiss – capitalized their earlier comparative advantages in craft manufacture through large-scale state supports. Increased labour demands in the 1950s and 1960s fuelled the immigration of South Asian textile workers (at first men, and later families). In mining

145

and port areas, West Indian immigration contributed to economic growth, labour market transformations and the social lives of towns.

Throughout the 1960s and 1970s, unemployment increased as textiles and light engineering lost markets and access to investment and operating capital waned. Within two decades, the initial hopes of social transformation had given way to regional decline, unemployment and intense spatial inequalities that compounded the already entrenched regional politics of North and South. Industrial towns throughout the region experienced large-scale immigration. Traditional male-dominated trade union power faded, particularly in mining. So-called "traditional" social values and practices organized around textile and mining labour elites were rapidly challenged by labour markets increasingly structured along lines of race and country of origin. The influx of new workers led quickly to corresponding changes in housing and neighbourhood segregation; white flight and Asian family reunion came to signify the immediate experience of people in places throughout the region, just as broader national and international economic policies and regional shifts further disadvantaged the ability of increasingly peripheral regions to adjust. Place came to matter even more as rapid structural and social change destabilized one form of life and gave rise to new forms of social and spatial organization. Defensive, racialized divides emerged across the towns of Lancashire, divides that were in turn mobilized to police the crisis of Thatcherism in other ways (Hall *et al.* 2013 [1978]).

It was in this context of rising social antagonism around Thatcherism that logics of economic regeneration around high-tech fantasies, innovation and growth emerged as an increasingly dominant set of stories framing the regional problem in ways that mischaracterized the causes and dynamism of declining regions (Massey 1973, 1978, 1979, 1987b). These fantasies were particular and partial, offering little to the older industrial areas of the North. Instead, they contributed further to notions of peripheral underdevelopment, conditions of lack, absence and lag, and deepened the regional crisis throughout much of the 1970s through the 1990s (and beyond). As Pigg (1992: 507) argued in the context of development studies: "As long as development aims to transform people's thinking, the villager must be someone who does not understand ... Hence the village becomes a space of backwardness – a physical space that imprisons people in what is considered an inferior and outmoded way of life" (see also Escobar 2001; Wainwright 2013).

Historical memory and relational geographies were lacking and it was these lacunae that her work sought to address with a wider concept of spatial relations and responsibility (Massey 1991c, 1999b, 2004a, 2007; Massey *et al.* 2009). This required a cartography of relationality to make clear the underlying geographies of responsibility. Places from the local to the global city constitute complex interwoven networks of relations and power.

CONCLUSION IN KILBURN

When Doreen asks how place can be understood in terms of the importance of rootedness and affiliation with locale without being defensive and reactionary, and how we can understand identity if we accept that borders are socially constructed, she is – in part – asking about how we can understand the class and gender politics of such places. But she is asking much more. In particular, she is asking how a neoliberal commitment to London's financial markets was solidified and normalized by large-scale effort and investment of resources to produce a new border of neoliberal commitments. That border consolidated the boundary lines around one set of political beliefs, economic opportunities and likely futures. Doreen saw this border – like all borders – as both ideological and material. In ways similar to the investments in science and society that had constructed and institutionalized a binary distinction between nature and society, neoliberalism has – in a few short decades – constructed a boundary policing the new normal. But, in a time when environmental and social problems are so deeply interwoven, such borders need to be deconstructed and challenged (Massey 2005). Concerned with pressing environmental and social problems, Doreen pushed for a thoroughgoing deconstruction of these normalized neoliberal "stories" (Hall *et al.* 1995; Massey 2005).

This complex relationality of the quotidian structures of life attracted both scholars and politicians alike to her work. Her articulation of the relationality of spatial divisions of labour, high-tech fantasies and the always uneven and combined nature of regional development processes made her analyses of regional growth and decline particularly powerful (Cochrane 2013). Eschewing the logics of "lack" that underpinned so much postwar British regional development, she repeatedly argued that regional decline was not first and foremost a problem of "lack", not a problem of the lagging region itself, but more importantly a result of policy choices that favoured particular fractions of capital, specific regions (particularly the hypergrowth of London and the southeast), and nexes of power that had come to dominate decision-making around increasingly taken-for-granted and "normalized" liberal and financialized models of the economy and individualized models of social life. The result was the growing dominance of the global city of London and the dynamic of its overheated property markets in shaping the national agenda and restructuring the geographies of opportunities across the country. Increasingly supported by investment, tax and credit policies aimed at solving the challenges of growth in London and the southeast, other regions experienced intense disadvantage and underdevelopment.

While much politics of the left and right were increasingly turning to highly technicist and managerial neoliberal conceptions of politics, Doreen remained committed to the possibilities offered by strong democratic practices and institutions (Massey 1995c; Amin, Massey & Thrift 2000). She worked regularly with working-class organizations around questions of education, housing and rights, driving such commitments and engagements to the heart of her work with the GLEB of the Greater London Council and her arguments with Ken Livingstone (Massey & Livingstone 2007; Wainwright 2013; Wills 2013).

In the on-going chapters of the *Kilburn Manifesto* these commitments to a spatial politics of responsibility became particularly clear (Hall *et al.* 2015). It was there that Doreen again made clear the political importance of a spatial imagination in changing the language we use: "The language we use has effects in moulding identities and characterizing social relationships. It is crucial to the formation of the ideological scaffolding of the hegemonic common sense. Discourse matters. Moreover it changes, and it can – through political work – be changed" (Hall *et al.* 2015: 3). And that is the challenge if the practices of regional and social inequality are to be addressed.

REFERENCES

For works authored and co-authored by Doreen Massey, please see Select Bibliography of Doreen Massey, beginning on p. 371.

Allen, J., D. Massey & A. Cochrane 1998. *Rethinking the Region*. Abingdon: Routledge.

Amin, A., D. Massey & N. Thrift 2000. *Cities for the Many Not the Few*. Bristol: Policy Press.

Anderson, B. 2008. "For Space (2005): Doreen Massey". In P. Hubbard, R. Kitchin & G. Valentine (eds) *Key Texts in Human Geography*. London: Sage Publications, 227–35.

Cochrane, A. 2013. "Spatial Divisions and Regional Assemblages". In D. Featherstone & J. Painter (eds) *Spatial Politics: Essays for Doreen Massey*. Oxford: Wiley-Blackwell, 87–98.

Escobar, A. 2001. "Culture Sits in Places: Reflections on Globalism and Subaltern Strategies of Localization". *Political Geography* 20 (2): 139–74.

Featherstone, D., S. Bond & J. Painter 2013. "'Stories So Far': A Conversation with Doreen Massey". In D. Featherstone & J. Painter (eds) *Spatial Politics: Essays for Doreen Massey*. Oxford: Wiley-Blackwell, 253–66.

Featherstone, D. & J. Painter 2013. "Introduction: There is No Point of Departure: The Many Trajectories of Doreen Massey". In D. Featherstone & J. Painter (eds) *Spatial Politics: Essays for Doreen Massey*. Oxford: Wiley-Blackwell, 1–18.

Grossberg, L. 2006. "Does Cultural Studies Have Futures? Should It? (Or What's the Matter with New York?)". *Cultural Studies* 20 (1): 1–31.

Grossberg, L. 2010. *Cultural Studies in the Future Tense*. Durham, NC: Duke University Press.

Grossberg, L. 2013. "Theorizing Context". In D. Featherstone & J. Painter (eds) *Spatial Politics: Essays for Doreen Massey*. Oxford: Wiley-Blackwell, 32–43.

Hall, S. 1991. "The Local and the Global: Globalization and Ethnicity". In A. King (ed.) *Culture, Globalization and the World-System*. London: Macmillan, 19–40.

Hall, S., D. Massey & M. Rustin 1995. "Uncomfortable Times". *Soundings* 1: 5–18.

Hall, S., C. Critcher & T. Jefferson 2013 [1978]. *Policing the Crisis: Mugging, the State and Law and Order*. London: Palgrave Macmillan.

Hall, S., D. Massey & M. Rustin 2015. *After Neoliberalism? The Kilburn Manifesto*. London: Lawrence and Wishart.

Harvey, D. 1987. "Three Myths in Search of a Reality in Urban Studies". *Environment and Planning D: Society and Space* 5 (4): 367–76.

Menendez, R. 2013. "The Socialist Transformation of Venezuela: The Geographical Dimension of Political Strategy". In D. Featherstone & J. Painter (eds) *Spatial Politics: Essays for Doreen Massey*. Oxford: Wiley-Blackwell, 224–34.

Mezzadra, S. & B. Neilsen 2013. *Border as Method, or the Multiplication of Labor*. Durham, NC: Duke University Press.

Mouffe, C. 2013. "Space, Hegemony, and Radical Democracy". In D. Featherstone & J. Painter (eds) *Spatial Politics: Essays for Doreen Massey*. Oxford: Wiley-Blackwell, 21–31.

Peck, J. 2005. "Economic Sociologies in Space". *Economic Geography* 81 (2): 129–75.

Peck, J. 2013. "Making Space for Labour". In D. Featherstone & J. Painter (eds) *Spatial Politics: Essays for Doreen Massey*. Oxford: Wiley-Blackwell, 99–114.

Peck, J., D. Massey, K. Gibson & V. Lawson 2014. "Symposium: The Kilburn Manifesto: After Neoliberalism?" *Environment and Planning A* 46: 2033–49.

Pickles, J. 2012. "The Cultural Turn and the Conjunctural Economy: Economic Geography, Anthropology, and Cultural Studies". In T. Barnes, J. Peck & E. Sheppard (eds) *The New Companion to Economic Geography*. London: Wiley-Blackwell, 537–51.

Pickles, J. 2018. "Place, Autonomy, and the Politics of Hope". In D. Billings & A. Kingsolver (eds) *Place Matters!* Lexington, KY: University of Kentucky Press, pp. 71–90.

Pigg, S. L. 1992. "Inventing Social Categories Through Place: Social Representations and Development in Nepal". *Comparative Studies in Society and History* 34 (3): 491–513.

Rustin, M. 2013. "Spatial Relations and Human Relations". In D. Featherstone & J. Painter (eds) *Spatial Politics: Essays for Doreen Massey*. Oxford: Wiley-Blackwell, 56–69.

Saldanha, A. 2013. "Power-Geometry as Philosophy of Science". In D. Featherstone & J. Painter (eds) *Spatial Politics: Essays for Doreen Massey*. Oxford: Wiley-Blackwell, 44–55.

Toft, D. 2012. "Occupy Kinder Scout: Remembering the Mass Trespass". *Red Pepper* 9 August. www.redpepper.org.uk/occupy-kinder-scout-remembering-the-mass-trespass/ (accessed 7 January 2018).

Wainwright, H. 2013. "Place Beyond Place and the Politics of 'Empowerment.'" In D. Featherstone & J. Painter (eds) *Spatial Politics: Essays for Doreen Massey*. Oxford: Wiley-Blackwell, 235–52.

Wills, J. 2013. "Place and Politics". In D. Featherstone & J. Painter (eds) *Spatial Politics: Essays for Doreen Massey*. Oxford: Wiley-Blackwell, 135–45.

CHAPTER 11

REFLECTIONS ON *CAPITAL AND LAND* BY MASSEY AND CATALANO

Richard Walker and Erica Schoenberger

Picking up this book again has been a curious form of time-travel. Doreen Massey was our friend, mentor and comrade for over thirty years, and our ideas and concerns evolved in tandem over those decades. Now to go back and confront some of Doreen's earliest work has been an eye-opener about where we all started, our relative naiveté in many respects, and the recurring need to reflect on the basic elements of one's intellectual edifice. The thinking at work in *Capital and Land* reflects a very different time and place from today, for better and worse.

One thing that jumps out from the pages of *Capital and Land* is the clarion voice of Doreen's authorship. The book is alight with good sense and careful reasoning, backed up by serious research. There is little to quibble with in how Massey and Catalano[1] go about their business of investigating capital and land in postwar Britain. The study is well grounded in empirical data about landownership and rents. The authors are extremely precise in their delineation of different categories of ownership, both empirically and theoretically, and they are determined to tease out in a rigorous way the modes of landownership pursued by various forms of capital (i.e. agrarian, industrial and financial).[2]

---

1. Alejandrina (Alex) Catalano joined the Centre for Environmental Studies in London as a research assistant in 1974, working with Doreen (and initially also Richard Barras and Andrew Broadbent) on a research project investigating the role of land in the economy; it evolved into *Capital and Land*. Alex has a BSc in Urban Studies from the University of California at Berkeley and an MSc in Urban and Regional Planning from the London School of Economics, but changed career in 1983, becoming a journalist specializing in real estate and finance. Although she and Doreen did not work together again after *Capital and Land*, they remained in close contact as friends.
2. They leave considerations of state ownership until the end, a choice dictated by their closeness to Labour party politics (see below).

151

One thing that surprises is how much empirical data on land ownership the authors were able to muster. If one were to try to update this work, one would still run into a wall of data silence. Property ownership continues to be a taboo subject in terms of published, publicly accessible data. There is a better chance of getting details on CIA black sites (Paglen 2009). In California, for example, landownership filings are masked by all sorts of false-fronts registered to corporate attorneys, untraceable entities and offshore holdings. This informational black hole is only likely to get worse in the present political conjuncture.

On the theoretical side, the authors offer a remarkable short course in Marxism, including the transition from feudalism to capitalism – which was still relevant to understanding the origins of certain categories of land ownership in the UK, even if their economic and political salience had been transformed. They base their theory solidly on Marx's analysis of rent, but do so without undue recourse to philosophical niceties or excursions into the marginalia of Volume 3 of *Capital*. At the same time, they are steeped in the Althusserian mode of the time, particularly the question of "class fractions". This goes hand-in-hand with their analysis of the specific kinds of landownership and whether they pose specific sorts of contradictions within capitalism.

Doreen Massey was always careful to place big theory into real world contexts of place, class and historical moment. Each of the empirical chapters meticulously analyses the different circumstances of particular groups and their differing stakes in landownership: who holds land for rental income, who holds it for capital appreciation, who holds it to develop, who holds it as a condition of production, and even who holds it out of a sense of *noblesse oblige*. Those kinds of social-spatial differences mattered to everything she did, and linked her form of theoretically sophisticated scholarship to her intense commitment to everyday politics.

Capital and Land is deeply rooted in postwar Britain in several ways. Let's start with the focus on large rural landholdings and the remnants of an old landed aristocracy. This is a pointedly British obsession, including that of the postwar Left, and for good reason: the landed aristocracy were never entirely displaced by the English Revolution (about which Christopher Hill wrote so incisively) and kept being replenished by the new bourgeoisie buying out the old landed estates for themselves (as Raymond Williams analysed so brilliantly) (Hill 1975; Williams 1975). The class struggles of the first half of the nineteenth century were primarily about capitalists versus landowners, and while the former triumphed in the Corn Laws of 1842, it took two world wars in the twentieth century to finally put an end to the latter as a significant political force.

Therefore, Massey and Catalano are determined to show that large landowning still persisted at the time they were writing, accounting for a roughly 30 per cent share of the total land area of Great Britain, which is an astonishing figure. But they also show that this was no longer the basis for a major landholder class, either in the feudal sense or as a class fraction of capital. It is not by chance that the data on rural landowning at that time were excellent, while the data on urban areas were flimsy, at best. Nor is it a simple predilection for the Frenchified Marxism of 1968 that led them to Althusser and Poulantzas to take up the question of class fractions – an idea that has virtually disappeared today.

Another way *Capital and Land* is rooted in its time is the chapter on "Industrial Land". In that discussion, one is transported back securely to a time when industrial capital ruled the land on both sides of the Atlantic. That world was about to be overthrown by the rise of international competition thanks to the revival of Japan and Germany, which led to falling profits on industrial capital in the UK and USA, and thence to the politics of Thatcher in Britain and Reagan in the United States. The gutting of industrial Britain was first and fastest, and it was followed by a full assault on the working class and on New Deal/Social Democratic protections. Massey would, of course, go on to consider deindustrialization and its geography, as well as what we all now call the Age of Neoliberalism (Massey 1984b).

An important element of the decline of postwar industry concerns land-holding, though this came a decade after Massey and Catalano were writing. That was the gutting of overblown, oligopolistic corporations by financial raiders in the 1980s. A significant dimension of the dismembering of both rotting and viable industrial companies was for the new owners to carve off their huge landholdings and sell them for a quick payoff (capitalized rent) – most notably in the cases of railroad yards and timber holdings.

A third postwar dimension of *Capital and Land* is its close relation to the British Labour Party and its critics on the radical Left. Massey and Catalano saw themselves as writing for the militant working class, and they assumed a politically sophisticated non-parliamentary Labour Party and Trades Union movement in dialogue with the state. In their conclusion, they take up the question of whether the Left should militate for nationalization of land – a real possibility in the postwar high tide of Social Democracy. On the whole, they accept that it would be better than not; but they are clear that such a move would not solve the problems of landownership, housing, rent and all the rest because it would not obviate capitalism and class power. Rather, the state as landowner would become the site of new political struggles with a bourgeoisie intent on manipulating the state into increasing its profits in whatever form they could get.

Well, that was the UK in 1978.[3] Writing from the US in 2017, one could fall over with envy. You mean there really was a solid, militant working class and its leadership that were ready to debate the finer points of Marxist analysis and urban policy? Actually, writing from the US, that would be a novelty at any time and practically treasonous for much of our lifetimes.

In taking up Marxian rent theory, Massey and Catalano were in keeping with a revived interest in rent on the Left in the 1970s. They were quite right to argue that rent matters, in two ways: as a slice of the total surplus value created by labour in industrial production and as a price signal that allocates land uses through the operation of the property market. This led them to produce a mercifully short discussion of absolute rent and a sensible treatment of differential rent. Their treatment of rent does not go as far as David Harvey's investigations of the subject taking form at the same time, but Massey and Catalano had different concerns that rested on prevailing political debates on the socialist Left in Britain (Harvey 1982).

On one side were those who felt that all rent on land is exploitation and that it should therefore be abolished in all its forms; on the other were those who thought that only absolute rent was egregious and that differential rent's allocative function should be retained (especially in cities). Massey and Catalano come down squarely between the two sides, explaining that differential rent does, in fact, play an important role in land use but that it is still exploitation that matters for the economics and politics of class relations.

What is odd, however, is that they do not look seriously at something that captivated Harvey: the pursuit of rent gains in land development and its implications for the process of urbanization. They recognize a segment of property capital that engages in rent-seeking, but their portrayal of that group of capitalists is surprisingly passive. There is no analysis of property owners and builders as a political force in the creation and reconstruction of cities. It's hard to know why property capital did not appear to them as a true sector of industry and as a major power in urban politics, but it undoubtedly had a lot to do with the national setting of postwar Britain, where developers were squeezed between the state and finance (see below).

Over in the United States, where Harvey settled circa 1970, property capital goes under the name of "real estate developers" and became a central topic in urban studies at the time. One thinks of Harvey Molotch's pathbreaking work on "urban growth coalitions", as well as Harvey's early research on property development in Baltimore, which led to his critical insights about

---

3. Consider how different was Manuel Castells' "consumptionist" take on *The Urban Question* (1977), written at the same time as *Capital and Land* and as part of the same series edited by David Harvey and Brian Robson.

land allocation in *Social Justice and the City* and the formation of cities in *The Urbanization of Capital* (Harvey 1973, 1985).

Harvey then went on to introduce his groundbreaking theory of the second circuit of capital, which tied industrial cycles to property development booms, such as the one that hit in the late 1960s and burst in the early 1970s. To some extent, Harvey was inspired by excellent studies of the London property sector and bubbles in the 1960s, such as the one by his former student, Bob Colenutt (Ambrose & Colenutt 1975). Massey and Catalano were undoubtedly familiar with that literature. But their main line of inquiry was still guided by working class-oriented debates over public housing and land planning, both still serious policies in Britain – unlike the disastrous pro-development policies of urban renewal and privatized suburbanization in the United States. As a result, the final chapter of *Capital and Land* is about housing policy, while Harvey's finale in *The Limits to Capital* was about the "spatial fix" as an outlet for overaccumulated capital – a much more abstract theory of the geography of capitalist expansion.[4]

Because of their interest in housing policy, the state is a presence throughout *Capital and Land*, but state politics around land development does not come in much for analysis except in the final few pages, and rather sketchily. Massey and Catalano could not know, of course, that in only two years the very idea of the state doing anything on behalf of the poor or working class was going to be trampled in the mud. If some enterprising graduate student wanted to update and deepen the book, s/he would be doing us all a favour by investigating how the various groups with stakes in the value of land, either as lenders or owners, fared under the neoliberal regime that followed Thatcher's counter-revolution.

Elsewhere in the book, Massey and Catalano pick up on a vital new tendency of the time, which has grown into a major phenomenon today: the move of finance capital into land and land development. On this front, they were remarkably prescient, in part, no doubt, because London was at the forefront of the rise of finance as a portion of capital and a force in the capitalist economy. Of course, that was before the major run-up of the stock markets, the ballooning of secondary mortgage activity, and the presence of such investment vehicles as Real Estate Investment Trusts (REITs); but it is valuable to recognize that the long arc of financialization did not begin with deregulation of the 1980s, as Americans often think, but with the Eurodollar market in London in the 1960s (Ryan-Collins *et al.* 2017).

---

4. For a critique of Harvey's much beloved and oft abused spatial fix theory, see Schoenberger (2004) and Walker (2010).

Financial institutions have a number of ways of participating in land ownership and property development, most importantly through long-term loans and through direct ownership. Banks may be less visible here because their loan horizons are generally more constricted, limiting them to shorter-term considerations. Property developers mainly use bridging loans to carry the operation through the construction stage, replaced by long-term mortgage financing once a building is completed. Insurance companies and pension funds, on the other hand, have evolved into important investors in land as constraints on their activities have been progressively removed.

The insurance industry became important after the Great Fire in 1666, but was mostly concerned with property and marine insurance; neither of these had investment time horizons that matched up with property ownership in an extremely illiquid land market. By the eighteenth century, however, insurers started to lend on land and speculative development projects. In the early twentieth century, the return on mortgages was attractive, especially compared to the fixity of rents which were rarely revised. Financial institutions found lending on property served their needs; ownership was not necessary.

After the Second World War, insurers began to commit sizeable resources to direct land ownership, principally in cities. But these were not the property and marine insurers of the past. The life insurance business had taken off, with two consequences. One is that since their investment horizons were long, they could invest in long-term assets. The second was that money was pouring into the sector and companies were scrambling to invest it all. Capital controls remained in effect in the UK until 1961, which meant that outlets for direct and portfolio investment had to be found locally. Even then, some controls on foreign direct investment remained in place for years (Bank of England 1967). For insurance companies, famously international, local land and property development seemed a perfect solution to their difficulties. This is what Harvey would call, a few years later, a spatial fix.[5]

As if this weren't enough, in 1955 pension funds were allowed to invest in property-based equities for the first time. The further development of property unit trusts and property bonds opened up a rich field for investment and the pension funds – which also had long investment horizons – were quick to take them up. This contributed substantially to the postwar property boom. At the same time, there was no noticeable reduction in the rate of growth of property-related *lending* such as mortgage lending. Banking and finance companies were both pushing money out the door. It would seem that the land surface of the UK was a sponge for capital.

---

5. This is a good reminder that the spatial fix does not always entail moving; a spatial fix can be accomplished by investing in long-lived, illiquid projects in place (Schoenberger 2004).

The mid-1960s saw another prompt for financial institutions' deeper engagement with land ownership. The 1965 Finance Act partially or wholly exempted insurance companies and pension funds from higher corporate taxes on investment income; by contrast, property development companies were hit especially hard given their high leverage. Hence, it became more interesting for the financials to invest directly in land to earn rents and see their capital appreciate than to lend capital to property companies. As Massey and Catalano stress, all of this capital pouring into land had the effect of raising prices and depressing yields (rents). The financial companies could live with this, counting on long-term appreciation. The property companies, dependent on loan financing, found it hard to maintain a sufficient level of profit to allow them to pay their investors in view of their less favourable rents and taxes.

The financial crisis of 1972–3 hit banks hard, but the insurers and pension funds emerged largely unscathed. Moreover, the government's efforts to stabilize the economy put the property companies in another unfavourable position. Business rents were frozen in 1972. Proposals for a tax on realized and unrealized development gains contributed to falling asset prices. Again, the financial firms could afford to wait this out; the property companies, not so much.

Of course, it mattered where and to what capital was flowing, and here the geographical and class distinctions are stark: London, and central London above all. Investors had lots of enthusiasm for office buildings, shopping centres and houses for the rich, but almost no interest in the working class and certainly none for the poor. These were not the local housebuilders and building societies of the past. Even those landowners who one would have thought to be guided by morality and a sense of duty – namely the Church of England and the Crown – divested from low-rent, long-lease properties and moved their capital into more lucrative places and properties. In the end, much of the problem of housing the unwealthy fell to the state, which was hoist by its own ideological petard since it had to pay market rates for the land, making public housing that much more expensive to build and unaffordable for the working class.

Massey and Catalano wanted to make a larger argument about the relationship between landed property and capital. The principle concern was the way that private property in land stands in conflict with capitalist development. Land is a necessary condition of production for capital, but because it siphons off rent from the surplus value available to industrial capital, it reduces the rate of accumulation. Similarly, to the degree that high rents hit the working class, they are likely to force an increase in wages and threaten profits from another direction. This concern has more in common, however, with David Ricardo, J. S. Mill and Henry George's nineteenth-century worries about the parasitism of the Landed Interest than with Leftist concerns with the dynamics of capital and the welfare of the working class.

In fact, capital has come up with many strategies for overcoming the barrier of rent, starting with the passage of the Corn Laws. Modern industrial capital mostly owns its own land to avoid sharing with landlords. The state has also been mobilized in various ways, such as making rent a tax-deductible business expense. The same goes for subsidies to workers' housing that help keep wages in check. Contrast this Ricardian worry about a lack of profits for accumulation with Harvey's late-twentieth-century thesis about using investment in land and buildings as a useful sink for overaccumulated capital.

Yet there is something that Harvey, Massey and Catalano, and the whole edifice of Marxist theory has been largely silent about. The exploitation of land rents is not just an accessory to the primary circuit of industrial capital; it has always been a fundamental dimension of capitalist profiteering, growth and expansion. Land rents, and the rents derived from the "free gifts" of the earth that go with land, have always been essential to the enrichment of investors, the dynamism of capital and the geographic expansion of world capitalism, as Jason Moore (2015) has argued forcefully. Moore, however, does not consider the urban facet of this massive rent-extraction, something that Lewis Mumford realized late in life. Given the importance of cities in the history of capitalism and their weight in the modern capitalist economy, rent-seeking has to be taken seriously by Marxism and Left political economy as a vital gear in the machinery of exploitation and accumulation (Brechin 1999).

There are some other silences in *Capital and Land* that reveal the time and concerns of the authors. For one, because Massey and Catalano remain laser-focused on land they never take up the question of the form of the built environment and the formation of cities atop that land. Needless to say, Massey would get around to that later, but even then, it is not her strongest suit – surprising given her deep affection for London and its politics (Massey 2007).

Similarly, there is no mention of land as nature and the relation of rent-taking to the environment. Massey and Catalano observe, for example, that owner-farmers run more capital-intensive agricultural operations than their renter counterparts, and increasing mechanization is mentioned in passing. But there is nothing said about the exploitation of the soil or the use of herbicides and pesticides that almost certainly accompanied intensification, and their ecological impacts (Walker 2004).

The complete silence on the subject of gender is another reminder of the hazards of time-travel, since Doreen Massey later became one of the premier voices of feminism in geography (1994; see also 2005). Presumably there were women involved in land ownership and development in postwar Britain besides the Queen. On the other hand, it is hard to see how you could winkle their status out of the already opaque data on landholding. It's one thing to do surveys and ethnographies of women in social housing, but quite another to figure out who exactly owns the ground under their feet.

Doreen Massey kept moving intellectually all her life, as this reflection back on a research project of the 1970s reveals. Hers was an endlessly restless and inquiring mind. Notably, she led the way for a thorough overhaul of industrial and economic geography in the 1980s, which we both had the good fortune to be part of thanks to her coming to Berkeley for a semester in the early 1980s, just after a similar visit by Bennett Harrison (Massey & Meegan 1982; Bluestone & Harrison 1982; Storper & Walker 1989; Schoenberger 1997). Doreen went on to revolutionize spatial theory in geography in the 1990s, and to renew her political project with the founding of the journal *Soundings* with Stuart Hall, her long-time friend and colleague, in the 2000s. Her early departure from our midst is a staggering loss for radical geography, but her work and her spirit will continue to inspire students of land, capital and cities – and much more – for many, many years to come.

REFERENCES

For works authored and co-authored by Doreen Massey, please see Select Bibliography of Doreen Massey, beginning on p. 371.

Ambrose, P. & B. Colenutt 1975. *The Property Machine*. Harmondsworth: Penguin.

Bank of England 1967. "The UK Exchange Control: A Short History". www.bankofengland. co.uk/archive/Documents/historicpubs/qb/1967/qb67q3245260.pdf (accessed 1 July 2017).

Bluestone, B. & B. Harrison 1982. *The Deindustrialization of America*. New York: Basic Books.

Brechin, G. 1999. *Imperial San Francisco: Urban Power, Earthly Ruin*. Berkeley, CA: University of California Press.

Castells, M. 1977. *The Urban Question*. London: Edward Arnold.

Harvey, D. 1973. *Social Justice and the City*. Baltimore, MD: Johns Hopkins University Press.

Harvey, D. 1982. *The Limits to Capital*. London: Edward Arnold.

Harvey, D. 1985. *The Urbanization of Capital*. Baltimore, MD: Johns Hopkins University Press.

Hill, C. 1975. *Change and Continuity in Seventeenth-Century England*. Cambridge, MA: Harvard University Press.

Moore, J. 2015. *Capitalism in the Web of Life: Ecology and the Accumulation of Capital*. London: Verso.

Paglen, T. 2009. *Blanks Spots on the Map: The Dark Geography of the Pentagon's Secret World*. New York: Dutton.

Ryan-Collins, J., T. Lloyd & L. Macfarlane 2017. *Rethinking the Economics of Land and Housing*. London: Zed Books.

Schoenberger, E. 1997. *The Cultural Crisis of the Firm*. Oxford: Blackwell.

Schoenberger, E. 2004. "The Spatial Fix Revisited". *Antipode* 36 (3): 427–33.

Storper, M. & R. Walker 1989. *The Capitalist Imperative: Territory, Technology and Industrial Growth*. Oxford and Cambridge, MA: Basil Blackwell.

Walker, R. 2004. *The Conquest of Bread: 150 Years of California Agribusiness*. New York: The New Press.

Walker, R. 2010. "Karl Marx Between Two Worlds: The Antinomies of Giovanni Arrighi's *Adam Smith in Beijing*". *Historical Materialism* 18 (1): 52–73.

Williams, R. 1975. *The Country and the City*. Oxford: Oxford University Press.

CHAPTER 12

THE ROAD TO BREXIT ON THE BRITISH COALFIELDS

Huw Beynon and Ray Hudson

SETTING THE SCENE

The late 1960s/early 1970s was a turbulent time. Although it may seem hard for those who weren't around at the time to believe it now, there was then quite a lively debate in geography about how best to understand uneven development and which theoretical positions gave most traction on this issue. The theories developed in geography and regional science in the 1960s effectively assumed away issues of uneven development in the real world; economic life was seen as conducted on a featureless isotropic plain. While initially seen as an exciting and important development in a discipline that had been – to say the least – theory-light, it soon became clear that such an approach to theory was more concerned with models of how the geometry of the economic landscape "ought to be" given a set of precise and restrictive assumptions (see Chorley & Haggett 1967) than with understanding the social processes that created uneven development registered in empirically observable differences between places in the real material world. The change was exemplified in the contrast between David Harvey's *Explanation in Geography* (1969) and his later books *Social Justice and the City* (1973) and more powerfully in *The Limits to Capital* (1982).

But this was not simply a debate within the narrow disciplinary confines of geography. The changes in geography also have to be situated in the context of broader changes that were then sweeping through the social sciences more generally, and in particular the rediscovery of Marxism and other strands of heterodox political economy. There had long been a recognition of the importance of uneven development at the global scale within Marxian political economy. Mandel's work beginning in the 1960s in Belgium (1963, 1968, 1970, 1975) was seminal in drawing attention to the significance of intranational as well as international uneven development to the accumulation process. For Mandel, the unevenness of development between different parts of a single country was an essential precondition for capital accumulation, the significance of which had been greatly underestimated in previous Marxian analyses. This was a key insight that was to have an important influence on the way in which geographers began to theorize uneven development and "the regional problem". This revived interest in Marxian approaches, and recognition of the importance of uneven development at a range of spatial scales was not restricted to geography. Issues of uneven development came onto the agenda in economics as well as in sociology and political science and in specialized centres like the Institute of Development Studies at Sussex University and the Institute of Latin American Studies in London. As a consequence, issues of "uneven development" were subjected to intense theoretical discussion and argument, setting the scene for a germinal phase of work that crossed both interdisciplinary and international boundaries. In the 1970s/1980s issues of the political economy of uneven development had become firmly established within the geography curriculum and central to Doreen's understanding of a research agenda. In the early 1970s Doreen had just returned from Pennsylvania and begun work at the Centre for Environmental Studies in London. She had gone to the University of Pennsylvania to seek to understand more about regional science and the sort of locational theories that had risen to prominence in geography in the 1960s. By the early 1970s, however, these approaches were coming under quite severe critique as geographers, Doreen included (although she was later to say that "…[I] found it very difficult to count myself as a Marxist" in Featherstone, Bond & Painter 2013: 253), along with other social scientists, rediscovered Marx and various versions of Marxism (in her case from her time in Pennsylvania, Althusserian) and of political economy. A group of leftist leaning geographers and social scientists, with Doreen as a prominent member, met regularly as the Conference of Socialist Economists "Regionalism" group throughout the 1970s as a UK forum for debate and discussion on these issues.

One focus of these debates was on the ways in which capital could both use existing differences between places and deliberately seek to produce spatial differentiation as a central element in its strategies to produce profits. This became a debate that crossed disciplinary boundaries (for example, see Bagnasco 1977; Harvey 1982; Lipietz 1977; Poulantzas 1978; Smith 1984). Neil Smith, for example, developed his "seesaw" theory of capital movement, which saw capital moving out of places as conditions for profitable production were eroded, then moving back in as conditions for profitable production were re-created, then moving out again as they were eroded … and so on, in a more or less automatic and mechanistic process. In this view, the production of uneven development tended to be oversimplified, reducing it to capital flows with no attention to the social and political structures of places or of national and local politics. In these accounts, labour – and people more generally – tended to be seen as passive, with no involvement in the production of the places in which they lived beyond their role as bearers of labour-power or in providing the conditions for its reproduction. Doreen vigorously challenged such perspectives and developed quite a distinctive position in these debates. She developed a theoretically more nuanced approach, and continued to refine her ideas on place in subsequent years (for example, see Allen *et al.* 1998 – although this essentially elaborated on her earlier theoretical insights). In her view, the economic success or failure of places was a result of the interplay between spatially specific attributes and social processes and wider systemic forces shaping flows of capital. In stressing that people can and do help shape the places in which they live and work, she was making both an acute theoretical observation and an important political point. In the context of the decline of many centres of industrial production in the 1980s and campaigns to defend places in the face of a neoliberal onslaught, this was an important intervention – one that retains its salience in the context of post-2008 depression in many parts of the world.

This is not to suggest that she was unaware of the power of capital, the rationale for and impacts of its diverse strategies and its influence on places and its involvement in creating the layers of investment and disinvestment that for better or worse shaped places – quite the contrary (as demonstrated in Massey & Meegan 1982, for example); and she was equally aware of the power of the state to shape places (although this perhaps did not always figure as prominently in her theoretical accounts as might have been expected). Put another way, the significance of both capital and state was taken as read. But she was equally insistent that people – particularly working people and their families – and their hard-won institutions in trades unions and political parties, also had a role in shaping their places and their trajectories of change. The political importance of this view was to be dramatically demonstrated in the coal districts of the UK in 1984.

1984 AND ITS LEGACIES

The importance of campaigns to defend place-based communities was demonstrated in spades in 1984, a truly momentous year. For one thing, it witnessed the publication of the first edition of *Spatial Divisions of Labour* (Massey 1984b), the most thorough and comprehensive account of Doreen's approach to that point, pulling together much of the seminal work she had carried out in the preceding years. Ironically the book by Piore and

Sabel, *The Second Industrial Divide*, that foretold of a positive post-industrial future[1] came out in the same year, the year that saw the beginning of the UK coal miners' strike centred on that very process. The single most important industrial dispute in the UK since the General Strike of 1926, the miners dramatically emphasized the political as well as the theoretical significance of local people and their activities in the creation and defence of place. The emphasis that Doreen had given to these processes was vindicated in the extent to which the year-long conflict revealed the deep attachment that people had to the places in which they lived and worked and the extent to which they were prepared to go to defend them. It was an attachment that was the product of a long process of struggle in settlements in which people were originally thrown together from diverse origins to provide labour-power in the pits. They created their own institutions as a way of getting by, of making their places habitable. Therefore this was not just about the defence of jobs in the pits, important though they were in single-industry settlements, but of a culture and a way of life. But the struggle to preserve a culture also helped to change it. One of the more positive effects of the strike was that women, who had typically been in the shadows in mining communities, became centrally involved in organizing support groups and politically challenging the state. Doreen was energized by this development, seeing it as a powerful corrective to stereotypical understandings of these coalfield places and the people who lived there. The strike was also a struggle that resonated with people well beyond the coalfields – often people in places in the Thatcherite heartlands that seemed unlikely sources of support. Doreen became involved in this, and when writing about this wider support, often with Hilary Wainwright, took issue with those who, on the basis of the social and geographical characteristics of the strike, dismissed it as:

> The last gasp of the old labour movement, in its decaying heartlands, isolated, sectional, macho, and with little resonance beyond its regions, its unions and – of course – what they call the "hard left". (Massey & Wainwright 1985: 150)

What actually happened they wrote "has been quite different". Support groups developed not only in the coal districts, but in urban centres and rural areas too. In the account of the development of these groups can be found

---

1. Although it is important to emphasize that their account was based upon extremely limited empirical evidence drawn from parts of the "Third Italy" that subsequently experienced a profound economic decline.

the beginnings of Doreen's later theorization of alliances built across space between people of various experiences and interests.

Doreen had an acute sense and awareness of class and of the relations between class and place. The miners' strike revealed clearly the extent to which the British state – through the Thatcher government – was pursuing a class-based strategy and the lengths that it was prepared to go to destroy the National Union of Mineworkers, the deep coal mining industry, and the coalfield communities that had been built around and through them. One expression of this was the police occupation of mining villages such as Easington in the summer of 1984, with police waving £10 notes to taunt striking miners and their families and running riot through peoples' houses and gardens, as Beynon (1984) graphically documented at the time – forget about the romanticized version subsequently shown in the film *Billy Elliot*. But it was above all the savage deployment of the repressive power of the state and the gratuitous physical violence by the police at "the battle of Orgreave cokeworks" that revealed the extent to which Thatcher was prepared to go to break the NUM and destroy the coal industry and coalfield communities. For all the Thatcherite rhetoric about the sanctity of markets, when push came to shove Thatcher was more than prepared to directly intervene and ruthlessly deploy the power of the state without limit to crush the miners and the NUM.

It is also worth remembering the extent to which the Thatcherites had assiduously prepared for this in the 1970s, following the miners' strikes of 1972 and 1974. Their campaign was a cold-blooded political calculation, driven above all by memories of 1972 and 1974 and a fierce determination to ensure that no future Conservative government would be defeated by trades union power. And especially not by the NUM. When in opposition in the 1970s, a series of committees and working groups was established to prepare policy agendas for a future government. One of these, the Economic Reconstruction Group (1977) coordinated a number of committees, one of which was chaired by Nicholas Ridley (a deeply committed neoliberal), and concerned with issues relating to the nationalized industries. Its report identified these industries as a major obstacle to the radical economic changes seen as needed, and presented a strategy through which their position as a key component of the national economy could be eroded and with it the trade union power that was based within them. These unions were seen as a key impediment to the free operation of market forces and symbolic of the centrality of the state and its relationship to organized labour and to the postwar settlement. As such, their power and influence had to be crushed. The Ridley Plan advised the future government to proceed with "stealth" and begin with softer targets like the steel industry. This was followed to the letter, as the government turned its eyes on the coal industry and the NUM: the greatest prize and the greatest challenge to the Thatcherite stripe of neoliberalism. Buoyed up by the military victory in the Falklands, Thatcher emerged prepared for battle with the coal miners, a group she was to define as "the enemy within".

Coal was thus seen as a much tougher, but strategically much more important, target than steel – if the NUM could be broken and defeated, then who could resist the thrust of Thatcherism? The door would be opened for a radical redrawing of the contours of political economy in the UK, breaking union power, redefining the role of the state and ending the postwar social democratic settlement. As such, a clash with the NUM was seen as pivotal, and requiring very careful preparation – not least building up coal stocks at the power stations and seeking to divide the mining labour force on spatial grounds.

People defending their places can have problematic and regressive as well as progressive effects – something Raymond Williams (1981) graphically described as the difficulties of bringing together "militant particularisms" into more broadly based and more generalized political movements. For defence of place can lead to a destructive competition between places for jobs and investment. With the emergence of the anti-strike Union of Democratic Mineworkers in the Nottingham coalfield, encouraged and supported by the government, with echoes of the earlier history of the 1926 General Strike, a deep schism was created within the mining labour force between those areas dominated by the NUM and supporting strike action and those dominated by the UDM where miners continued to mine coal.

The government was acutely aware of the importance of the UDM for coal production and this contributed significantly to the strike being called off after a year in March 1985. The end of the strike then served as the prelude to savage cuts in the industry: almost 200 mines closed in the ten years that followed, and with them almost a quarter of a million people directly employed in the industry lost their jobs. All this effectively signalled the end of deep coal mining in the UK and the de facto destruction of the NUM as an industrial union. That the Nottingham pits that had kept working during the strike were closed along with the rest, and the UDM rendered irrelevant, was of little consolation to those in areas that that had supported industrial action through 1984 and 1985. Former coalfield places, in all former coalfields, became blighted by problems of poverty, high levels of unemployment across generations and ill-health and ill-being.

THE LOCALITIES DEBATE AND ITS AFTERMATH

Following the publication of *Spatial Divisions of Labour*, Doreen's work became very influential in geography and more generally in the social sciences. Her appointment to the Chair at the Open University (where she developed a close collaboration with Stuart Hall) followed by her membership of the Economic and Social Research Council's Research Grants Board, sealed her position within UK geography. Her influence was seen in the decision by the ESRC to fund a major research initiative under the rubric of the "Localities Programme" in 1985. Given the experiences of the miners' strike and the Thatcher government's continued programme of industrial restructuring, it says much about Doreen's reputation that the ESRC (itself under government scrutiny) was prepared to provide the funding for research into this highly sensitive area. The programme was built around a core of seven case studies and the fact that none of the bids from coalfield areas, or from Scotland and Wales, were selected caused more than a little comment.[2] We were involved in this programme and carried out one of the case studies based on Teesside. Our book (with David Sadler, 1994) can be read as an extension of Doreen's analysis in the context of a deindustrializing conurbation. At the same time, and in an attempt to remedy what we thought of as a significant defect in the programme, we also found a way to work a study of the effects of colliery closures in east Durham into our case study (see Beynon *et al.* 1991). To us, it seemed perverse – to say the least – that a programme dedicated to a comparative and grounded analysis of economic and political restructuring in the UK in the 1980s should ignore the very places where these processes were most dramatically evidenced. That some of the reasoning for the absence echoed the very assumptions that Doreen had argued against ("smokestack industries", old-fashioned, the past not the future etc.) made matters worse. As did the fact that the coalfield areas provided important examples of the ways in which places and peoples were changing in the context of a dramatic shift toward neoliberalism with its rhetorical emphasis upon the market and entrepreneurialism.

With the disappearance of mining jobs, the political emphasis switched to a rhetoric of regeneration – of creating and bringing new jobs to former coalfields. Place still mattered, but in a very different way than in the struggles of the past. In part, this involved increasing competition among former coalfield areas for inward investment in new manufacturing plants – a "race to the bottom" exemplified by the "beauty contest" among potential destinations for the Nissan car assembly plant in the mid-1980s (it eventually located at Washington, on the former Durham coalfield). The other dimension of

---

2. However the severest criticism was reserved for the overall approach which was seen to be deeply empiricist and lacking any coherent theoretical underpinning to the ongoing debates on uneven development within the discipline (see the varied contributions in *Antipode* and *Society and Space* in the latter years of the 1980s; Cooke 1989).

regeneration centred on self-employment and the creation of new small firms. This was typically wrapped up in the Thatcherite language of enterprise and of people creating their own forms of work via entrepreneurialism and embracing an enterprise culture. Not surprisingly, in the depressed former coalfield economies, such as those in Durham and south Wales, this cut little ice – there were clear limits to the numbers of taxi drivers and hairdressers that such economies could support.

While the coalfield areas had little expectation of support from a Tory government, the historical links between the NUM and the Labour Party might have been expected to lead to a more supportive policy stance from a Labour government, especially after Deputy Leader John Prescott had been given responsibility for improving things on the coalfields. He set up the Coalfields Taskforce (1998) and its description of the coalfield districts had a ring of Massey about it with its emphasis upon their unique mix of problems, social isolation and "concentrated joblessness". But the political solution it offered was light years away from the visionary possibilities written about in the previous decade. For all of its emphasis on "newness", the solutions of New Labour were the old-fashioned ones based on competing to attract inward investment and to secure grant aid for small local projects.

In the 1990s, however, competition was tougher and most likely to be for warehouses, distribution centres or the back offices of banks and financial services firms, typically offering poorly paid jobs with precarious contracts. In essence, the former coalfields were left to decline as the New Labour government continued with its prioritization of finance and the City of London, in practice a policy to widen the already yawning chasm between the former coalfields (and indeed many former manufacturing areas) and those parts of the southeast within the sphere of influence of the City.

In these ways, many of the coalfield communities were set adrift. The linkages made during the strike, and which Doreen had hoped would create a new pattern of engagement across and beyond place, had been weakened by the competition for funds and jobs, a competition encouraged by and central to the new state policies. Left to their own devices, however, these places, through the husk that remained of their trade unions, still managed to find ways to resist. Unions took legal action aimed at proving the National Coal Board negligent in relation to a range of occupationally specific illnesses and diseases, first to the incidence of vibration white finger and then emphysema and bronchitis amongst ex-miners. Successful in both cases, this action by the miners' unions provided much needed financial support. It also provided a revealing example of how the capacities of places are not reducible to capital flows or their economic structure. Any full account of place needed to include the historically established institutions of its people.

BREXIT AND WHAT OF THE FUTURE?

The collapse of the financial bubble economy after the crash of 2007/8 vividly revealed the fragility of the economy encouraged by New Labour's economic policies. Chancellor of the Exchequer Gordon Brown's claims to have evened out the process of boom and bust inherent to capitalist economies were graphically revealed as fatally flawed. It was a sign of the times that having abandoned Clause 4 of Labour's constitution in 1995, which had committed it to nationalization of the commanding heights of the economy, in the wake of the collapse of financial markets the New Labour government de facto nationalized major banks and building societies – or more precisely nationalized their debts and financial obligations while leaving control in the hands of the private sector.[3] In so doing, the party performed a 180 degree turn that made it very clear that the state was indeed a capitalist state, and that ultimately ensuring the viability of capitalist interests took precedence over any other concern. The subsequent imposition of deep austerity policies – seen as the solution to the excesses revealed by the financial crash – simply further confirmed that the former coalfields and coalfield communities, and indeed many other former industrial areas, were seen as politically marginal and as such faced a grim future. Their regeneration became even less of a political and policy priority. Neither of the major political parties, nor the coalition government formed following the 2011 general election, had an answer to their problems. Indeed, they failed to feature in their policy priorities and as such they were left to decay and decline further. In short, the already marginalized former coalfields and other deindustrialized places outside the southeast were left to suffer the effects of austerity policies. Uneven development was still more deeply inscribed across the UK's economic landscape with no prospect of this being addressed in any meaningful way.

Given all this, the turn away from formal politics, the rejection of the mainstream political parties, the rise of support for the British National Party (BNP) and then the UK Independence Party (UKIP) and the rejection of the EU registered in the Brexit vote in the former coal districts and more generally in places devastated by industrial decline hardly came – or at least should not have come – as a surprise. Low – at best – on the priorities of Conservative, New Labour and coalition ConDem neoliberal governments alike, people in these places began to look for other sources of support for their concerns and interests. Seeking to defend their local interests

---

3. As the former Governor of the Bank of England put it: "In terms of its balance sheet, the banking system had been virtually nationalised without collective control over its operations. That government rescue cannot be conveniently forgotten. When push came to shove, the very sector that had espoused the merits of market discipline was allowed to carry on only by dint of taxpayer support" (King 2016: 4).

took precedence over any concern as to where that support was coming from. Support for UKIP and rejection of the EU became generalized across deindustrialized swathes of peripheral England and Wales, those places that had been stripped of their industrial economy from the 1980s and in which many people saw themselves with nothing more to lose. It was no surprise that the "remain" appeal to vote to stay in the EU – to avoid losing what they had – failed to resonate; for these were people in places that had already lost all they had, some time ago, and had been left to suffer the consequences by the political elite. Was it any wonder they stuck a proverbial two fingers up at the political establishment in the Brexit vote?

At a different scale, attachment to place continued to matter in a different way that vividly demonstrated the significance of cultural and political differences against a background of similar economic circumstances and a trajectory of decline. For Scotland followed a different political trajectory, because for many of its residents, remaining in the EU was linked to the nationalist cause and prospects of an independent Scotland. While swathes of Scotland had suffered the same trajectory of decline as the coalfields and industrial areas of England and Wales, the nationalist promise of independence within the EU held out hope and prospects for the future in a way that was not the case in England and Wales. As a result Nairn's (1975) vision of the "break up of Britain" may well be a step nearer.

What would Doreen have made of all this? We do not think she would have been at all surprised at the way place and the difference between places impacted on the Brexit vote. We do think she would have been both angered and saddened by the outcome. She was encouraged by the changes in the Labour Party, with the engagement of young people around the election of Jeremy Corbyn as leader and the leftward shift to more socialist policy positions. She thought that there was much more to do if the Party was to engage effectively with these new voices. Forever the optimist, in her meetings around the Kilburn Manifesto she was already thinking of ways to build a more progressive way forward, and given the mess we're in now, the absence of her voice is a significant one.

REFERENCES

For works authored and co-authored by Doreen Massey, please see Select Bibliography of Doreen Massey, beginning on p. 371.

Allen, J., D. Massey & A. Cochrane 1998. *Rethinking the Region.* London: Routledge.
Bagnasco, A. 1977. *Tre Italie: La Problematica Territoriale dello Sviluppo Italiano.* Bologna: Il Mulino.
Beynon, H. 1984. "The Miners' Strike at Easington". *New Left Review* 148: 104–15.

Beynon, H., R. Hudson & D. Sadler 1991. *A Tale of Two Industries: The Contraction of Coal and Steel in the North East of England.* Milton Keynes: Open University Press.

Beynon, H., R. Hudson & D. Sadler 1994. *A Place Called Teesside: A Locality in a Global Economy.* Edinburgh: Edinburgh University Press.

Chorley, R. & P. Haggett 1967. *Models in Geography.* London: Edward Arnold.

Coalfields Task Force 1998. *Making a Difference: A New Start for England's Coalfield Communities.* London: Great Britain Department of the Environment, Transport and the Regions.

Cooke, P. 1989. "Locality Theory and the Poverty of 'Spatial Variation.'" *Antipode* 21 (3): 261–73.

Duncan, S. & M. Savage 1989. "Space, Scale and Locality". *Antipode* 21 (3): 179–206.

Economic Reconstruction Group 1977. "Final Report of the Nationalised Industries Policy Group". www.margaretthatcher.org/document/110795 (accessed 3 May 2015).

Featherstone, D., S. Bond & J. Painter 2013. "'Stories So Far': A Conversation with Doreen Massey". In D. Featherstone & J. Painter (eds) *Spatial Politics: Essays for Doreen Massey.* Oxford: Wiley-Blackwell, 236–66.

Harvey, D. 1969. *Explanation in Geography.* London: Edward Arnold.

Harvey, D. 1973. *Social Justice and the City.* London: Edward Arnold.

Harvey, D. 1982. *The Limits to Capital.* London: Edward Arnold.

King, M. 2016. *The End of Alchemy: Money, Banking and the Future of the Global Economy.* London: Little, Brown.

Lipietz, A. 1977. *Le Capital et Son Éspace.* Paris: Maspero.

Mandel, E. 1963. "The Dialectic of Class and Space in Belgium". *New Left Review* 20 (Summer): 5–31.

Mandel, E. 1968. *Marxist Economic Theory, Vols. 1 and 2.* London: Merlin.

Mandel, E. 1970. *Europe Versus America; Contradictions of Imperialism.* New York: NYU Press.

Mandel, E. 1975. *Late Capitalism.* London: New Left Books.

Nairn, T. 1975. *The Break-up of Britain.* London: New Left Books.

Piore, M. & C. Sabel 1984. *The Second Industrial Divide.* New York: Basic Books.

Poulantzas, N. 1978. *State, Power, Socialism.* London: New Left Books.

Smith, N. 1984. *Uneven Development: Nature, Capital and the Production of Space.* Oxford: Blackwell.

Warde, A. 1989. "Recipes for Pudding: A Comment on Locality". *Antipode* 21 (3): 274–81.

Williams, R. 1981. "The Forward March of Labour Halted?" In M. Jacques & F. Mulhern (eds) *The Forward March of Labour Halted.* London: Verso. Reprinted in R. Williams 1989. *Resources of Hope.* London: Verso, 247–55.

CHAPTER 13

INDUSTRIAL RESTRUCTURING AND SPATIAL DIVISIONS OF LABOUR: UNDERSTANDING UNEVEN REGIONAL DEVELOPMENT IN THE UK

Richard Meegan

INTRODUCTION

As this collection so amply demonstrates, Doreen was a complete geographer whose relational approach to understanding space and places transcended intradisciplinary divides. Her initial path-breaking contribution to the discipline was in economic geography, however, initially through research on the geography of industrial restructuring and the book that the editors of a recent collection of essays on research design and methodology in the subdiscipline claimed "transformed the field ... and triggered one of the sharpest paradigmatic shifts in contemporary economic geography" (Barnes *et al.* 2007: 2) – *Spatial Divisions of Labour* (Massey 1984b).

UNDERSTANDING UNEVEN DEVELOPMENT: INDUSTRIAL RESTRUCTURING AND SPATIAL DIVISIONS OF LABOUR

Doreen's early work on the geography of industrial reorganization laid the intellectual foundations for *Spatial Divisions of Labour* with its overarching argument that space is not passive in processes of industrial restructuring. Early applications of the approach focused on the industrial restructuring of particular sectors (Massey & Meegan 1979). The title of one of the papers from this early research – "Industrial Restructuring versus the Cities" – captures its basic theoretical argument: shifting explanation of employment change away from a purely locational explanation (the characteristics of cities and their residents) to the macro-economy and process of capital accumulation (Massey & Meegan 1978). The same argument was powerfully reiterated in an early use of her notion of "spatial divisions of labour" to understand the UK's "regional problem" (Massey 1979).

Later work attempted to formalize some of the initial findings on production reorganization through an extension of the empirical research to a range of industries cutting jobs in the UK in the late 1960s and early 1970s (Massey & Meegan 1982). A trinity of forms of production reorganization ("rationalization", "intensification" and "investment and technical change"), each with potentially distinct uses of space, was identified and used to support the argument that spatial outcomes – the geography of job loss – needed to be understood in the broad causal chain emanating from the macro-economy and process of capital accumulation.

Spatial Divisions of Labour took the "industrial restructuring" argument to another level. Its subtitle – "Social Structures and the Geography of Production" – captured the essence of its theoretical arguments, which related to:

> the way in which we conceptualise economic and to some extent more general social space, about how to understand the differences and structures of inequality within it, and about how we might begin to think of the economic identities of unique places.
>
> (Massey 1995b: xii)

Drawing on Erik Olin Wright's class analysis she came up with a framework relating occupational groups to functional positions in production relations and class location. She then identified "spatial structures of production" of leading firms and sectors in which the process of production and its attendant social – and power – relations (of ownership, control, function and status) are distributed over space. Spatial hierarchies of ownership and control, she argued, have increasingly had overlain on them spatial hierarchies of production as new sectors in which production can be functionally and spatially separated have grown in relative importance. The interweaving of hierarchies of control and production in different sectors creates particular spatial structures of production – identifying three important, but not meant by any means to be exhaustive, types: "locationally-concentrated", "cloning branch plants" and "part-process". These and other spatial structures of production combine to form overall "spatial divisions of labour" that are constructed and reconstructed over time as rounds of investment in the economic landscape are superimposed on and interact with previous rounds. This process not only creates new economic structures and new geographies of production – and class relations – but also has the potential for transforming the social composition of affected localities; with, for example, increased female labour force participation in particular areas being accompanied by a restructuring of gender relations not only in the local labour market but also in households.

The notion of social relations being stretched out over space is at the core of Doreen's concepts of "spatial structures of production" and "spatial divisions of labour":

> new spatial divisions of labour (forms of economic uneven development) are thorough re-workings of the social relations which construct economic space (for divisions of labour themselves are conceptualised as constructed through social relations). They are more than just new patterns of employment, a kind of geographical reshuffling of the same old pack of cards. They represent whole new sets of relations between activities in different places, new spatial forms of social organisation, new dimensions of inequality and new relations of dominance and dependence.
>
> (Massey 1995b: 3)

For Doreen, then, the concept of "spatial divisions of labour" was all about thinking relationally – specifically in spatializing the relations of capitalist production – and from which all her subsequent arguments about the critical relations between space and power developed (Massey *et al.* 2009: 405).

SPATIAL DIVISIONS OF LABOUR: THE RADICAL CRITIQUE

As Michael Rustin argues in his chapter in this collection, it is Doreen's relational approach to understanding space that sets her apart from historicist and temporally focused Marxist approaches and it was no surprise that criticism *of Spatial Divisions of Labour* came from that quarter. David Harvey and Neil Smith were particularly forthright:

> Every sentence in *Spatial Divisions of Labour* is so laden down with a rhetoric of contingency, place, and the specificity of history, that the whole guiding thread of Marxian argument is reduced to a set of echoes and reverberations of inert Marxian categories.
>
> (Harvey 1987: 373–4).

> Spatial divisions of labour did not, in the end, transcend its illustrative case studies and failed to paint a picture of the general movement involved in contemporary geographical restructuring. The entirely necessary shift to a greater concern with locational difference has proceeded in practice by jettisoning many of the theoretical insights and frameworks that would allow the general movement to be comprehended. (Smith 1989: 154)

Doreen was unrepentant. She had no difficulty in relating to "capital" as a concept or to theories addressing the "general movement involved in con-temporary geographical restructuring". The whole thrust of the notion of "spatial divisions of labour" – and of the "industrial restructuring approach" before it – had been precisely that, to relate space to the imperatives of capital accumulation. Where she remained resolute, however, was in challenging the argument that it is possible to explain concrete instances of spatial change solely by reference to such high-level abstractions:

> Harvey said the book was "loaded down with a rhetoric of contin-gency, place and the specificity of history", which it certainly was and which I meant it to be! (Martin *et al.* 1993: 71)

In the second edition of *Spatial Divisions of Labour*, she restated her concerns over metanarratives, arguing that her concept of "spatial divisions of labour" recognizes the existence of multiple and non-totalizing broad structures but these, crucially, are not assumed to be inexorable in their operation. The aim of the approach is conceptualization and understanding, not "prediction" of pre-determined outcomes. In this sense, it is very much poststructuralist (Graham 1998).

The locality debate: much ado about something

Doreen's work on *Spatial Divisions of Labour* was partly funded by a research grant from the UK's Social Science Research Council and it inspired the latter, by then politically rebadged as the Economic and Social Research Council, to sponsor a research programme exploring the "Changing Urban and Regional System in the UK". The research was, in a sense, a UK-based theoretical and empirical testing of "spatial divisions of labour". It was not surprising, therefore, that the radical critique featured heavily in the "locality debate" that the research provoked – principally, the alleged retreat from theory to empiricism. Huw Beynon and Ray Hudson revisit the "locality debate" in their chapter in this collection. I want to concentrate here on Doreen's intervention – as, it should be said, a somewhat bemused bystander not having being directly involved in the research – over understanding the "local". She meticulously dismantled the argument that theory is unable to address the unique and specific, highlighting the mistaken conflation of the local with the concrete and the confusion of geographical scale with abstraction that infused the "locality debate". As she so forcefully argued, theory needs to be able to address the unique and specific. The local is just as concrete an entity for study as the global and studying localities does not, as some critics claimed, amount to fetishizing them (Massey 1991c).

Broadening the analysis

In the second edition of *Spatial Divisions of Labour*, Doreen reflected on debates that the book, in part, provoked. In relation specifically to her conceptualization of space, she accepted that "spatial structures of production" needed more explicitly to link intranational with international divisions of labour and to integrate inter- with intrafirm relations of production. She did not attempt to do this in further empirical research, however, and moved on to developing her more wide-ranging relational understanding of space, place and gender.

Economic geography has also moved on from the "paradigmatic shift" that *Spatial Divisions of Labour* triggered – but the influence of the approach persists, even if, as time passes, not always explicitly acknowledged. Labour research, for example, has been directly inspired by it (Peck 2013). So too, as Linda McDowell argues in her chapter in this collection, has feminist geography. Its influence can also be detected in the various strands of heterodox "economic geography" that have developed, partly in opposition to the limited "new economic geography" of orthodox economists (Sheppard *et al.* 2012). There are clear sounding echoes of "spatial divisions of labour" in institutional and evolutionary economic notions of "path dependency", "regional lock-in" and "regional systems of innovation". The "regional worlds", "new industrial spaces" and "industrial clusters" of "new regionalism" research, I would argue, have the understanding of "spatial structures of production" at their very core and, dare I say it, are sometimes "locality studies" in all but name. What this research tends to underplay, however, are the social relations of production that constitute these geographies – the social relations on which "spatial divisions of labour" are constructed.

UNDERSTANDING UNEVEN DEVELOPMENT IN THE UK: SPATIAL DIVISIONS OF LABOUR AND SOCIAL RELATIONS OF PRODUCTION

Understanding uneven development

Spatial Divisions of Labour was published in the mid-1980s when there had been a reversal of the narrowing of regional inequality in the UK that the postwar "golden age" had brought. The argument was that central to understanding uneven development were the economy's changing relationship with the international division of labour, the spatial structures of production and overall spatial division of labour configured by the changing sectoral structure of the economy and changing balance between dominant forms of technological and industrial organization and political interpretation of how to promote capital accumulation (Massey 1988b).

This argument, I would contend, still holds for understanding uneven spatial development in the UK, which has become even more marked in the three decades since publication of *Spatial Divisions of Labour* – expressed most tellingly in a sharp and widening divide between London and the south of England and the rest of the country (Gardiner *et al.* 2013).

The UK economy is even more embedded in an extended international division of labour in which knowledge and technology flows within corporate structures dominate global production chains (McCann 2016) – and around which, key elements of the national spatial division of labour are even more tightly configured than they were when *Spatial Divisions of Labour* was published.

When *Spatial Divisions of Labour* appeared, there were still just over 5 million jobs in UK manufacturing, around 19 per cent of total employment. There are now 2.9 million, 9 per cent of total jobs. A "spatial divisions of labour" perspective clearly needs to be even more sensitive to understanding the spatial structures of service provision – in which four-fifths of the workforce are now employed. This applies to both private and public consumer services, with the former shaped by capital accumulation imperatives and the latter by state-directed, population-related provision. Analysis also needs to accommodate the growth in importance of producer services, notably in "knowledge economy" sectors. It is these sectors where the new industrial districts identified in "new regionalism" research appear to be forming – structured around the co-location of research-intensive global corporations, variously networked small and medium-sized enterprises, public scientific research organizations and science and technology based university research centres.

Governments of different political persuasions have come and gone in the UK since *Spatial Divisions of Labour* was published. Yet, while there have been some differences in political positions, notably in relation to social welfare policies and devolution under New Labour and the latter's brief flirtation with Keynesian intervention in immediate response to the Great Recession of 2008, the economic emphasis throughout has been fundamentally "neoliberal". Political intervention in production has been limited accordingly.

London – "World City"

Nothing captures the prevailing national spatial division of labour more truly than London's privileged position in it. It is where capital has accumulated exceptionally. When *Spatial Divisions of Labour* was published, the city's population was falling but, since the early 1990s, it has increased by over a quarter – from 6.8 million in 1984 to 8.7 million now. Over the same period, its workforce has grown by just over a third – from 4.1 million to 5.5 million.

Investment statistics are not available at regional level, but recently released, admittedly highly qualified, estimates of regional gross fixed capital formation for the period 2000 to 2014 provide an indication (ONS 2017a). Over those years, London, on average, accounted for around 15 per cent of the total – but 32 per cent of the figure for financial and insurance activities, 27 per cent for information and communication activities, 24 per cent for distribution, transport, accommodation and food services and 19 per cent for business service activities. Much of this investment originated from overseas. Between 2011 and 2016, for example, London – measured against the relative size of its workforce – accounted for two and a half times its share of foreign direct investment projects and just under double its share of the new jobs associated with them (Department for International Trade 2016).

When Doreen was an advisor on the Greater London Enterprise Board's Industrial Strategy, London still had around 476,000 manufacturing jobs. At the last count, it was down to 134,000, a 72 per cent decline. New rounds of investment have been predominantly in service sectors and London is now a service-sector economy writ large and one that currently positions it in a uniquely advantaged position in the spatial division of labour – both national and global, as Doreen forensically showed in *World City* (Massey 2007). It is the spatial embodiment of the financialization of the UK economy, operationalized in the established spatial cluster of firms in the City of London, and the new cluster in east London's former docklands in Canary Wharf, the latter an iconic symbol of the transformation of "locality" by rounds of invest-ment. Its role as a global financial centre, a role that Allan Cochrane explores more fully in his chapter in this collection, is one that sets it distinctively apart from the financial centres in second-tier cities – like Edinburgh, Leeds and Manchester – in the rest of the country. It is also a role that has been funda-mentally bolstered by the state. State intervention – from "Big Bang" deregu-lation in the 1980s to the assistance (in the shape of "quantitative easing") in the wake of the Great Recession crash – has been critical in the consolidation of its global financial role. It has also benefitted disproportionately from state infrastructure spending, notably on transport infrastructure.

Ownership and control functions remain concentrated in it, with head office activities and management consultancy the fastest growing professional services sector in employment terms in recent years. A recent count shows 40 per cent of the world's largest 250 companies have their European headquarters located in it. With government and public administration functions also centralized in it, it embodies the spatial concentration of power at the core of interregional relations in the UK (Amin *et al.* 2003).

High status professional, scientific and technical activities are now the city's largest employer, accounting for 14 per cent of employment and 22 per cent of workplaces (ONS 2016). Many are linked to the city's financial role while others are central to the "knowledge economy" sectors based around publishing and communications and digital information technology: the former structured around large national media corporations and clusters of small firms and self-employed freelance workers and the latter in "innovation districts" of global IT companies and small firm networks with growing links with universities – as symbolized in "Tech City" and "Silicon Roundabout" in east London (Hanna 2016).

London's scale and growth have also generated investment in private and public consumer services including, in the former, the rapid growth of a part-time and "self-employed", "precariat" "gig economy". Health and social work activities are the second largest sector in terms of jobs. Its labour market attracts workers from across the country but also globally – over a third (37 per cent) of all migrants to the UK live in London, providing a key source of both high- and low-skilled workers.

In aggregate, London's workforce is polarized between high-status, high-paid managerial and professional jobs and low-status, low-paid routine employment with workers in the latter, as Doreen was careful to emphasize, essential to the functioning of London as *World City* but increasingly marginalized within it (Massey 2007). The high status managerial and professional jobs are the ones that express London's intranational and global power relations. It is difficult to contradict Philip McCann's (2016) argument that London's global positioning – revealed in its specialization in financial services, its attraction of multinational companies, foreign direct investment and international migrants – has resulted in its effective "decoupling" from the rest of the UK economy.

The southern "sunbelt"

Doreen saw London sitting at the centre of what she described as the southern "sunbelt" – the "swathe of tamed rurality" stretching from the southwest, through the southeast to the east of England – that has been transformed by rounds of investment in "hi-tech" manufacturing and professional, scientific and technical service sectors. It is here where the separation in production of conception from execution that Doreen emphasized is most apparent. In 2015, Berkshire, Buckinghamshire and Oxfordshire in the southeast of England and the east of England (principally Cambridgeshire) had shares of employment in scientific research and development, for example, three and a half times greater than their share of all jobs nationally (ONS 2017b).

The clustering of "knowledge economy" sectors has recently been documented in the state-sponsored "science and innovation audits" that are designed to inform a belated "industrial strategy" (Department for Business, Energy and Industrial Strategy 2016, 2017). The audits show a clustering of research-based activity in advanced engineering, digital technologies and life sciences stretching from Bristol and Bath in the southwest, through Berkshire and Oxfordshire in the southeast to Cambridge's "Silicon Fen" in the east of England. The spatial structures are a mix of globally connected corporations, local small and medium-sized firms and public-sector research and innovation organizations. Cambridge's "Silicon Fen" is the quintessential example of these "new industrial districts" based around a mix of large externally controlled companies and a clustering of small and medium-sized firms – some university "spin-offs" – operating alongside publicly funded research facilities and with easy access to London and locally based venture capital (including in the latter, sourced from the university itself).

The role that the "sunbelt" plays in the national and global division of labour can be characterized by some examples of corporate restructuring. The Anglo-Swedish pharmaceutical company, Astra Zeneca, recently transferred its R&D facilities from Cheshire in the northwest of England to its newly established headquarters in Cambridge. ARM, also based in Cambridge, is a quintessential high-tech, knowledge-based "producer service" company. It concentrates on microchip design and development and sells its designs to IT producers around the world. Spatially separated from direct production, it is also now externally controlled, having recently been acquired by Japan's Softbank. Dyson, the technology-based household appliances company, has its headquarters and research and development activities in Wiltshire in the southwest end of the "sunbelt" but has shifted production to Malaysia and Singapore.

The ongoing restructuring of "old industrial Britain"

The successive rounds of investment in London and the southeast of England need to be set against the ongoing erosion of the old spatial division of labour – with "deindustrializing" job losses still disproportionately impacting on industrial cities, towns and coalfield areas in the "old industrial Britain" of the north and midlands in England and in western Scotland and Wales (Beatty and Fothergill 2016). Workers have been pushed into economic inactivity and reliance on welfare benefits while the new jobs that have been created in service sectors have been routine and low-paid, as reflected in the much greater claims for in-work benefits in these areas than in parts of southern England (*ibid.*). The places affected are being transformed. In *Spatial Divisions of Labour*, for example, Doreen explored the impact of the decentralization of jobs "for women" on the male-dominated labour market and gender relations in the coalfield areas. One outcome of the subsequent remaking of these areas has been women facing increasing competition with men for jobs – previously viewed as "women's jobs" – and a greater proportion of new jobs taken by men (Beatty 2016).

There have been selective rounds of investment in "advanced manufacturing" that have retained a core of manufacturing jobs. A few, like the car and aerospace manufacturing plants in the Midlands, have had investment in R&D activities alongside production; but most have generally reinforced the "cloning branch plant" and "part-process" spatial structures that Doreen identified, now even more tightly integrated into global production networks. These externally controlled plant structures in manufacturing have their counterparts in private service sectors – such as wholesale and retail with their spatial structures visible in the out-of-town warehouse complexes and department stores and supermarkets located in town-centre and suburban locations. With their relatively cheap and available labour reserves, northern regions have also been the principal locations of new investment in, for example, contact consumer service ("call") centres. Public education and health services remain important sources of jobs in both number and quality, albeit with investment currently constrained by post-recession austerity.

In Scotland, the balance of investment has shifted perceptibly from the "old industrial" west to the east and northeast of the country – in Edinburgh's financial services and services linked to its devolved governance role and digital industries; and in Aberdeen's oil-related and more recent R&D based energy and electronics sectors.

The legacy of "deindustrialization" in "old industrial Britain" has constrained but not completely discouraged investment in "knowledge economy" sectors, as again recorded in recent "science and innovation audits" (Department for Business, Enterprise, Energy and Industrial Strategy 2016, 2017). These sectors include medical and digital technologies in Leeds and Manchester, advanced manufacturing in Sheffield, life sciences and material technologies in Liverpool and advanced transport technologies in the Midlands. Emerging clusters of digital technology and broader "creative" industry sectors in northern second-tier cities appear to be breaking from "old industrial path dependency" to an extent, but, overall, not yet on anything like the scale of their counterparts in London and the southern "sunbelt" (Mateos-Garcia *et al.* 2018).

Spatial divisions of labour and social relations of production

What all the rounds of disinvestment and investment in these different areas are producing is a spatial reconfiguration of the social relations of production that, for Doreen, lie at the core of the spatial division of labour. Capturing this class restructuring process with official statistics is difficult but the National Statistics Socio-Economic Classification does provide a measure of employment conditions and relations. Being occupation-based, it excludes members of the "elite" with no occupation who derive income from their economic assets but for those in work it does attempt to categorize employment relations distinguished in status from control and supervision and conception to directed and routine participation in the labour process.

Applying the "three-class" categorization – which distinguishes higher managerial, administrative and professional from intermediate and routine and manual occupations – to data from the last Census of Population – provides a snapshot that validates the broad pattern of employment relations that Doreen suggested was taking shape three decades ago (Massey 1988b). The three classes have a distinct geography. While present in all areas, they are very much weighted differentially across them. Figure 13.1 shows the marked concentration of higher managerial, administrative and professional occupations in London and "sunbelt" areas. It should be noted that the "three-class" categorization combines the "higher" and "lower" professional categories that are separately identified in the finer "eight-class" version. The data for the "higher" category alone show an even greater spatial concentration of these high-status jobs in London and the "sunbelt". Figure 13.2 shows the contrasting picture for "old industrial Britain". In none of the areas, summarized in the figure by region, is the share of higher managerial, administrative and professional occupations greater than the national average. In contrast, routine and manual occupations are very heavily concentrated in them. The different positions of London and areas of the "sunbelt" and "old industrial Britain" in the national and global division of labour are revealed in this spatial stretching of employment relations.

As Doreen would also be at pains to accentuate, these different geographies of employment conditions and relations have a – socially contingent – gender dimension, with men nearly twice as likely to be in higher managerial,

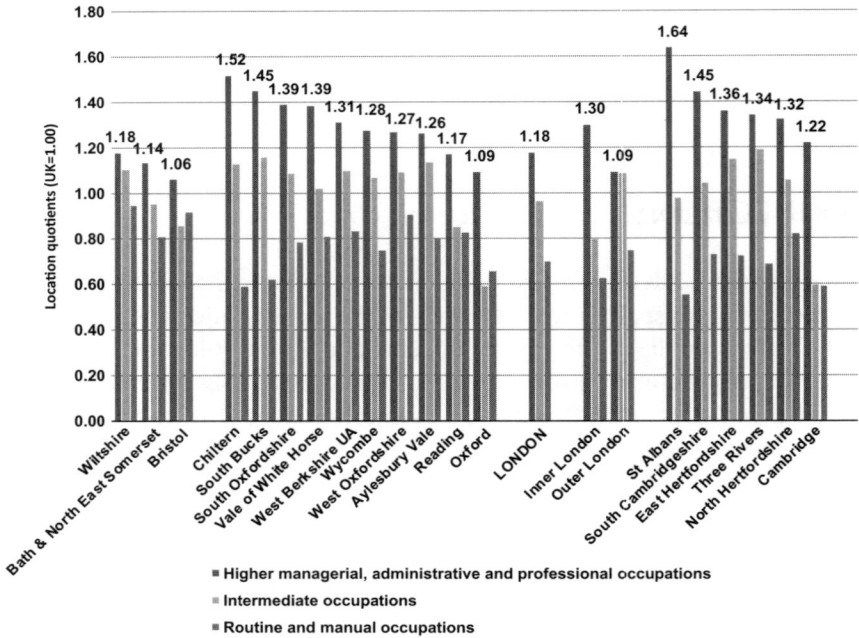

Figure 13.1 Relative concentrations of socioeconomic classes, London and "sunbelt" areas, 2011: National Statistics Socio-Economic Classification "3 classes" Location Quotients UK = 1.00

Note: Location quotients are used to measure the geographical concentration of socioeconomic positions on the basis of the National Statistics Socio-Economic Classification's broad "3-class" categorization. A location quotient of 1.0 indicates that the area's share of a particular category is equal to the national share. A location quotient greater than 1.0 indicates a relative concentration of the category in the area – the higher the location quotient, the higher the relative concentration.
Source: Calculated from Census 2011 data (ONS).

administrative and professional occupations and women nearly three times as likely to be in intermediate occupations.

CONCLUSION: "SPATIAL DIVISIONS OF LABOUR" REVISITED

Ron Martin has criticized "new regionalism" research on the grounds that the plethora of concepts it has generated only offer a partial explanation of uneven spatial development by losing sight of the bigger picture of large processes and structures – of the fundamental forces and logics of capital creation, circulation and accumulation (Martin 2015). He makes a powerful plea

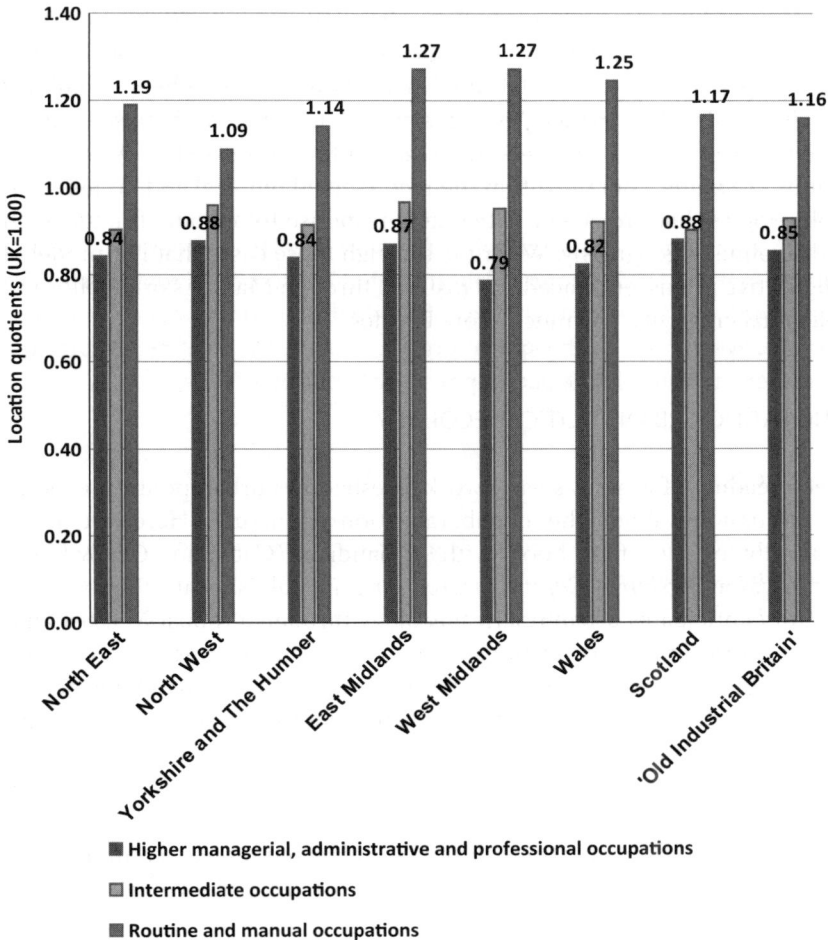

Figure 13.2 Relative concentrations of socioeconomic classes in "old industrial Britain" by region, 2011: National Statistics Socio-Economic Classification "3 classes" Location Quotients UK = 1.00

Note: See Note to Figure 13.1.
Calculations for areas of "Old Industrial Britain" and summarized regionally are based on the local authority area definitions in Industrial Communities Alliance (2015).
Source: Calculated from Census 2011 data (ONS).

for an evolutionary-historical, geographical political economy approach to understanding uneven development – an approach that can analyse both the spatially biased cumulative causation of economic growth and the evolving political-institutional context in which it operates.

185

I would argue that Doreen's notion of "spatial divisions of labour" still offers much to such an approach. It is historical and evolutionary, seeking to understand the spatial unevenness of capital accumulation over time. It recognizes that spatial uneven development is systemic in – unregulated – capital accumulation and that political interpretation of the requirements of capital accumulation influence the form that uneven development takes. It recognizes the uniqueness and specificity of place and that the economy of any locality is a complex outcome of the locality's succession of roles in wider national and international spatial divisions of labour. What it also brings to analysis is its foregrounding of the social relations of production in understanding uneven regional development – the spatial organization of the relations of production rather than simply the geographical distribution of jobs.

REFERENCES

For works authored and co-authored by Doreen Massey, please see Select Bibliography of Doreen Massey, beginning on p. 371.

Amin, A., D. Massey & N. Thrift 2003. *Decentering the Nation: A Radical Approach to Regional Inequality*. London: Catalyst.

Barnes, T., J. Peck, E. Sheppard & A. Tickell 2007. "Methods Matter: Transformations in Economic Geography". In A. Tickell, E. Sheppard, J. Peck & T. Barnes (eds) *Politics and Practice in Economic Geography*. London: Sage, 1–24.

Beatty, C. 2016. "Two Become One: The Integration of Male and Female Labour Markets in the English and Welsh Coalfields". *Regional Studies* 50 (5): 823–34.

Beatty, C. & S. Fothergill 2016. *Jobs, Welfare and Austerity: How the Destruction of Industrial Britain Casts a Shadow Over Present-day Public Finances*. Sheffield: Sheffield Hallam University Centre for Regional Economic and Social Research.

Department for Business, Energy and Industrial Strategy 2016. *Science and Innovation Audits: Wave 1 Summary Reports*. London: Department for Business, Enterprise, Energy and Industrial Strategy.

Department for Business, Energy and Industrial Strategy 2017. *Science and Innovation Audits: Wave 2 Summary Reports*. London: Department for Business, Enterprise, Energy and Industrial Strategy.

Department for International Trade 2016. Foreign Direct Investment (FDI) Projects by UK Region (Tax Year 2011 to 2012 to Tax Year 2015 to 2016). www.gov.uk/government/publications/foreign-direct-investment-projects-by-ukti-regions-201011-to-201415/foreign-direct-investment-investment-projects-by-uk-region-201011-to-201415 (accessed 8 January 2018).

Gardiner, B., R. Martin, P. Sunley & P. Tyler 2013. "Spatially Unbalanced Growth in the British Economy". *Journal of Economic Geography* 13 (6): 889–928.

Graham, J. 1998. "Review of Spatial Divisions of Labour: Social Structures and the Geography of Production 2nd edition, by D. Massey". *Environment and Planning A* 30 (5): 942–3.

Hanna, K. 2016. *Spaces to Think: Innovation Districts and the Changing Geography of London's Knowledge Economy*. London: Centre for London.

Harvey, D. 1987. "Three Myths in Search of a Reality in Urban Studies". *Environment and Planning D: Society and Space* 5: 367–76.

Industrial Communities Alliance 2015. *Whose Recovery? How the Upturn in Economic Growth is Leaving Older Industrial Britain Behind*. Barnsley: Industrial Communities Alliance.

Martin, R. 2015. "Rebalancing the Spatial Economy: The Challenge for Regional Theory". *Territory, Politics, Governance* 3 (3): 235–72.

Martin, R., A. Markusen & D. Massey 1993. "Classics in Human Geography Revisited: *Spatial Divisions of Labour*". *Progress in Human Geography* 17 (1): 69–72.

Mateos-Garcia, J., J. Klinger & K. Stathoulopoulos 2018. *Creative Nation: How the Creative Industries Are Powering the UK's Nations and Regions*. London: NESTA.

McCann, P. 2016. *The UK Regional-National Economic Problem: Geography, Globalisation and Governance*. Abingdon: Routledge.

Office for National Statistics 2016. *Earning, Learning and Business Churning: Revealing London's Industrial Economy in 2015*. London: Office for National Statistics. www.ons.gov.uk/employmentandlabourmarket/peopleinwork/employmentandemployeetypes/compendium/earninglearningandbusinesschurning/revealinglondonsindustrialeconomyin2015 (accessed 8 January 2018).

Office for National Statistics 2017a. *Regional Gross Fixed Capital Formation, 2000 to 2014*. London: Office for National Statistics. www.ons.gov.uk/economy/grossvalueaddedgva/adhocs/006749regionalgrossfixedcapitalformation2000to2014 (accessed 8 January 2018).

Office for National Statistics 2017b. *The Spatial Distribution of Industries in Britain: 2015*. London: Office for National Statistics. www.ons.gov.uk/employmentandlabourmarket/peopleinwork/employmentandemployeetypes/articles/thespatialdistributionofindustriesingreatbritain/2015 (accessed 8 January 2018).

Peck, J. 2013. "Making Space for Labour". In D. Featherstone & J. Painter (eds) *Spatial Politics: Essays for Doreen Massey*. Oxford: Wiley-Blackwell, 99–114.

Sheppard, E., T. Barnes & J. Peck 2012. "The Long Decade: Economic Geography, Unbound". In T. Barnes, J. Peck & E. Sheppard (eds) *The Wiley-Blackwell Companion to Economic Geography*. Chichester: Wiley-Blackwell, 1–24.

Smith, N. 1989. "Uneven Development and Location Theory: Towards a Synthesis". In R. Peet & N. Thrift (eds) *New Models in Geography: The Political-Economy Perspective, Volume One*. London: Unwin Hyman, 142–63.

CHAPTER 14

WHERE IS LONDON? THE (MORE THAN) LOCAL POLITICS OF A GLOBAL CITY

Allan Cochrane

This book is a celebration of Doreen Massey's life of political action and geographical thinking. For her, of course, it was not just that these two aspects informed each other but that they were inseparable, neither able to exist without the other. Politics without geography was impossible, but it was equally impossible to conceive of geography without politics.

The notion of dialogues that frames the book is fundamental to this chapter. The last time I saw Doreen Massey, we had begun (yet another) discussion about contemporary political possibilities, one which we never had the chance to complete. As a result, Doreen is still present in my head – always arguing, and somehow still winning all the arguments. In other words, it feels as if am in a continuing dialogue with her and her ways of thinking.

In what follows, I want to engage with the sadly absent yet still present Doreen in the context of a debate about the UK's political geography. Doreen Massey consistently highlighted the significance of uneven development for the UK's economic geographies, for the grounded specificities of its social relations (of gender, race and class) and for the dominant practices of British politics. She approached these issues from a number of different angles over the years, but my focus is on just one aspect of her writing – namely the position of London and the London city-region within the increasingly uneasy political and economic settlement constituted by the United Kingdom of Great Britain and Northern Ireland (Allen & Massey 1988; Allen *et al.* 1998; Amin *et al.* 2003; Massey 2007). Doreen never directly engaged with the political geography of Brexit – the UK's vote to leave the European Union – because the vote took place some months after her death. But it would have come as no surprise to her. Her work on uneven development, and the role of London in particular, was highly prescient, indeed almost prophetic. In the context of that vote I draw on that work to reflect on some of the geographies of the vote and their implications.

THE (PECULIAR) CASE OF LONDON

The tensions associated with those places identified as global city regions are continuing ones: are they part of the nation states in which they find themselves or better understood through their connections within wider networks? And these questions are still more intense when the city – like London – has been the metropolitan centre of an imperial project, remains a capital city and is at the centre of an extensive urban (mega) region. The UK's recent decision to leave the European Union and the potential breakup of the UK itself merely reinforce the need actively to interrogate the many over-lapping geographies of the London city region. Some have begun to argue that London needs to be understood as a political territory in its own right, positioned within a global network, even as others have complained about its role in stunting development possibilities elsewhere in the UK.

In one sense, of course, the question posed in the title of this chapter is easily answered. London has a clear and material expression in the south-east of England, with a widely distributed suburban and exurban hinterland that stretches around the metropolitan core. It is a node for major transport networks (road, rail and air) all of which generate significant economic activities and distinctive settlement patterns. The city region has a population of 20 million or so, depending on where the boundaries are drawn, while the more narrowly defined metropolitan area has one of nearly 9 million. The housing and infrastructure associated with such populations is apparent not only to those living in and passing through it, but also in all those maps and familiar images constructed to show energy usage, connectivity etc. And, yet, all that inescapable solidity of place only begins to open up wider questions about the London city region's position within national and inter-national urban hierarchies and networks – its relations with elsewhere.

For Doreen Massey, the region in question was never fixed or finalized. What mattered were the dynamics of region-making, rather than the fixing of the "region" as a specimen on the page (or even on a map). Over the last quarter of century, London and its city region have been defined through growth. This was widely understood to be the UK's "growth region" (see, e.g., HM Treasury 2007). But, as Doreen and others (Allen *et al.* 1998) noted it was a particular sort of growth – often presented as self-generated, but in practice underpinned by significant state infrastructural investment. This was a region presented politically in the heyday of Thatcherism as a triumph of neoliberal market led growth, rooted in the City of London's position within globalized networks of finance, but it was equally emblem-atic of the extent to which neoliberalism relied on the state to deliver not just its authoritarian disciplinary framing (Hall 1985) but also the material underpinnings of success.

In *Rethinking the Region* (Allen *et al.* 1998) Doreen and others emphasized both the holes in the growth region and the significance of its connections to elsewhere. In defining the region as a (neoliberal) growth region (and indeed the region whose apparent success was mobilized politically to highlight the failure of the UK's other regions) this made it possible to identify the areas which had been bypassed by growth, left as islands, products of some previous round of growth (like Luton and Dagenham as all too obvious emblems of Fordism, as well as the coastal towns whose tourist industry has dried up). In other words, the region is defined through unevenness and disconnection within it as well as between it and other regions. But it is also defined through its connections to elsewhere within and beyond the UK.

Within the UK, London's tentacles stretch out to incorporate economic and social actors (particularly in the financial sector) in many other cities, as well as many of the quasi rural enclaves of privilege to be found dotted around those cities. They were connected into and defined by London's growth. Beyond the UK, as Doreen goes on to argue so forcefully in *World City* (2007) and elsewhere (Massey 2004a), not only were the globalized networks of finance critically important, but London's success relied on the more mundane contributions of migrant labour and production, generating often unacknowledged geographies of responsibility. Within London, this also found another reflection as Greater London became an international city, in the sense not only that many of those living in it identified more with London than with any national formation but also that the white British population as defined by the Census became a minority (those identifying as white British declined from 60 per cent to 45 per cent between 1991 and 2011). Connections to elsewhere are increasingly significant, not only among the elite but also among the most exploited of the city's population.

LOCATING A GLOBAL CITY REGION IN A NATIONAL CONTEXT

In some discussions of London's position it feels as if it is somehow floating free in a globalized world, only touching down reluctantly because of the crude expectations and requirements of material existence. Yet the sets of relations through which the mega region is defined are stubbornly connected through a range of spatial practices. So, for example, it is necessary to recognize both the diversity of London's economy (which is not reducible to finance, advanced producer services or even the cultural industries) and the importance of its economic linkages to the rest of the United Kingdom. London's role as a world city (or with a particular position within world city networks) does not mean that its economy (and society) are divorced from the rest of the UK. In this context, Doreen Massey notes that "London's main export market is in fact the 'rest of the UK'" (Massey 2007: 38), even for the financial-services products which are generally assumed to be the most trad-able and traded in global markets.

The recognition of London's role in the national economy is not the end of the story, however. Some see London as driver of the national economy – a net contributor though its taxes to less fortunate places. Not surprisingly, this is a position that has been endorsed by a series of London mayors, from Ken Livingston, through Boris Johnson to Sadiq Khan. From another perspective, however, the matter has been understood quite differently. London's position within the UK's space-economy reflects a deeply unequal set of social and economic relations – there is a bigger gap between the wealth and economic prosperity of London and the more disadvantaged regions and territories of the UK than there is in most advanced economies. As Philip McCann (2016: 1) puts it: "the economic geography of the UK nowadays increasingly reflects the patterns typically observed in developing or former-transition economies rather than in other advanced economies".

McCann (2016) argues that the UK now has to be understood as being made up of three economies: London and what he call its hinterland (the London city region as it stretches to the southwest of England); the North and Midlands of England, Northern Ireland and Wales; and Scotland. Almost half of the UK's population, he says, now live in regions where productivity is similar to or below that of many regions in the post-communist economies of central and eastern Europe (including East Germany). While Scotland is now more prosperous than many English regions, he suggests that the core of the UK economy has "gone south" towards the region around London and decoupled from the rest of the UK.

In this context, Doreen and others (Amin *et al.* 2003) have gone on to argue economic policy in the UK, is overly influenced by the state of the regional economy in London and South East, with steps being taken to restrain the economy when the region is "overheating", even when the rest of the country still has significant capacity for growth. In other words, the cen-tralization of power in London means that a "significant element of national policy making effectively functions as an unacknowledged regional policy for the South Eastern part of England" (Amin *et al.* 2003: 17).

Doreen Massey (2007) goes further to argue that the emphasis on London's global role is a political strategy (as much as an economic reality) because of the way in which it reinforces particular ways of thinking (which she iden-tifies as neoliberal). The "geographical concentration" of the very wealthy in London and the southeast, "into a self referential echo chamber reinforces their distance from the rest of us" (Massey 2007: 66), and serves to reinforce a policy agenda which includes a commitment to deregulation, an emphasis on the "untouchability" of the financial sector, and a drive to privatization of various sorts (including "competitive individualism and personal self-reliance") (Massey 2007: 38–40. See also Atkinson *et al.* 2017 for a powerful discussion of the role of wealth elites in London). The "global" is mobilized precisely to reinforce the city's national dominance to the extent that the "Reinvigoration of London … represents the rise of a new elite, and the culture in which it is embedded" (Massey 2007: 49).

Despite a rhetorical shift in the language of national policy in England towards other regions identified for growth (such as the Northern Powerhouse or the Midlands Engine) and towards forms of devolved government in the UK's nations, the logics of development continue to reinforce the centrality of the Greater South East in public policy practice (Martin *et al.* 2016). The recent history of megaprojects certainly points in this direction. So, for example, the success of London's bid for the 2012 Olympic Games was a reflection of London's position as a world city (previous attempts by UK cities were said to have failed because of their more lowly position in urban hierarchies) and, of course, the Olympics itself is a globalized and globalizing phenomenon. The bidding process was a national initiative, with London at its core, and the bid was underpinned by the promise to transform (regenerate or remake) a significant area of east London. The infrastructural investment that followed has been reinforced since then (in the form of Crossrail as well as parkland and support for the building of new residential areas and new forms of commercial development). Major infrastructural development associated with Crossrail (cutting across London from east to west) and a new high speed railway line (HS2) providing faster connections between London and the northwest of England also provide the basis for major state sponsored and private sector led development around two new rail terminals in a previously run-down part of the city.

National priorities and national investments seem to have a continuing role in reinforcing London's position within the national space economy, even when they do not have any specific geographical framing. As Ian Gordon puts it, one consequence of London's ability to draw upon the generosity of the national state is that "bail-outs, implicit subsidy and quantitative easing … have been translated specifically into employment/spending power within London and overseas rather than elsewhere within the UK" (Gordon 2016: 336).

BREXIT AND BEYOND

In many respects, Massey's argument is a powerful one and in some respects, the referendum vote on 23 June 2016 undoubtedly highlighted aspects of the disconnection between the London city region and the rest of England.

193

All of England's regions, outside London, recorded a vote to leave the EU, although a majority in most of the larger (more cosmopolitan) cities voted for the UK to remain a member. A majority also voted to leave in Wales, although there were majorities to remain in Northern Ireland and Scotland, where the politics were very different. Of course (as one would also expect on the basis of Massey's writing) the regions and nations of the UK should not be seen as homogenous entities and there was significant variation between them and within them. In England, a majority voted remain in London and several other big (and some not so big) cities – Liverpool, Manchester, Bristol, Newcastle, Leeds, Cambridge, Oxford, York, Exeter, Brighton. It also confirms that even as most of the suburban Home Counties of the southeast of England (Oxfordshire, Surrey and Sussex) voted remain, other parts of that supposedly prosperous region (including much of Kent and Hampshire) voted to leave. Even in London, several of the boroughs on the outer east of the city (including Barking and Dagenham) voted to leave.

But what matters here is to recognize that the pattern of voting did reflect patterns of uneven development within the UK. Paradoxically, perhaps, that helped to explain the outcome in Northern Ireland and Scotland, where a remain vote reflected the linkages to another economic and political entity in the case of Ireland, and a readiness to develop a more autonomous (if not independent) relationship with Europe that went beyond the UK in the case of Scotland. In England (and Wales) in a sense the vote can be understood as a rejection of the existing set of arrangements, even if the nature of the alternative was not clear.

In many of the older industrial (and deindustrializing) regions, the European Union and its precursors had been active participants in the process of restructuring and consolidation that characterized their changing regional economies (Allen & Massey 1988), and the payments made through various regional development schemes hardly helped to compensate for those shifts. Indeed even those payments can be seen as part of the process of managing that restructuring. To put it at its most modest, people living in these regions had little reason to feel positively about the European Union. The referendum effectively threw up a strange alliance between those in the Conservative heartlands (for whom traditional forms of social and political security were fading in a post-imperial age) and those in the deindustrialized regions for whom the promise of "Europe" was always tarnished.

In this context the evidence that they are the ones most dependent on trade with Europe, or indeed most reliant on grants from Europe becomes irrelevant. This has been captured in popular and journalistic language in the notion of the "left-behind" – the argument has been that those who have been "left-behind" by globalization and the patterns of growth associated with it, were those who voted to leave. Once, however, it is recognized that uneven development is a process which actively repositions places and people through forms of economic restructuring – in other words a process in which there is a continuing relationship between the geographies of "growth" and "decline" (success and failure, moving on and being left-behind) – then a rather different set of conclusions can be drawn.

The vote can be seen as a reaction to the process by which forms of spatial and social inequality are generated and maintained. London may be less reliant on the EU than some other regions of England, precisely because it is connected into much more extensive global networks, but voting to leave the EU was also a way of voting against the effects of uneven development driven through an economic and political system focused on London and the needs of its elites. Incidentally, this also raises some fundamental questions for those who see the English vote as representing the rise of a new English nationalism – it may do, but it also highlights the extent to which "England" itself is fundamentally divided (not just between London and the rest, but also between the cities and the rest), rather than united around some clear-cut nationalist agenda.

The weight of London and the southeast of England at the heart of the UK's political settlement is not a new phenomenon, but its particular contemporary form certainly is, because of the way it is linked into wider global circuits of finance and service industries. The repositioning of London is a response to living in a post-imperial world – no longer at the centre of an empire, even a declining and fading one, but one node (one world city) among others. But – as Tom Nairn recognized long ago – this also means that some of the other assumptions about London as capital city of the UK as imperial state no longer hold as they once did – cities like Glasgow and Belfast may once have been deeply embedded in the imperial project, supplying ships and heavy engineering (as well as people) but they no longer are.

BACK TO LONDON: LIVING IN A WORLD CITY

But this is where the dialogue with Doreen becomes still more challenging and perhaps still more productive. The prime focus of *World City* is on the way in which London's role as financial centre has shaped its relationship with the rest of the UK, but – of course – London is more than that. In some respects Doreen recognizes that, particularly in her discussion of the inequalities and exclusions that go alongside the defence of privilege in the city and its region. As Doreen Massey's writing emphasizes, any story of uneven development and, indeed, of the making of a world city is necessarily incomplete. Uneven development is about much more than the generation of inequality – although, of course, that is one of its consequences. From the perspective of capital, uneven development may generate new opportunities for investment and profitable production (as Doreen emphasized in *Spatial Divisions of Labour* (Massey 1984b)), and from the perspective of labour, it may provide a basis for forms of social solidarity and political action, whether in those areas suffering decline or in those where growth is taking place.

London is a world city, but it is also the national capital. It is a diverse and multicultural city, a feature celebrated by local (and even national) politicians. It has a growing middle class population and workforce but at the same time it is a deeply divided city, characterized by growing inequality. A place within which there are concentrations of poverty and one which relies on a low paid (often migrant) workforce to service an economy dominated by (often migrant) highly paid and wealthy others. It has a nationally significant economic role, to the extent that its continued success is frequently identified by government as a necessary condition for the maintenance of the UK's prosperity.

Some residents of the city, such as those Taylor (2004: 214) identifies as the "network bourgeoisie", may be defined through their involvement in transnational networks but Castells suggests that the networked, interconnected world is one with significant "holes" in it. He identifies "multiple black holes of social exclusion throughout the planet" which are "present in literally every country, and every city, in this new geography of social exclusion" (Castells 1998: 164). In this context, it is perhaps not surprising that there is a strong tradition in academic writing on London, which suggests that it is a "dual" or "divided" city (Sassen 2001; Fainstein *et al.* 1992). There is a sharp disjunction between the (connected) urban elites and the new migrants who have a rather different position in global networks and have the task of servicing the business operations of the elites as well as supporting their leisure and consumption activities (May *et al.* 2007; Wills *et al.* 2010).

London cannot solely be understood in terms that identify it as a privileged space, or a place whose residents all benefit from the arrangements that give it such a central position in national hierarchies of power and wealth. It is a fundamentally divided city in which poverty and wealth grow alongside each other. Although, as Doreen Massey (2007) notes, these different – and coexisting – aspects of London are often presented in terms of paradox, particularly because the growing wealth of the city and some of its residents exists alongside continuing poverty for many, this is not some paradoxical outcome but, rather, an inherent part of the development process.

She suggests that the process by which inequality in London is not only reproduced but actively produced is a direct consequence of the way in which the city is imagined as a global city – that is, one in which the priorities of global financial markets are taken for granted and the need to keep and attract the financial institutions and the staff associated with them is assumed. But this also means, says Massey, that "London's poor ... and those without higher level skills, are caught in the cross-fire of the city's invention" (Massey 2007: 64). Most of the new jobs, she points out, are not ones for which those who are unskilled or with traditional skills are well-suited and those which are created for the lower end of the labour market may be taken on by labour drawn in from around the globe (Massey 2007: 64–5). In 2015 it was estimated that 18 per cent of all workers earn less than the London Living Wage, as identified by the Living Wage Foundation flowing from a calculation of the basic cost of living in the city (Linneker & Wills 2016: 761).

TOWARDS A MORE COMPLEX POLITICAL GEOGRAPHY: POSSIBILITIES AND PROSPECTS

This emphasizes the danger of understating – or perhaps better, failing sufficiently to recognize – the ambiguity of London's position. It is not enough just to point to the role of its elite in political and economic decision-making in the UK, significant though that is. In doing so, two other aspects of change may be underplayed. The first is simply to recognize that even the industrial sectors dominated by the elite (particularly business services, higher education, the media, publishing and tech industries) require a workforce that is not reducible to that elite. London draws in young people from across the UK and beyond to work in the new post-industrial industries that dominate within it, even as they face dramatically increased living costs in doing so. They may not be poor, but nor are they (yet) part of the elite.

Second, non-elite transnational aspects of the London experience may be downplayed in this narrative. As indicated above, in recent years, London and the southeast has been the region with the highest proportion of migrants from inside and outside the EU and in London itself those who identify as "white British" are now in a minority. The linkages and connections to elsewhere that are implied by such a population highlight what it means to imagine a global city region from below as well as above. London is one of the places within which living with difference is a taken for granted rather than an exceptional experience (Neal *et al.* 2013). Stuart Hall's warning that London is made up an intricate lattice of differences, which may in turn help to generate dramatic lines of divisions (expressed, for example in the 2011 riots) can also be interpreted as highlighting another range of possibilities as links are made across those lines of division (Hall 2006: 20–51).

In other words, while the outcome of the Brexit vote was undoubtedly a reaction to some of the concerns identified by Massey and others, it is also important to recognize the extent to which the vote in London and the southeast was itself a reflection of the emergence of forms of transnational and differently cosmopolitan politics – in which those who live in London often stress a London identity above a national one. These are not necessarily members of a global elite (the divisions within London and the wider city region make it impossible to identify them in this way) which may begin to open up the possibility of a different sort of political identity, within and beyond the nation.

In a sense the Brexit vote produced some very strange, if necessarily, temporary alliances. On the one hand, as I have already indicated, the geography of the leave vote brought together the relatively privileged in the Home Counties coping with the long echo of imperial decline, with the rather different experience of those who had suffered the experience of uneven development and economic restructuring, which lay waste to the UK's traditional industries. On the other, the geography of the remain vote seemed to bring together the cosmopolitan elite, the masters of financial capital, with newly insecure populations working in emergent industrial (creative, cultural etc.) sectors as well as migrant and transnational populations operating in still less secure labour markets.

The challenge now, of course, is how any of this might translate into a positive politics of connection rather than one of division and disconnection. And that is a challenge Doreen Massey would have welcomed as she actively sought to develop new alliances and new political potential. Just a year after the referendum, it was possible to identify some messages of hope in the UK General Election which saw a Jeremy Corbyn-led Labour Party (for those not obsessed with UK politics, that is a Labour Party with a left-wing leader) make significant gains, leaving the previously confident Conservatives without a majority and having to do deals with Northern Ireland's Democratic Unionist Party. In principle, at least, it looked as if the temporary accommodations of the Brexit vote had crumbled in response to a more radical political programme seeking to build alliances for change, rather than retreat. It is always dangerous to draw too many conclusions from one political moment (as some of those who did so in the wake of the referendum discovered). But I feel confident that Doreen would have been actively engaged in looking for ways of building on the incomplete success of the 2017 General Election, bringing her insights to bear in seeking to challenge the inequalities associated with uneven development – and even reconnecting London more positively into the politics of the UK.

REFERENCES

For works authored and co-authored by Doreen Massey, please see Select Bibliography of Doreen Massey, beginning on p. 371.

Allen, J. & D. Massey (eds) 1988. *Restructuring Britain: The Economy in Question*. London: Sage.

Allen, J., D. Massey & A. Cochrane 1998. *Rethinking the Region*. London: Routledge.

Amin, A., D. Massey & N. Thrift 2003. *Decentering the Nation: A Radical Approach to Regional Inequality*. London: Catalyst.

Atkinson, R., S. Parker & R. Burrows 2017. "Elite Formation, Power and Space in Contemporary London". *Theory, Culture and Society* 34 (5–6): 179–200.

Castells, M. 1998. *End of Millennium, Volume III of The Information Age: Economy, Society and Culture*. Oxford: Blackwell.

Fainstein, S., I. Gordon & M. Harloe (eds) 1992. *Divided Cities: New York and London in the Contemporary World*. Oxford: Blackwell.

Gordon, I. 2016. "Quantitative Easing of an International Financial Centre: How Central London Came So Well Out of the Post-2007 Crisis". *Cambridge Journal of Regions, Economy and Society* 9 (2): 335–53.

Hall, S. 1985. "Authoritarian Populism: A Reply to Jessop *et al*". *New Left Review* 1 (151): 115–24.

Hall, S. 2006. "Cosmopolitan Promises, Multicultural Realities". In R. Scholar (ed.) *Divided Cities: The Oxford Amnesty Lectures 2003*. Oxford: Oxford University Press.

HM Treasury 2007. *Review of Sub-National Economic Development and Regeneration*. London: HM Treasury.

Linneker, B. & J. Wills 2016. "The London Living Wage and In-Work Poverty Reduction: Impacts on Employers and Workers". *Environment and Policy C: Government and Policy* 34 (5): 759–76.

Martin, R., A. Pike, P. Tyler & B. Gardner 2016. "Spatially Rebalancing the UK Economy: Towards a New Policy Model? *Regional Studies* 50 (2): 342–57.

May, J., J. Wills, K. Datta, *et al*. 2007. "Keeping London Working: Global Cities, the British State and London's New Migrant Division of Labour". *Transactions of the Institute of British Geographers* 32 (2): 151–67.

McCann, P. 2016. *The UK Regional-National Economic Problem: Geography, Globalisation and Governance*. Abingdon: Routledge.

Neal, S., K. Bennett, A. Cochrane & G. Mohan 2013. "Living Multiculture: Understanding the New Spatial and Social Relations of Ethnicity and Multiculture in England". *Environment and Planning C: Government and Policy* 31 (2): 308–23.

Sassen, S. 2001. *The Global City: New York, London, Tokyo*. 2nd edition. Princeton, NJ: Princeton University Press.

Taylor, P. 2004. *World City Network: A Global Urban Analysis*. London: Routledge.

Wills, J., K. Datta, Y. Evans, *et al*. 2010. *Global Cities at Work: New Migrant Divisions of Labour*. London: Pluto.

CHAPTER 15

FINDING PLACE IN THE CONJUNCTURE: A DIALOGUE WITH DOREEN

John Clarke

In this chapter I explore a set of persistent questions about the practice of conjunctural analysis concerning the ways in which time and space are constitutive elements of a conjuncture. Although time seems the most obvious, if not foundational, element since the conjuncture is about the "present moment", the spatial dimensions are equally critical, if harder to diagnose. Starting from a simple question – where does the conjuncture take place? – I consider how Doreen Massey's profoundly relational view of space illuminates the question of the conjuncture by enabling us to think about the condensation of multiple spatial dynamics and relations.

In exploring some of these problems and possibilities of conjunctural analysis, this chapter carries on an unfinished conversation with Doreen Massey, following up on a promise to "talk more" about how to think about the conjuncture when we last met. The topic had been a running thread in both public and private conversations for some years: we were both convinced about the importance of thinking conjuncturally as a method of getting to grips with the present and its troubles. We shared with Stuart Hall and others a sense that conjunctural analysis was a way of avoiding the short-circuiting tendencies of reductionism by demanding attention to the multiple tendencies, forces and contradictions that made up the present moment. We had participated in a number of events, panels, and discussions considering both the current conjuncture and the way of thinking that conjunctural analysis represented. With a bitter sense of irony, I recall that the last of these encounters was in a panel on conjunctural analysis at a conference at Goldsmiths' College celebrating the life and work of Stuart Hall (see Henriques & Morley 2017). The orientation – and the stakes – can be seen in this characteristic exchange between Stuart and Doreen in the pages of *Soundings*, discussing the importance of resisting the temptations to treat the current crisis as "economic":

Massey: The other thing that's really striking is the importance of thinking of things as complex moments, where different parts of the overall social formation may themselves, independently, be in crisis in various ways. So although we see this moment as a big economic crisis, it is also a philosophical crisis in some kinds of ways – or it could be, if we got hold of the narrative. So it's really important that we don't only "do the economy", as it were.

Hall: Absolutely not. It is not a moment to fall back on economic determinism, though it may be tempting to do so, since the current crisis seems to start in the economy. But any serious analysis of the crisis must take into account its other "conditions of existence" …

But we must address the complexity of the crisis as a whole. Different levels of society, the economy, politics, ideology, common sense, etc, come together or "fuse." The definition of a conjunctural crisis is when these "relatively autonomous" sites – which have different origins, are driven by different contradictions, and develop according to their own temporalities – are nevertheless "convened" or condensed in the same moment. Then there is crisis, a break, a "ruptural fusion."   (Hall & Massey 2010: 38)

I take from this not just the imperative to think about the "current crisis" as complexly constituted, and the drive to avoid simply identifying it as "economic", but also the less developed but troubling aspects of its framing: one concerns the "when" of the conjuncture (note the point about different temporalities); the other brings us to the "where" of the conjuncture. Where is this "current crisis" taking place? Where is this "society" with its different levels? This, after all, was Doreen's "home ground", where she taught me – and many others – about why "geography matters" (Massey & Allen 1984). In the first section of the chapter, I will briefly consider the issue of time and temporalities. Secondly, I will ask more extensively about the place(s) where the conjuncture "takes place". Thirdly, I will consider the tangled relationship between a Massey-inspired understanding of space and place, the political formations of the current conjuncture's spaces, and the spatial imaginaries through which proposed resolutions of the current crisis are being constructed. This brings into view the insurgent nationalisms of contemporary populist politics that promise to rescue the nation – as place and people – from a dangerous world.

TELLING THE TIME: THE CONJUNCTURE AS CONDENSED TEMPORALITIES

There are three temporal issues worth thinking about in considering the present conjuncture. The first, and most difficult, concerns when this conjuncture emerged: is there a definite moment at which one can identify a conjuncture taking shape? There is a range of possibilities that identify the timing and character of the conjuncture with different tendencies. These might range from the long duration of neoliberalism's arrival and accumulation of triumphs, failures and new antagonisms since the mid-1970s, to a medium-term focus on the profoundly unsettling effects of the global financial crisis that began in 2007, through to a much more immediate framing around the new nationalisms and populisms of the last few years (Brexit, Trump, Le Pen, etc.; see Müller 2016). But I think this view of temporality is problematic precisely because it reduces the conjuncture to a matter of periodization, rather than dealing with the different temporalities that come together and are condensed in the conjuncture.

This is the second temporal issue: how is the conjuncture produced by the articulation of different temporalities. For example, Brexit certainly emerged out of the long trajectory of neoliberalization – the economic and social dislocations that it unevenly distributed and the political alienation that the various forms of "managing neoliberalism" engendered (see Beynon and Hudson, this volume). But these political strategies and (de-)mobilizations also have a distinctive rhythm of their own, not just of party transformations and the problems of "leadership" but also the British version of the crisis of social democracy or Labourism (see, for example, Hall 2003; Streeck 2016). There was also the very immediate temporality of individual and party political calculation associated with the EU referendum and the political futures of David Cameron, Michael Gove, Boris Johnson, Theresa May and the "trickster" Nigel Farage (thanks to Paul Stenner for that image). Yet the question of the Nation, posed so directly in the moment of Brexit, combines equally complex temporalities – the very *longue durée* of postcolonial melancholia (Gilroy 2005), the more immediate crises of mobility and migration, together with the contradictory position of the nation-state as a site for the management of a globalizing neoliberalism. It is precisely in the conjunctural *entangling* of different dynamics that we can find the conditions in which these different dynamics and their distinctive temporalities come together in complex articulations as they groom, condition, interrupt and unsettle one another. What I wish to underline, however, is the importance of thinking of this presence of different temporalities as taking the form of active and intense condensation rather than a passive or indifferent coexistence. This issue has also been addressed by Michele Filippini in his recent book on Gramsci, in which he argues that Gramsci developed a distinctive view of multiple temporalities but that:

> This temporal plurality should not be confused, however, with an objective, eternal condition that sees fragmentariness as a value in itself, and which consequently expresses a politics that tends to incorporate these diverse temporalities into one "harmonious plurality".
>
> (2017: 109)

Instead, he suggests, Gramsci saw this multiplicity as the locus of a form of hegemonic struggle in which one temporality became accepted not just as dominant but as normal, subordinating or even suppressing other temporalities in the process. Might this be understood as one aspect of the present conjuncture? That is, not just a question of the intersection and condensation of different temporalities, but the conjuncture as a moment in which the normalized hegemonic temporality – the rhythm of capitalist time itself – is at risk, or has become unsettled as other temporalities assert themselves? These might include the varieties of "speed up" that we currently encounter, from the intensified speed of capitalist calculation, financial flows or the digitally enhanced sense of immediacy; or the sense of slowing down or even stasis attached to the lives of those (in the current pacifying metaphor) "left behind" by globalization/neoliberalism/progress. There might be a tension between the apparent urgency of planetary time in the face of ecological disaster and the foot-draggingly slow time of "business as usual". We might also consider the temporal dramatization of populist political choices and their apparently immediate effects (Brexit as "independence day") and its strained encounters with the slow pace of "governmental time", as treaties are negotiated, laws drafted, discussed and ratified, policies formed and reformed or judicially tested (Clarke & Newman 2017).

Such tensions between different – and differently located – temporalities manifest themselves in political struggles over the capacity to "tell the time". At stake is the power to articulate the present, and its relationships to past and future; to identify our proper attachments to them; and to define the appropriate rhythms of economic, social and political life (and to denigrate or exclude those who fail to move in step with them). Struggles over "telling the time" form the third and final issue of temporal dynamics and dimensions of the conjuncture that are significant. But I must move on, given the urgency of addressing questions of space.

Taking place: the spaces of the conjuncture

Let me return to a question that I posed at the start: where is the current crisis taking place? So much of contemporary social and political analysis remains transfixed by the imagined geography of nations and nation-states that is hardly surprising that crises – and conjunctures – are typically understood as national events, peculiar to the time-space of a particular place. The typical escape route from the straitjacket of this conceptual and methodological nationalism has been to substitute a methodological globalism, shifting scales while avoiding the challenge of rethinking place and space. By contrast, Doreen Massey's work has provided a vital resource for the possibilities of thinking about the relational production of place in ways that make conjunctural analysis simultaneously richer – and more difficult. Later, I will argue that the current crisis takes distinctive national *forms*, while being profoundly structured by transnational forces, tendencies and relationships whose significance demands that we think about the "nation" (and its associated state) as a (shifting) relational entity. In an interview with Andrew Stevens, Doreen gave a beautifully compressed version of her view of relationality:

> For me, places are articulations of 'natural' and social relations, relations that are not fully contained within the place itself. So, first, places are not closed or bounded – which, politically, lays the ground for critiques of exclusivity. Second, places are not 'given' – they are always in open-ended process. They are in that sense 'events'. Third, they and their identity will always be contested (we could almost talk about local-level struggles for hegemony).
>
> (Massey & Stevens 2010)

It is this understanding, centred on places as articulations of multiple relations (and disconnections), that enables us to think about where the conjuncture takes place in ways that escape the national versus global binary. Instead, the conjuncture articulates multiple spatial relations, such that politics come to play out on a terrain that combines and condenses multiple sites – the local (the deindustrialized city or region); the national, the regional (embodied in the EU, for example) and the global, whilst recognizing that all of these are folded into one another. It is, surely, her insistence on the concept of articulation in the above extract that points to how we might think multiple places (and scales) together, rather than as separate entities (ordered by borders or nested in scalar hierarchies). And "articulation" forges many connections between Massey and Hall, reflecting long running shared intellectual and political commitments.

However, what may be distinctive about *this* conjuncture is the way that it brings one of those spatial formations to the fore as the focus of political contestation. The nation has come to be the focus of conflict and mobilizations, not least in the proliferation of fantastic imaginings of the "way forward" requiring the restoration of the nation. The recurrent failures and perverse effects of a globalizing neoliberalism have been felt, most directly, in national forms, not least because one of the chosen engines for managing neoliberal development was the nation-state (alongside a variety of transnational apparatuses). So, to take one instance within the present conjuncture, the strange case of "Brexit" took place on nationalized terrain, through the strained apparatuses of a nation-state, and involved the deployment of powerful nationalist imaginings and discourses. But it was hardly "national" in its causes, its conditions or its consequences, even as it promised the restoration of a liberated and sovereign nation. This Britain was marked precisely by the production of a divided, unequal and contradictory place through complex relational dynamics: *economic* (deindustrialization combined with financial hegemony); *social* (the degradation of waged work; familial responsibilization and urban decay) and *political* (anti-welfarism, anti-statism and antagonisms around migration). These dynamics were condensed in lived experiences of dislocation, dispossession and despair that came to be articulated in a multiplicity of forms of political disaffection (Gilbert 2015). Many of these dynamics are shared with other places; many of the lived experiences of dislocation are common, but the specific forms of political articulation have taken distinctive national shapes and have articulated distinct and divisive nationalisms. That is to say, far from the nation being the natural location for these experiences, or nationalism being the normal political response, the nation is given specific salience and significance conjuncturally by the ways in which the contradictions, forces and dynamics have been articulated in and around the nation and nation-state. Nationalism, then, needs to be understood as a conjuncturally particular formation, rather than a generic political disposition. In the following two sections, I take this view further: first by considering the contradictory place of the nation-state in these processes and then by reflecting on the revitalization of nationalism as the focus of populist mobilizations in this conjuncture.

THE UNSETTLED HYPHEN: NATION-STATES AND THE MANAGEMENT OF NEOLIBERALIZATION

The contemporary struggles over the nation and the state are a tribute to the persistence of the nation-state form in the era of neoliberalism and its globalizing tendencies. Abstractly, the nation-state form might seem ill-suited to managing the dynamics of globalization, juggling with the (different) insertions of national economies into the global, and combining economic development with international beauty contests to win the affections of a footloose and mobile capital seeking fertile ground for capital accumulation (see for instance, Massey on London as "World City", 2007). Neither the political processes nor the bureaucratic apparatuses of the nation-state system seem well-adjusted to such challenges – and it is hardly surprising that they have both been the focus of endless "reform" initiatives since the 1970s to make them precisely more business-like as well as more business friendly. Yet such states have continued to play a leading role in managing the conditions, consequences and contradictions of neoliberalization within specific national spaces (and the spaces in-between through such institutions as the WTO, IMF and EU). Their role has been partly to drive the processes of capital accumulation (not least through the transformation of public resources into private ones), partly to stabilize the surprisingly frequent "wobbles" of the economy (especially during the recurrent financial crises), and also to enthusiastically naturalize the "economic" as the fundamental human condition, disposition and way of life (Newman and Clarke 2009, chapter 4; Elyachar 2005 and many others, including Massey 2013a).

These nation-states have also had a critical role in creating and maintaining the political, social and cultural conditions for neoliberalization and renewed capital accumulation: constructing consent, policing crises and managing contradictions (even while dismembering particular state apparatuses). At the core of this process has been the welfare state and Claus Offe's famous contradiction: "The contradiction is that while capitalism cannot co-exist *with*, neither can it exist *without*, the welfare state" (Offe 1984: 153; emphasis in original). As I have argued elsewhere (Clarke 2013), we have certainly seen forty years of constant innovation – varieties of welfare reform and state reform – attempting to resolve this contradiction, creating new welfare apparatuses that are simultaneously diminished and more disciplinary. Such reforms have developed models of "corporate welfare" while making social welfare increasingly antisocial (responsibilizing welfare, "do-it-yourself" welfare and welfare as surveillance and scrutiny, for example; see Lawson and Elwood, this volume). Despite this, the contradiction persists, not least because of the glaring failures of neoliberalization, including growing inequality, social dislocation and the inability to meet the most basic human needs. As recent national-populist political movements have indicated, people still look to nation-states to provide support and well-being, even if these movements have constructed those desires in nationalist/nativist terms – welfare for "our people" (see, *inter alia*, Guia 2016 and Walsh 2015 on Trump). The unsettled political conditions emerging in a variety of places have made visible the increasing disaffection of publics from the conventional practices and institutions of political representation and the emergence of new political projects (or at least attempts at new political projects – their effects and stability remain open questions). I borrow the notion of "disaffection" from Jeremy Gilbert's (2015) brilliant discussion of the politics of neoliberal consent in the UK but intend to make it work to cover multiple forms of disaffection that form the terrain of popular politics in this moment. And, to return to my theme, these politics increasingly take place around the question of the nation – imagined as both place and people (whose unity should be restored). One of the consequences of the nation-state playing such a central role in managing neoliberalization is that the role has engendered new stresses and strains, unsettling both the state form and the way the nation is inhabited and imagined. The shifting relationship between the nation and state is nicely captured in Akhil Gupta's conception of the "hyphenated entity":

> That this curiously hyphenated entity, the nation-state, does not evoke constant surprise is a testimony to its complete ideological hegemony. Scholarly work has tended to underestimate seriously the importance of that hyphen, which simultaneously erases and naturalizes what is surely an incidental coupling …
>
> (Gupta 1998: 316–7)

In his discussion of the "postcolonial condition", Gupta views this hyphenation of nation and state – and its destabilization – as critically important. He argues that the unsettling of the hyphen reveals the contingently constructed coupling of nation and state, such that they can be more easily recognized (and contested, both analytically and politically) as separable. At the same time, he sees the unsettling of the hyphen as bringing new social tensions and forces into play:

> The hyphen between nation and state holds together a particular bundle of phenomena that are increasingly in tension. It is this that makes the "postcolonial condition" different from the order of nation-states brought together by colonialism and nationalism.
>
> (1998: 327)

I would want to suggest that the era of multiple and divergent neoliberalizations also brings new destabilizing forces to bear on the hyphen (and the formations that it couples). Debates about the rescaling of government, governance and the state; the proliferation of forms of anti-statism and the contemporary resurgence of nationalism swirl around that unsettled hyphen. For example, we might argue that "Brexit" emerged from this turmoil – a profoundly unsettled domestic politics, an unstable spatial and scalar relationship between nation-state and governing, and a profoundly affective re-imagining of the nation as a unity of people, place and sovereignty (Clarke 2017). This unstable mix of space, scale and sovereignty and their contested imaginings embodies Doreen Massey's view of why "space" had to be central to critical intellectual work. Refusing the subsumption of space into time (the singular narrative of modernity and progress), she argued that:

> "Recognising spatiality" involves (could involve) recognising coevalness, the existence of trajectories which have at least some degree of autonomy from each other (which are not simply alignable into one linear story) … On this reading, the spatial, crucially, is the realm of the configuration of potentially dissonant (or concordant) narratives. Places, rather than being locations of coherence, become the foci of the meeting and the nonmeeting of the previously unrelated and thus integral to the generation of novelty. (2005: 71)

This argument guides me in thinking about the return of place in the conjuncture, both as a relational formation and potent political imaginary. Brexit (and its contemporary echoes in other places such as the USA, Hungary and France) invokes that impossible object of desire – the nation in all its glory. Celebrating the Britain-to-come after Brexit, the new Prime Minister and leader of the Conservative Party, Theresa May told the Conservative Party conference in 2016 that:

> A truly Global Britain is possible, and it is in sight. And it should be no surprise that it is. Because we are the fifth biggest economy in the world. Since 2010 we have grown faster than any economy in the G7. And we attract a fifth of all foreign investment in the EU. We are the biggest foreign investor in the United States. We have more Nobel Laureates than any country outside America. We have the best intelligence services in the world, a military that can project its power around the globe, and friendships, partnerships and alliances in every continent. We have the greatest soft power in the world, we sit in exactly the right time zone for global trade, and our language is the language of the world. (May 2016)

Brexit has enabled the renewal of imperial fantasies – a mercantilist Britain, capable of leading the world as a sovereign power, can both direct and benefit from a new phase of Free Trade. This is a profound spatial and scalar imaginary that seeks to remake (some of) the spatial relationships that have produced the nation. The question of whether such relationships can be governed and directed in this manner remains open: there is, at least, a suspicion that the constitutive "power-geometries" might prove intractable.

"TAKING BACK CONTROL": INSURGENT NATIONALISM AND THE PROMISE OF POWER

To repeat, the politics of this conjuncture draw on, and are animated by, rich imaginaries of space, scale and place – and how they might be "properly" aligned by the restoration of "national sovereignty". Such contested imaginings of space, scale and place reflect the ways in which the nation-state has been exposed to new pressures as it tries (and fails) to manage the crises and contradictions of globalizing neoliberalism. The "unsettled hyphen" of the nation-state form has been destabilized by both the recurring efforts to "retool" the state to make it fit for its new purposes and by the return of the national question. Both the place of the nation and the peopling of the nation have become increasingly politicized as the strains of neoliberalism are lived in the everyday. The spatial formation of the nation (and its attendant state) has been challenged by the flows of money, power, goods and people across national borders. The last of these – more precisely, people in the form of refugees, asylum seekers and migrants – have emerged as the overarching symbol of the Border (whether it be Trump's projected Wall, the Fidesz government's fences at the edge of Hungary or the UK government's "taking back control of its own borders"). It is, of course, not accidental that it is the human flow (rather than capital or goods) whose mobility has emerged as the symbolic content of the Border question. Such flows of non-business class people disrupt the imagined link between nation-as-territory and nation-as-people that has been central to the articulation of a range of right-wing populisms in recent years. Defending the distinctive "character" of the (imagined) nation has been one of the crucial political-discursive strategies linking Brexit, Trump, Orbán, Wilders, Le Pen and many more (including Modi's version of Hindu nationalism, see, *inter alia*, Anand 2011). A variety of terms swirl around these movements and their representations of the divide between the people and its others – nationalism, xenophobia, nativism, racism – but they point to the centrality of the nation/people as the articulating point for these new (or revitalized) movements.

Such insurgent nationalist mobilizations have tried to articulate a variety of disaffections and desires. It may be worth distinguishing between the multiplicity of disaffections and desires and the limited range of their *articulation* within these new populisms (rather than treating such movements as "reflections" of popular experiences and sentiments – see, for example, Judis, 2016). Such a distinction is analytically and politically important, since reflection metaphors collapse the politics of selection, silencing, voicing and ventriloquism that are at stake in processes of articulation. Keeping such a space open also marks the site of possibility of alternative articulations, rather than assuming that left or even liberal alternatives must also "reflect" popular feelings in a direct or unmediated way (on this theme, see the thought-provoking discussion of Gramsci's view of the relationship between common sense and hegemony in Crehan 2016). Nevertheless, such movements have been recently very successful at identifying senses of loss (of livelihoods and ways of life, of relative privilege, of conceptions of the future as progress, of stability) and feelings of disempowerment, disenchantment, abandonment and betrayal as fertile grounds that could be brought to voice around a cluster of key themes – restoration (Make America Great *Again*; putting the Great *back* into Great Britain etc.); authenticity (the people versus the cosmopolitan elite); unity (the people versus its others) and, in the Brexit moment explicitly, the promise of "taking back control". The promise of taking back control raises uncomfortable questions about when "we" had control, about what sorts of control may be at stake and, perhaps most troublingly, what happens when such promises are not fulfilled: what happens *after* insurgent nationalism?

At the core of these promises of restoration, authenticity and control are the foundational liberal fantasies of the sovereign individual, family and nation (and their economic shadows – the enterprising self, the self-reliant family and the autonomous corporation). So, it seems strange to read that such insurgent nationalisms mark a crisis of liberalism and neo-liberalism (e.g. Fraser 2017). Rather, I would suggest that we remain locked in a conjuncture that is framed by the presumptions (and fantasies) of liberalism and neoliberalism in all their contradictory and antagonistic glory. However, their instabilities (their contradictions, crises and antagonisms) have generated new responses and potential new lines of political force and movement. On the one side, new populisms and nationalisms try to resolve these contradictions in profoundly authoritarian and regressive directions while also trying to secure the rule of national and international oligopolies. In the same moment, alternative imaginings of political futures continue to emerge that disrupt and challenge such regressive politics and offer the prospect of new alliances and solidarities being built (see, for example, David Edgar's discussion of Labour in the UK, 2017).

Gramsci once observed (in a typically complicated and subtle phrase) that "the life of the state can be conceived as a series of unstable equilibria" (1971: 182). I remain fascinated by this paradoxically dynamic image and find it helpful for thinking about the processes of settling and unsettling of social and political formations. We may well be at a new point in this series – a point of disequilibrium – in which the state (and its national form) have been profoundly unsettled by the accumulating contradictions and crises of a globalizing neoliberalism. Each of the nation-states caught up in this moment of disequilibrium is striving to find a new point of balance, often seeking a new settlement that will enable an effective return to "business as usual". As Doreen knew only too well, the difficulty may be that it is precisely "business as usual" that is the problem.

REFERENCES

For works authored and co-authored by Doreen Massey, please see Select Bibliography of Doreen Massey, beginning on p. 371.

Anand, D. 2011. *Hindu Nationalism in India and the Politics of Fear*. Basingstoke: Palgrave Macmillan.

Clarke, J. 2013. "Widersprüche des Heutigen Wohlfahrtsstaates". *Sozialwissenschaftliche Literatur Rundschau* 67: 68–81.

Clarke, J. 2017. "Imagining Scale, Space and Sovereignty: The United Kingdom and 'Brexit'". Unpublished manuscript.

Clarke, J. & J. Newman 2017. "'People in this Country Have had Enough of Experts': Brexit and the Paradoxes of Populism". *Critical Policy Studies* 11 (1): 101–16.

Crehan, K. 2016. *Gramsci's Common Sense*. Durham, NC: Duke University Press.

Edgar, D. 2017. *Cosmopolitans, Communitarians and the New Fault-Line: How to Renew the Traditional Labour Alliance*. London: Compass, Thinkpiece 87.

Elyachar, J. 2005. *Markets of Dispossession: NGOs, Economic Development, and the State in Cairo*. Durham, NC: Duke University Press.

Filippini, M. 2017. *Using Gramsci: A New Approach*. London: Pluto Press.

Fraser, N. 2017. "The End of Progressive Neoliberalism". *Dissent*, 2 January. www.dissentmagazine.org/online_articles/progressive-neoliberalism-reactionary-populism-nancy-fraser (accessed 8 January 2018).

Gilbert, J. 2015. "Disaffected Consent: That Post-Democratic Feeling". *Soundings* 60 (Summer): 29–41.

Gilroy, P. 2005. *Postcolonial Melancholia*. New York: Columbia University Press.

Gramsci, A. 1971. *Selections from the Prison Notebooks*. London: Lawrence and Wishart.

Guia, A. 2016. *The Concept of Nativism and Anti-Immigrant Sentiments in Europe*. Florence: European University Institute Max Weber Programme Working Papers. http://cadmus.eui.eu/handle/1814/43429 (accessed 8 January 2018).

Gupta, A. 1998. *Postcolonial Developments*. Durham, NC: Duke University Press.

Hall, S. 2003. "New Labour's Double Shuffle". *Soundings* 24: 10–24.

Hall, S. & D. Massey 2010. "Interpreting the Crisis". In R. Grayson & J. Rutherford (eds) *After the Crash: Re-Inventing the Left in Britain*. London: Soundings, Social Liberal Forum and Compass, 37–46.

Henriques, J. & D. Morley (eds) 2017. *Stuart Hall: Conversations, Projects, Legacies.* London: Goldsmiths Press.

Judis, J. 2016. *The Populist Explosion: How the Great Recession Transformed American and European Politics.* New York: Columbia Global Reports.

May, T. 2016. "Brexit Speech to Conservative Party Conference". *The Independent* 2 October. www.independent.co.uk/news/uk/politics/theresa-may-conference-speech-article-50-brexit-eu-a7341926.html (accessed 19 January 2018).

Müller, J.-W. 2016. *What is Populism?* Philadelphia, PA: University of Pennsylvania Press.

Newman, J. & J. Clarke 2009. *Publics, Politics and Power.* London: Sage Publications.

Offe, C. 1984. *Contradictions of the Welfare State.* London: Hutchinson.

Streeck, W. 2016. "Social Democracy's Last Rounds". *Jacobin* 25 February. www.jacobin mag.com/2016/02/wolfgang-streeck-europe-eurozone-austerity-neo-liberalism-social-democracy (accessed 8 January 2018).

Walsh, J. 2015. "Donald Trump's Nixonian Populism: Making Sense of his Grab Bag of Nativism and Welfare Statism". *Salon*, 31 August. www.salon.com/2015/08/31/donald_trumps_nixonian_populism_making_sense_of_his_grab_bag_of_nativism_welfare_statism (accessed 8 January 2018).

CHAPTER 16

LAMPEDUSA IN HAMBURG AND THE "THROWNTOGETHERNESS" OF GLOBAL CITY CITIZENSHIP

Matthew Sparke and Katharyne Mitchell

Drawing on Doreen Massey's arguments about the politics of "throwntogetherness" we return here to the city of Hamburg, out of which she drew a series of lessons about the global politics of place (Massey 2005). These lessons, we argue, are all the more salient today amidst intensifying refugee reception crises worldwide. Citizens of cities such as Hamburg are being challenged to reflect anew on the material meaning of global city citizenship as they variously reject and accommodate those thrown together by border-crossing forms of violence. And refugees themselves are actively resisting exclusionary encodings of citizenship. In these kinds of context, we want to suggest that Massey's attention to the time-space complexity of urban throwntogetherness offers a useful way of conceptualizing the resulting opportunities for the remaking of global city citizenship.

The contested character of citizenship in Hamburg has recently been re-articulated with special attention to the global composition of local throwntogetherness by a refugee protest and solidarity group called Lampedusa in Hamburg. Comprised largely of sub-Saharan African workers turned into refugees by the war in Libya, the group was thrown together in seeking asylum on the Italian island of Lampedusa. Many of them subsequently found their way to Hamburg after the Italian authorities offered money for the journey, along with a one-year humanitarian permit for the European Union (EU) Schengen area. Owing to the EU's Dublin regulations, however, the refugees were not considered legal asylum claimants or workers outside of Italy. This negation of rights has formed the basis for their subsequent claims and protests. As we demonstrate in what follows, Lampedusa in Hamburg raises critical questions about global city citizenship – questions that we believe highlight Massey's concern with how such cities pose – both physically and politically–"the question of our living together" (Massey 2005: 151).

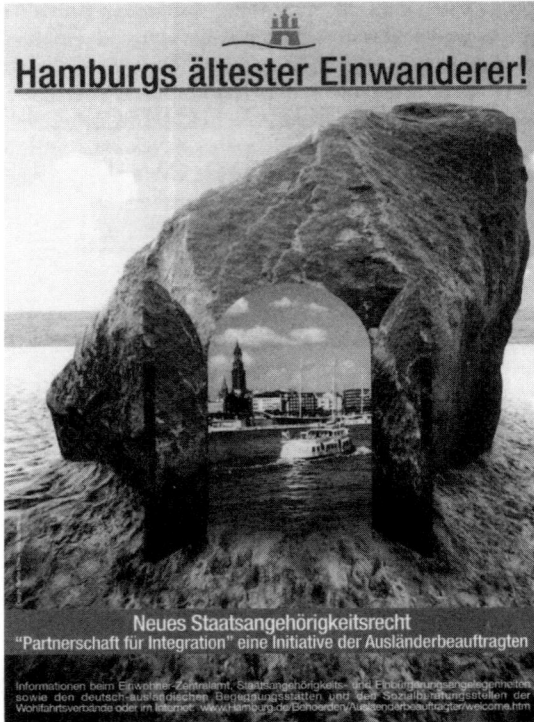

Photo 16.1 "Hamburg's Oldest Immigrant"
(reproduced courtesy of Steffan Boehle who holds the copyright)

Massey's arguments about "throwntogetherness" – a neologism that deliberately combines words in a way that is meant to metaphorize the local jumbling together of global human geographies – begins with a series of reflections on the migration politics of a poster (see Photo 16.1).

The poster shows an image of a rock in Hamburg harbor that had been pushed south and deposited by a glacier thousands of years earlier. With its emphatic accompanying statement, *Hamburgs ältester Einwanderer!* (Hamburg's oldest immigrant!), and open archway, the image served to rework urban boosterism *vis-à-vis* the city being a welcoming "gateway to the world". For Massey, the poster illustrated a basic geomorphological argument about local geographies including global moving parts. But as a form of physical geography, she was also fascinated with the rock's symbolic representation of intersecting global-local human geographies. It provocatively recoded global city citizenship by renaturalizing migration as a fundamentally physical piece of Hamburg's status as gateway to the world. In Massey's words "[t]he poster [thereby] addressed a challenge to established German citizens to make this logo (this already existing [gateway] self-image) meaningful in another way, to take it at its word and press it home" (Massey 2005: 149). At the same time, she argued, the poster was also "an invitation to migrants to find out more" (Massey 2005: 149).

The challenge turned invitation of the poster was, for Massey, one more compelling case for why we need to rethink place, and the politics of place more generally, as being inherently founded on the throwntogetherness of converging time-space trajectories. The opening gateway cut through the rock, she suggested, invoked "place as a constellation of trajectories, both 'natural' and 'cultural' where if even rocks are on the move the question must be posed as to what can be claimed as belonging" (Massey 2005: 149). Taking up this embodied concern with belonging in the global city, we want to suggest that the solidarity and organizing work of Lampedusa in Hamburg today is further radicalizing Massey's interest in unsettling the givenness of place, and posing new questions about "our living together" in contemporary Europe.

The etymology of Lampedusa itself – thought to derive from ancient Greek words for rock (Λαπαδούσσα/*Lopadoússa*) and navigational lamps (λαμπάς/ *lampás*) – would itself seem to indicate a newly transnational trajectory of rocky throwntogetherness turned illumination by the work of Lampedusa in Hamburg. We are interested, in this way, in how the "our" in "our living together" is itself now being thrown together in ways that invite further reflection on who is given place, who is not, and how this relates to what we might call, following Massey's geomorphological arguments, the landscaping of global city citizenship.

First, there is the immediate urban landscape of throwntogetherness to which Massey's writing on the global city draws our attention and through which she developed her much discussed global sense of place (see Cochrane, this volume; Leitner and Sheppard, this volume). But following the interventions of Lampedusa in Hamburg, we wish to take Massey's arguments in two further directions beyond the urban landscape itself. The first of these is the *geopolitical throwntogetherness* of the international relations and tensions that have so violently forced refugees together and sent them across multiple borders into cities such as Hamburg. The second is what we describe as the *geosocial throwntogetherness* of the politics of refugees and their allies in the global city. This is a border-crossing but also interpersonal aspect of politics that embodies a series of affective engagements and experiences of transnational activism and networking that radically complicate geoeconomic gateway depictions of the global city.

Augmenting Massey's arguments we argue that the throwntogetherness of place conceptualized through the geopolitical-turned-geosocial landscapes of Lampedusa in Hamburg offers new insights into the profoundly uneven landscaping of global city citizenship. These include differential degrees of disenfranchisement and the ways in which these are embodied in varying levels of citizenship and subcitizenship amidst the austerities of the contemporary global city (see also Sparke 2017a). The activism of Lampedusa in Hamburg underlines the error in explaining the situation of the dispossessed, precarious and marginalized as one of bare life bereft of agency (see also Ramsay 2017). Instead, it shows how the agency of immigrants themselves can inform our theories of global city subcitizenship and the inequalities and asymmetrical power relations *within* throwntogetherness.

LAMPEDUSA IN HAMBURG AND URBAN THROWNTOGETHERNESS

Lampedusa in Hamburg began organizing in Hamburg in 2013 and has gone on subsequently to become a model and collaborator for similar solidarity groups throughout Germany and beyond. Underlining their resistance to deportation from the city, the group's repeated reworking of an anchor logo in various posters and wall murals invokes Hamburg's harbour heritage and, like the poster highlighted by Massey, also invokes the port's gateway tradition to make the case for refugee belonging (see Photo 16.2).

Lampedusa in Hamburg's insistence on the right to stay has already been documented by other scholars (e.g. Borgstede 2017). It has also been highly generative and inspiring for local and international activists. Over the four years of its existence it has spawned conferences, a website, a Facebook page, an award winning documentary – Lampedusa in Berlin, a public informational "tent", refugee strikes, hunger strikes, public art, and numerous media articles and events.

At a recent event on migrant asylum that we attended at the Friedrichshain-Kreuzberg Church of the Holy Cross in Berlin in November 2016, a representative from Lampedusa in Hamburg was featured as one of the guest speakers. In addition to insisting on the work and residence rights of refugees, the church speaker also invited participants at the event to join upcoming protests. These included one in Berlin aimed at the embassies of North African countries that were then considering EU inducements to implement offshore border controls of refugees. It is this concern with broader forms of political activism, combined with the group's geopolitically innovative name and geosocially networked ties to other refugee solidarity efforts, that reframes the politics of living together with such radical attention to the inequalities and opportunities within throwntogetherness.

Lampedusa in Hamburg was composed initially of roughly 300 refugees who arrived in Hamburg after staying in Lampedusa and other Italian refugee

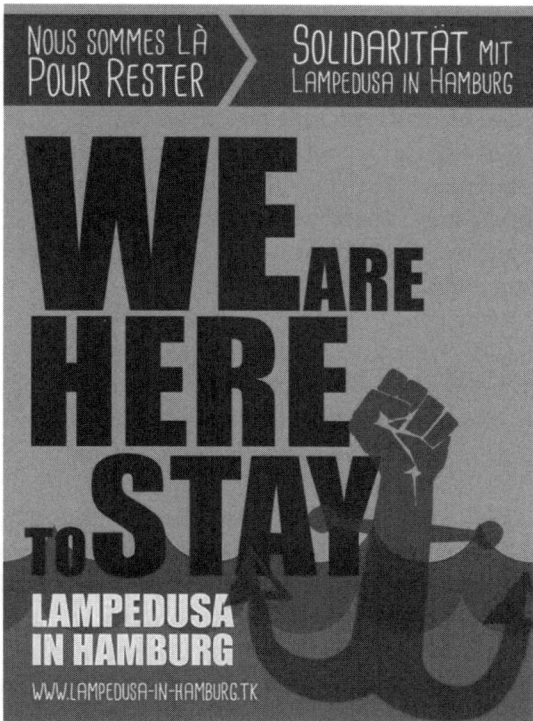

Photo 16.2 "We are here to stay"
(reproduced courtesy of Lampedusa in Hamburg)

camps, boardinghouses and hotels, where they were housed through the Emergency North Africa Program after fleeing the civil war in Libya. Most of the men (they were nearly all young men) were originally from sub-Saharan Africa. Thrown into motion by the legacies of structural adjustment and other forms of structural violence in countries ranging from Burkina Faso and Mali to Somalia and Sudan, they were also thrown together as labourers in Libya prior to the war. During and after the war, however, they were perceived as mercenaries of the Gaddafi regime and forced to escape across the Mediterranean. Their subsequent regrouping in Hamburg was then made possible by the funding provided by Italian authorities, funding which German authorities have, in turn, condemned as a way of deliberately dodging Italy's asylum-assessment obligations under the Dublin regulations (BBC 2013). Owing to the Dublin principle that the country where immigrants first arrive should be responsible for asylum processing, German officials do not consider members of Lampedusa in Hamburg to be legal asylum claimants and, for the same reason, they do not allow them to work. The net result of these

219

experiences has since been well summarized by the group themselves on their website:

> We worked and lived a good life in Libya until the unjust NATO-war in 2011 forced us to flee. We lost everything and were forced to flee across the Mediterranean. The ones that survived were accommodated in camps in Italy. These camps were closed in early 2013 and we were pushed out into the streets. As a group we are united by this common fate ... We don't understand why our Italian papers are not recognized in Germany and why we are not even allowed to work. Here in Hamburg we have fought nearly two years for our right to finally rebuild our life. We are not allowed to work, in order to survive we depend on the solidarity of Hamburg's citizens. (Lampedusa in Hamburg 2015a)

The urban throwntogetherness of Lampedusa in Hamburg is manifested in the bodies and identities of the refugees, as well as the conditions of their urban experiences, and their demands for a right to the city. "We are all part of this city", they underlined at a meeting of the Green Party in 2015, "we all have a right to the city" (Lampedusa in Hamburg 2015b). Coming from diverse countries and traditions, many of the migrants arrived in Hamburg only after arduous journeys by train, bus, plane and on foot. Initially housed through a winter assistance programme, the government support ended in April 2013 and most of the men were ejected onto the streets in the hopes that they would leave the city permanently. But due to their joint experiences of labour in Libya, surviving the war, travelling under duress, dealing with bureaucratic red-tape, inertia and indifference along the way, and then dealing with rejection and violence in Hamburg, the journey led ultimately to solidarity and resistance. The bonds holding these disparate workers together were, in this sense, strengthened by their throwntogetherness.

As a protest movement, Lampedusa in Hamburg became the voice and the representation of these refugees' collective biographies, their embodied journeys and experiences, and their unwillingness to passively accept the politics of rejection or victimhood. They took up their case as a collective unit, asking for housing, and the right to work and to integrate fully into the life of the city. In December 2013 they marched behind a red banner, declaring in English, German and French: "We did not survive the NATO war in Libya, to die on the streets of Hamburg" (BBC 2013). Their associated demands, in turn, became the basis of their ongoing campaigning for basic city citizenship rights in the following years (Lampedusa in Hamburg 2015b):

A work permit for everybody who lives in Hamburg!

The right to stay for all people who live in Hamburg – no more deportations!

The right of housing in humane living conditions – flats for everybody instead of *Lager*!

"We want to become part of Hamburg society", a spokesperson summarized. "We cannot and do not want to go back to misery, neither in Italy nor in an African country."

(quoted in Meret & Della Corte 2014)

With these political actions, then, Lampedusa in Hamburg became a new and illuminating rock in the gateway global city – a material reality and representation of embodied difference both temporary and immovable, a "provocation to new thinking" (Massey 2005: 159). For the same reason, the solidaristic movement became a provocation to a new articulatory politics – in the terms Massey adapted from Stuart Hall, Chantal Mouffe and Ernesto Laclau. "Long live the Green Party", participants declared in 2015, "long live the citizens of Hamburg and long live Lampedusa in Hamburg, because we are here to stay!" (Lampedusa in Hamburg 2015b).

Those involved in Lampedusa in Hamburg were thrown together as individuals, with biographies of forced movement, labour and temporary lives lived under threat; they were thrown together as a collective unit, with different languages and traditions, various religious beliefs, diverse food preferences, dissimilar habits and dispositions; and they were thrown together in a city that itself was formed as a diverse and changing landscape, where "place" itself is always a temporary constellation. Symbolizing and embodying this constellatory urbanism, Lampedusa in Hamburg erected a temporary tent for distributing information in the city, a tent that, despite often being removed, has also reasserted refugee agency and a kind of time-out from the dominant temporalities of "crisis", "waiting" and "limbo" that are so often used to discipline and depict those seeking asylum (Brun 2016; Ramsay 2017). It has thereby indexed a kind of disjunctive protest time or what Massey evoked when she described how a "radical difference in temporalities emphasises more than cities ever can that a 'constellation' is not a coherent 'now'" (Massey 2005: 160).

LAMPEDUSA IN HAMBURG AND GEOPOLITICAL THROWNTOGETHERNESS

The changing spatiality of the throwntogether city with its throwntogether people is also clearly connected in the case of Lampedusa in Hamburg to broader geopolitical formations and representations. "We are Africans", they explained to the Green Party meeting in 2015, "and we are fathers, mothers, brothers and sisters who lived and worked in Libya until the unjust NATO war that paved the way for mass exodus of migrants to the shores of Lampedusa and Europe as a whole" (Lampedusa in Hamburg 2015b). Their reasons for migration thereby connect the local to international relations, with both manifesting uneven and contingent tendencies of openness and closure relating to specific African identities.

For members of Lampedusa in Hamburg the opportunity of working as bricklayers or painters in Libya was an opening and a chance for a decent living, but one necessitated by the global politics of colonial relations, neo-colonial capitalism and climate change, as their own livelihoods were increasingly compromised at home. After the NATO war in Libya, the migrants were again compelled by broader geopolitical forces to move on, finding some initial assistance after 2011. But when this support was ended by the Monti government in February 2013, the Italian authorities cast them out and pushed them onward and northward, forcing them to confront the realities of European connection and disconnection once again.

This geopolitical throwntogetherness of Lampedusa in Hamburg is clearly not simply an experience of "the refugee crisis" and it has certainly not led to an experience of bare life in the Agambenian sense. Instead, it exposes the harsh realities of the EU refugee reception crisis by bearing witness to its evolving geopolitical formations and deformations. Lampedusa in Hamburg statements have made this explicit in a variety of ways. For example, the Dublin accords, which were both a response to geopolitical conflicts and a source of ongoing tension between EU member states, have also come to mediate the geopolitics of racist rejectionism. Much legal discourse about the accords avoids saying this, but Lampedusa in Hamburg has not shied away from doing so. They called instead on the Green Party to advance "human dignity in order to stop the racist European refugee policies and the Dublin III treaty" (Lampedusa in Hamburg 2015b). Likewise they have more recently sought to highlight the racial hierarchies that belie claims about the global city and Europe more generally being welcoming of refugees.

> The Lampedusa Tent in has become the symbol of the inequality and degrading values in our so called "welcoming" society. The gates have been open to some but remain closed to others. How can it be that for over three years now, the African-War-Refugees from Libya are still referred to as "illegals"? Black Africans are at the bottom of the refugee hierarchy that has developed in Hamburg.   (Lampedusa in Hamburg 2016)

Then there are the double standards of war itself on which the group was especially critical in its statement against the G20 meetings in Hamburg in 2017.

> The leaders of the of G20 countries who would meet in Hamburg are all war criminals who have waged wars against the world and produced groups of displaced people and immigrants to control their countries and loot their wealth. These dirty countries that committed many immoral crimes have exterminated nations and peoples and today they speak of human rights and moral principles. (Lampedusa in Hamburg 2017)

And more recently, in the context of the shift towards more extreme xenophobic forms of reactionary populism in Europe and associated efforts to ramp up detentions and deportations, Lampedusa in Hamburg's Facebook posts have taken note, becoming a running log of disputed detentions and deportations.

• Sudanese refugees in Hannover attacked by police, 27 April 2016
• Police forcing refugees out of church in Münster, 14 August 2016
• Deportations from Austria to Nigeria, 18 September 2016
• Deportations to Nigeria, 18 October 2016
• Deportations to Afghanistan, 14 December 2016
• Deportation to Greece from Germany, 14 January 2017

Connecting their alter-geopolitical contestation of EU rejectionism with wider resistance to the global rise of authoritarian governments, Lampedusa in Hamburg has more recently begun to articulate solidarity with the Black Lives Matter movement in the US. This certainly reflects an alter-geopolitical sensitivity to the ways in which racism connects reactionary forms of populist hypernationalism on both sides of the Atlantic (see also Bessner and Sparke 2017). But more than this we want to suggest it represents a form of geosocial solidarity-building as well. This is another kind of throwntogetherness that has been articulated with the global city citizenship claims of Lampedusa in Hamburg, linking their struggle with wider webs of relations in the EU and worldwide. From Ghana to Libya, Libya to Lampedusa, and Lampedusa to Hamburg, participants are joined through innumerable threads and entanglements: those of history, geography, economics and activist politics. These connections are what we are calling *geosocial* (see Mitchell and Kallio 2017; Sparke 2017b). We see the geosocial throwntogetherness of Lampedusa in Hamburg bringing together local and transnational alliances of migrant activism, ally-ship and solidarity, and it is to this geosocial work we now turn.

LAMPEDUSA IN HAMBURG AND GEOSOCIAL THROWNTOGETHERNESS

Lampedusa in Hamburg is a collective political action that insists on the agency of all, embracing the human interlocutors who speak, document, laugh, play, write, cook and march together, as well as the non-human yet productive forces of streets, tents, kitchens, football fields, churches and public squares where these activities emerge and merge. It involves the refugees struggling for their right to stay and to work and integrate in the city; it also involves the solidaristic actions of many others both near and far.

The main slogan of the campaign, "We are here to stay!" is emblematic of a demand that extends beyond Hamburg itself. It encompasses the experiences and insistent demands of all postcolonial migrants in Europe – from post Second World War "guest" workers to those contemporary refugees fleeing war and climate-induced famine and poverty. The link between their own conditions of precarity and European actions both past and present is made in numerous speeches and manifestos, including the ubiquitous poster, "We did not survive the NATO war in Libya to come and die in the streets of Hamburg." These types of sentiments and actions manifest the distinctions – and what Massey might have called the articulatory antagonisms – that Lampedusa in Hamburg consistently makes vis-à-vis dominant discourses of geopolitics and geoeconomics in the global city. They work in this way to show how consciousness and organization against exploitation and precarity crosses boundaries to unite migrant workers and refugees with others rendered surplus in the globalized urban economy. As Martin Bak Jørgensen notes, "Their banners claim belonging to a broader united precariat" (Jørgensen 2017: 66; see also Photo 16.3).

Lampedusa in Hamburg was successful in uniting refugees not just with other members of the precariat, but also with activists, faith-leaders, football coaches and fans, academics, volunteers, city residents and trade unionists, among others. Many of these actors came to support the refugees through the shared values of close-knit neighbourhoods. For example, the St Pauli church was one of the first institutions that welcomed many of the first refugees into their community, providing them with shelter and food for several months. This type of social support was continued by the St Pauli football club, which helped establish the FC Lampedusa Hamburg football club, coached by women involved with football and other social projects. Football at FC Lampedusa Hamburg has a clear political mandate, it "wants to create awareness and draw attention to the evils of the European

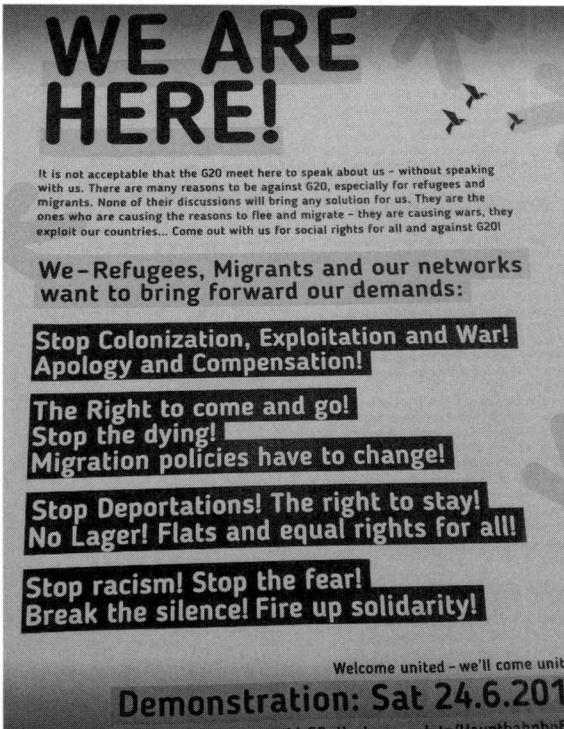

Photo 16.3 "We fight for our right"
(reproduced courtesy of Lampedusa in Hamburg)

refugee policy and the situation of refugees in Hamburg, Germany and the European Community". Its ongoing efforts on behalf of the refugee-players has continued long beyond the end of media interest in the Lampedusa in Hamburg phenomenon.[1]

The refugees' insistence on integration and a "right to the city" reverberated positively with many of the St Pauli residents and others who felt that Hamburg, Germany and Europe itself should be held accountable for their plight and responsible for their wellbeing. They demanded these rights in a series of protests that grew rapidly in scope and size beginning in May 2013. On 1 May the refugees appeared publicly at a church event for the first time; on 3 May they sat on the steps of the Rathaus and asked the mayor to speak with them; in June they were joined by 600 people at a demonstration; in June and August they connected with the "right to the city" movement and with many St Pauli district residents to cook and eat together in the city park; in July they were joined by trade unionists from

---

1. See, for example, FC Lampedusa St. Pauli (2017).

Photo 16.4 Wall painting
(reproduced courtesy of Lampedusa in Hamburg)

Ver.di, and also opened an exhibition "We want our lives back" at the St Pauli Church; and on 17 August, they were accompanied by 2500 people in another demonstration.[2]

By the end of autumn these events and demonstrations were so numerous and had articulated with so many disparate actors that it was soon evident that Lampedusa in Hamburg had transcended the specificities of the local refugees and their case for urban justice. Graffiti, wall paintings, exhibitions, conferences, theatre, music, film and ongoing demonstrations across the nation manifested the growing awareness of the plight of refugees in Germany that had been galvanized by the Lampedusa in Hamburg movement. Their mural on a physical wall in Hamburg as global city thereby articulated with struggles far beyond Hamburg too – also being notably featured on Lampedusa in Hamburg's Wikipedia page (see Photo 16.4).

Moreover, because of their own actions and demands, and the struggles against geopolitical formations past and present, the refugees opened up the space of awareness and discussion about the multiple historical ties, uneven geographies and ongoing inequities between Africa and Europe. All this awareness and articulation around geosocial throwntogetherness has nevertheless evolved in ongoing tension with forces of rejectionism. Some of these we have already addressed above in terms of the violence of geopolitical throwntogetherness against which much of Lampedusa in Hamburg's

---

2. This chronology of events is laid out well in "Lampedusa in Hamburg", in Wikipedia: https://de.wikipedia.org/wiki/Lampedusa_in_Hamburg.

resistance and solidarity work has been organized. But in conclusion, and in the spirit of Massey's cautions against concluding that the politics of throwntogetheness can ever be fully bounded in time and space, we want to highlight how the geosocial throwntogetherness of Lampedusa in Hamburg also challenges the geoeconomic promotion of the city as a global logistics hub for the global business class and its business.

COMPLICATING CONCLUSIONS ABOUT GLOBAL CITY CITIZENSHIP

Our goal here in closing this chapter is to reflect on how the urban, geo-political and geosocial throwntogetherness of Lampedusa in Hamburg complicates the ongoing geoeconomic promotion of Hamburg as a trad-itional global city. They do so, in part, by offering a competing counter-narrative that contests the global city as a natural home for the global business class. As Massey's own work on London made clear, one of the repercussions of neoliberal globalization is that cities today operate in conditions of constant competition. As David Harvey (1989) also identified early on, this competitive order has led from the more managerial functions of times past, to an increasingly entrepreneurial urbanism that involves competing for business opportunities and resources in a global climate of scarcity and austerity. And as individual cities attempt to attract capital and attention worldwide, many have turned to the promotion of cosmopolit-anism and tolerance – branding themselves as attractive for investment because of an enlightened and enjoyable lifestyle, population and urban landscape (Sparke 2011). This is part of a broader strategy endorsed by pundits and academics such as Richard Florida (2004), who argued that cre-ating the conditions of a socially and financially liberal metropolis will lead to the rise of a "creative class" of knowledge workers, which will, in turn, galvanize the post-industrial city and lead it to new financial and techno-logical heights. But what Lampedusa in Hamburg's competing articulations of global city citizenship urge us to consider are the limits and exclusivity of this creative class cosmopolitanism.

Cosmopolitanism is supposed to signal greater acceptance of the for-eigner, and as Massey underlined, the wandering rock in the harbour was represented thus in the Hamburg poster as a welcome guest. Hybridity, multiculturalism, cosmopolitanism, welcome, acceptance and tolerance for difference were all meant to be part of Hamburg's campaign to recog-nize the fundamental openness of the city as global shipping hub. But what Lampedusa in Hamburg has illuminated instead is the limits of this place-promoting discourse. The rhetoric of urban openness turns out to be not an abstract universal offer or claim, but rather occurs in what Massey herself

227

described as "the ever-mobile power-geometries of space-time". While it might present as openness to *all* forms of difference, in fact, the invitation is quite specific, contingent on Hamburg's position and place in business class worldviews of global relations and geoeconomic visions and rankings of global cities. Massey's points about the uneven power relations of place are all the more salient in this respect.

> The issue is one of power and politics as refracted through and often actively manipulating space and place, not one of general "rules" of space and place. For there are no such rules, in the sense of a universal politics of abstract spatial forms; of topographic categories. Rather, there are spatialised social practices and relations, and social power.   (Massey 2005: 166)

Massey's concern with the specificities of power-geometries helps highlight how citizenship in a city such as Hamburg is contoured by precisely the kinds of power hierarchies illuminated by the struggles of Lampedusa in Hamburg. As the second busiest port and prime logistics centre for Europe, Hamburg really is a political-economic "Gateway to the World" (*Tor zur Welt*). Its key role as a transport and logistics hub manifests the central economic importance of optimized commodity circulation in the "new" global city. But as Deborah Cowen (2014) has underlined, such competitive economic optimization strategies are always bound up with violent politics of inclusion and exclusion. The increasingly fast-paced and complex functions of logistical global cities such as Hamburg mean that any potential impediments or obstructions to the free flow of commodities and capital can be quickly assessed as obstacles to ongoing geoeconomic promotion – which is also why boundary-drawing geopolitical visions and violence remain widely entangled with such geoeconomic vistas of cross-border market integration (Sparke 2018). Thus while hybridity and the foreign are proclaimed in a rhetorical declaration of openness and welcome, it is a cosmopolitan invitation only applicable to certain bodies. Immigrant bodies perceived as potentially slowing or derailing the smooth circulation of goods, systems of capital management, allocation of resources or representations of urban jouissance, health and creativity are delayed, deferred and detained rather than welcomed with open arms (Mitchell and MacFarlane 2016). Indeed, it is not accidental that the poster of the rock in the harbour shows an image of a *ship* passing through the boulder, not a person. It is also no accident that Lampedusa in Hamburg has, in response, turned an image of a ship's anchor into a raised fist of resistance. And it should likewise come as no surprise that their information tent in downtown Hamburg featured posters and flyers announcing anti-neoliberal opposition to the Transatlantic Trade and Investment Partnership (TTIP).

As Lampedusa in Hamburg has made clear, racism and antiracism are at work in all of this too. As black Africans racially coded as different than the "local" German population, their foreignness cannot be hidden from view. This has been used by some to criminalize the group – portraying them as deviant and ungovernable, a "foreign" element that is essentially unassimilable and must be violently expelled from the body politic.[3] Newspaper reports indicate that the Hamburg housing market is notoriously unwelcoming to foreigners, even those with papers, and the unemployment rate is far higher for black Africans than for the general population (Heinrich 2013; Schmeller 2013). The treatment of the Lampedusa in Hamburg migrants can be viewed in this way as an illustration of how racism continues to disenfranchise people from global city citizenship in cities such as Hamburg. It continues with contemporary efforts to detain, detach and deport those migrants perceived as siphoning off the precious resources of the nation and infiltrating and slowing the circulation of tourists, trains, ships, goods, money and all else that must flow smoothly and quickly through the slipways and freeways of the logistical city.

Massey begins *For Space* with a discussion of the colonial relations between the Aztecs and the Spanish. She introduces the book with this historical tale as a way of indicating how "conceiving of space as in the voyages of discovery, as something to be crossed and maybe conquered, has particular ramifications" (Massey 2005: 4). In the case of Cortés and how the story of "discovery" is often narrated, this spatial imagination of crossing and taking space renders Moctezuma and the Aztecs immobile and passive, deprived of their histories and trajectories.

Massey asks us to "reorientate this imagination", and to question our habits of thinking about space as a surface to be crossed. As we have sought to underline in this chapter, Lampedusa in Hamburg has done this in ways that turn a very different kind of voyage of discovery into the basis of geosocial opposition to geopolitical relations and geoeconomic representations of the global city. In doing so their resistance and solidarity work underlines the importance of Massey's argument that the global city is always already a "meeting-up of histories" rather than an inert space to be traversed. This landscape of throwntogetherness, marked as it is by extreme variations in citizenship and subcitizenship, is also at the same time a place of "intersecting trajectories", one that, as Massey argued so inspiringly, can provide the opportunity for an open-ended place of articulatory politics.

---

3. See Lampedusa in Hamburg (2013) and Dolzer (2013).

REFERENCES

For works authored and co-authored by Doreen Massey, please see Select Bibliography of Doreen Massey, beginning on p. 371.

BBC 2013. "Hamburg Blames Italy Over 300 Homeless African Refugees". *BBC News* 23 May. www.bbc.com/news/world-europe-22694022 (accessed 8 January 2018).

Bessner, D. & M. Sparke 2017. "Nazism, Neoliberalism and the Trumpist Challenge to Democracy". *Environment and Planning A* 49 (6): 1214–23.

Borgstede, S. 2017. "'We Are Here to Stay': Reflections on the Struggle of the Refugee Group "Lampedusa in Hamburg" and the Solidarity Campaign, 2013–2015". In P. Mudu & S. Chattopadhyay (eds) *Migration, Squatting and Radical Autonomy: Resistance and Destabilization*. London: Routledge.

Brun, C. 2016. "There is No Future in Humanitarianism: Emergency, Temporality and Protracted Displacement". *History and Anthropology* 27 (4): 393–410.

Cowen, D. 2014. *The Deadly Life of Logistics: Mapping Violence in Global Trade*. Minneapolis, MN: University of Minnesota Press.

Dolzer, M. 2013. "Afrikaner Underwünscht". *Junge Welt* 17 October. www.jungewelt.de/loginFailed.php?ref=/2013/10–17/045.php (accessed 19 January 2018).

FC Lampesdusa St. Pauli 2017. "A visit to FC Lampedusa St. Pauli's central midfielder in deportation detention". http://fclampedusa-hh.de/?p=568 (accessed 9 June 2017).

Florida, R. 2004. *Cities and the Creative Class*. New York: Routledge.

Harvey, D. 1989. "From Managerialism to Entrepreneurialism: The Transformation in Urban Governance in Late Capitalism". *Geografiska Annaler Series B, Human Geography* 71 (1): 3–17.

Heinrich, D. 2013. "Foreigners Not Welcome: Racism in Germany's Housing Market". *DW*, 13 November. www.dw.com/en/foreigners-not-welcome-racism-in-germanys-housing-market/a-17223748 (accessed 8 June 2017).

Jørgensen, M. B. 2017. "The Precariat Strikes Back – Precarity Struggles in Practice". In C.-U. Schierup & M. B. Jørgensen (eds) *Politics of Precarity: Migrant Conditions, Struggles and Experiences*. Leiden: Koninklijke Brill, 52–77.

Lampedusa in Hamburg 2013. "Statement on current issues regarding our politics – Discussion on violence". *Karawane* 19 December. http://thecaravan.org/node/3993 (accessed 19 January 2018).

Lampedusa in Hamburg 2015a. "We Demand the Recognition of Our Italian Residence and Working Permit!" *Lampedusa in Hamburg blog* 19 February. http://lampedusa-hamburg.info/recognition-of-the-italian-residence-and-working-permit/ (accessed 8 January 2018).

Lampedusa in Hamburg 2015b. "Greeting Words in the Talk with the GREEN Party". *Lampedusa in Hamburg blog* 12 February. http://lampedusa-hamburg.info/greeting-words-in-the-talk-with-the-green-party/ (accessed 8 January 2018).

Lampedusa in Hamburg 2016. Facebook post, 23 May. www.facebook.com/lampedusainhamburg/posts/1025460787489188 (accessed 8 January 2018).

Lampedusa in Hamburg 2017. Facebook post, 26 May. www.facebook.com/lampedusainhamburg/posts/1345701725465091:0 (accessed 8 January 2018).

Meret, S. & E. Della Corte 2014. "Between Exit and Voice: Refugees' Stories from Lampedusa to Hamburg". *Open Democracy* 22 January. www.opendemocracy.net/can-europe-make-it/susi-meret-elisabetta-della-corte/between-exit-and-voice-refugees-stories-from-la (accessed 6 May 2017).

Mitchell, K. & K. P. Kallio 2017. "Spaces of the Geosocial: Exploring Transnational Topologies". *Geopolitics* 22 (1): 1–14.

Mitchell, K. & K. MacFarlane 2016. "Crime and the Global City: Migration, Borders, and the Pre-Criminal". In K. Mitchell & K. MacFarlane (eds) *Oxford Handbooks Online*. New York: Oxford University Press.

Ramsay, G. 2017. "Incommensurable Futures and Displaced Lives: Sovereignty as Control over Time". *Public Culture* 29 (3): 515–38.

Schmeller, J. 2013. "Immigrants Struggle in German Job Market". *DW* 12 September. www.dw.com/en/immigrants-struggle-in-german-job-market/a-17083761 (accessed 8 June 2017).

Sparke, M. 2011. "Global Geographies". In M. Brown & R. Morrill (eds) *Seattle Geographies*. Seattle, WA: University of Washington Press, 48–70.

Sparke, M. 2017a. "Austerity and the Embodiment of Neoliberalism as Ill-Health: Towards a Theory of Biological Sub-Citizenship". *Social Science and Medicine* 187: 287–95.

Sparke, M. 2017b. "Situated Cyborg Knowledge in not so Borderless Online Global Education: Mapping the Geosocial Landscape of a MOOC". *Geopolitics* 22 (1): 51–72.

Sparke, M. 2018. "Globalizing Capitalism and the Dialectics of Geopolitics and Geoeconomics". *Environment and Planning A* 50 (2): 484–9.

CHAPTER 17

HEGEMONIES ARE NOT TOTALITIES! REPOLITICIZING POVERTY AS RESISTANCE

Victoria Lawson and Sarah Elwood

INTRODUCTION

In 2014, Doreen Massey discussed the Kilburn Manifesto at the American Association of Geographers Meeting. "The Kilburn" is a project she co-organized with Stuart Hall and Michael Rustin that is at once powerful public scholarship and potent political intervention.[1] The room felt like a political meeting – packed with people sitting on the floor and standing in the doorways, all in rapt attention. She ended by noting that the current project of government is one in which "we must *not* be allowed to know that there are alternatives – we must *not* know that hegemonies are not totalities" (Massey 2014, authors' emphasis). She argued that in the face of widely circulating ideology scripting the current neoliberal conjuncture of financialized and globalized capitalism as natural, inevitable and ideal, the Left must narrate it instead as a complex economic, philosophical, moral and political crisis. That night, as in countless other interventions throughout her career, Doreen insisted that cracking open hegemonies is urgent political work that requires new modes of thought, action, and solidarities (see also Hall & Massey 2010).

We take up this charge in relation to poverty. Poverty is a central site of politics, resistance and struggle over "common sense" that shapes how members of a society relate to each other. We respond to Doreen and her co-authors' call in the Kilburn Manifesto to repoliticize poverty, and turn to Hall and Massey's (2010) conjunctural analysis as one avenue towards doing so. Conjunctural analysis traces the social, political, economic and ideological contradictions that are held together in particular space-times and makes visible the techniques by which these contradictions are rendered invisible, unremarkable and off limits for intervention. We argue that persistent poverty is one such contradiction that must be repoliticized because its very persistence

---

1. The Kilburn Manifesto is available at: www.lwbooks.co.uk/soundings/kilburn-manifesto.

challenges economistic narratives of "progress" and "fairness". Repoliticizing poverty stands to crack open the hegemony of neoliberal market fundamentalism, imagine alternatives and build new political solidarities.

In this chapter, we repoliticize poverty. This involves two moves: first, we trace "thinkable politics" that stabilize the political-economic and sociocultural processes of impoverishment. Conceptualizing poverty governance as thinkable politics illuminates multidimensional systems of dominance that frame structural causes of impoverishment as "individual choices" and circulate "common sense" that makes racialized, classed, gendered, bordered econocentric rationalities appear natural, necessary and unquestionable. Second, we explore "unthinkable politics" (Cacho 2012: 31) that exist in dialectical tension with thinkability and that challenge existing social orders to repoliticize poverty in the current conjuncture. "Unthinkable poverty politics" mark the outsides of existing power hierarchies and are unrecognizable under its normative and ideological terms. The examples we trace are unthinkable because they involve actions by people whose lives, theory and politics are framed as irrelevant, because they arouse the threat of state-sponsored violence, and because they involve alliances forged across difference. As Doreen insisted, repoliticizing concepts like poverty opens the door to new narrations of the current economic, moral and political crisis and creates new possibilities for solidarities.

POVERTY AS ONE SITE FOR BUILDING ALTERNATIVE POLITICS

Our analysis centres on poverty politics under late liberal capitalism and its associated projects of governance (Schram 2000; Goode & Maskovksy 2001; Giles 2018; Schram 2015). Doreen Massey and Stuart Hall's conjunctural analysis brings into view political-economic and sociocultural processes of impoverishment *and* the ways they interrelate to secure particular social, political and economic arrangements. Hall and Massey (2010) build a conjunctural analysis in the UK, tracing the articulation of shifts in capital accumulation and its labour politics with cultural productions that drive a wedge between working people on the basis of race, gender, citizenship, sexuality and other differences. More broadly, late liberal capitalism in North Atlantic States has brought deindustrialization, financialization and globalized capital accumulation, with attendant practices that uphold them: offshoring of profits, downward harmonization of wages, workfare, union busting (Peck 2002; Peck & Theodore 2012). In parallel, powerful state, economy and media actors double down on discourses of individualism, naturalization of markets as "fair", and poverty as personal failure or the result of bad choices (Massey 2014; Campbell 2014). This political and cultural discourse thoroughly naturalizes "economic forces" and excises any critical analysis of capitalism. Hall (1988, 1994) theorizes this move to cleanse popular discourse of concepts like inequality, poverty, public ownership, social good, or even society as an ideological erasure that works to silence dissent and leaves exploitive neoliberal capitalism and elite power unchallenged.

This conjuncture of economic, discursive and social processes produces "common sense" about poverty that makes collective and structural poverty politics unspeakable. The need for redistribution, the need to rebuild the public sphere, the possibility of collective care for those disadvantaged, the ways that finance capitalism and taxation structures produce impoverishment: All this lies outside the bounds of hegemonic imaginaries and interventions. Race, gender, immigration and sexuality politics enacted through cultural productions of difference and boundary-making drive wedges between working people. In the face of these interlocking forces, Massey and Hall argue that the Left must grapple with the complexity of the current conjuncture as a whole. This involves two moves: refusing to reduce the problem of poverty to one of economism alone, while simultaneously refusing popular claims of an ideological crisis – such as anti-immigrant and racist politics – that are supposedly separate from changes in the economy and deepening poverty.

We are living through precisely this separation of an ideological crisis from economic analysis in the US today. The labour politics of the current moment are in contradictory relation with cultural productions that mitigate against solidarities among working people. Historical forms of class solidarities have not effectively challenged neoliberal austerity and racial capitalist violence. Union-busting has been reinvigorated post-2008 through "right to work" state legislation and attacks on public sector unions (Censky 2011; McCartin 2016). Beyond unionism, class itself is thoroughly depoliticized with most people identifying as middle class despite extreme, widespread poverty and inequality (Žižek 2000; Zweig 2000; Lawson *et al.* 2015). Simultaneously, social difference is produced and deployed to powerful political effect, dividing low-income people from one another along lines of race, citizenship status, gender and more. Throughout the 2016 presidential election campaign, right-wing politicians, TV, talk radio and print and social media successfully mobilized a discourse in which immigrants and people of colour are responsible for the impoverishment of working white men by "stealing" jobs and overwhelming the welfare system. Donald Trump's promise to build "the Wall" on the US–Mexico border constructs political common ground between the white working class and right-wing wealthy elites through appeals to national security, white identity and neoliberal narratives of economic "fairness". These narratives obscure the ways in which poverty and inequality are driven by obscene wealth accumulation by elites exercising political influence on tax codes, exemptions for unearned income, deregulation of work/environment and more. For example, Donald Trump's presidential campaign doubled down on a racist politics of fear and division and explicitly endorsed state-sponsored violence against African-Americans, Latinos and Muslims, setting the stage for possessive investments in whiteness and white supremacy that make class solidarities unthinkable.

These politics effectively erase poverty from view, focusing instead on nationalist populism, racialized "security" concerns, and a politics of defending white privilege. Refusing this ideological shift, we repoliticize poverty as a site of analysis and resistance. As Doreen says in the Kilburn Manifesto: the need to repoliticize keywords like poverty seems so obvious – but we must argue it, and understand the political possibilities it raises. Massey and Rustin (2015: 213) argue that "the struggle over how society is organized, how its members are to relate to each other, and what will emerge as its central values and symbolic representations, needs to take place in a multiplicity of locations". A primary and vital location for political struggle is inequality and poverty. Poverty has always been a problem for neoliberal capitalist states – precisely because its persistence belies narratives of progress and fairness. Persistent poverty is a contradiction that must be depoliticized. This is accomplished through arguments about the fairness of markets, personal responsibility for poverty, and people's own bad choices and failures.

The work of repoliticizing poverty begins from relational poverty analyses that explore how poverty is produced by capitalist, white supremacist, sexist and heteronormative institutions and rules; by political-economic processes and through identity-making by middle classes, elites, policy-makers, religious or sectarian nationalists. Theorizing impoverishment as produced through intersecting power relations of race, caste, class, identity, coloniality, ability and gender counters theoretical closures in much poverty knowledge. A relational analysis theorizes impoverishment within racist histories of dehumanization, capitalist dispossession, exploitation, cultural erasure and linked oppressions of ability, sexuality, gender and citizenship. Poverty is produced in the relations between these processes and geo-histories (Schram 2000; Hickey & du Toit 2007; De Hert & Bastiaensen 2008; Elwood *et al.* 2016).

Repoliticizing poverty entails two moves: first, recognizing that neoliberal governance and identity projects that individualize poverty are deeply political. These projects produce both knowledge and actions that solidify neoliberal claims about who is poor and why; second, repoliticizing poverty is a project of recognition of multiple transgressive and rebellious politics that lie outside the hegemonic liberal social order and its imaginaries. In the sections that follow, we trace thinkable and unthinkable poverty politics in order to imagine new narrations of the current economic, moral and political crisis and think about possibilities for new solidarities.

THINKABLE POVERTY POLITICS

Thinkable politics are what Doreen and Stuart refer to as "speakable politics". They are projects of government and culture that identify problems, justify interventions and frame solutions in ways that stabilize dominant forms of economic and political power (Rose & Miller 1992; Roy & Crane 2015). Thinkable poverty politics arise in relation to the current conjuncture of globalized and financialized capitalism and are secured by ideological projects that (re)produce the "common sense" modes of thought and action of neoliberal hegemony (Hall & Massey 2010). These projects of government focus on incorporating individuals into existing structures of culture and economy, advancing the notion that poverty can be addressed through personal improvement "solutions" like getting a job, becoming a better parent or learning about household budgeting. This ideology produces deserving poor subjects understood as flawed but reformable, as well as undeserving poor subjects seen as deeply threatening and in need of discipline – a poverty politics that works against solidarities by differentiating people.

Thinkable poverty politics are enacted through a range of site and practices, such as the bureaucracies of poverty management. US poverty assistance policies ushered in 20 years ago through the "Personal Responsibility and Work Opportunity Reconciliation Act" (PRWORA) are a powerful example. This welfare reform bill dramatically intensified governance of impoverished people, including home visits to confirm that single parents receiving Temporary Assistance for Needy Families (TANF) cash benefits have no other adults in their homes, drug testing, real-time review of personal bank accounts, and required participation in job training, work, parenting classes and other "self-improvement" activities. Along with this intrusive oversight came a five-year lifetime limit on cash benefits, and punitive sanctions on those who cannot meet its requirements. The US poverty management bureacracy at every turn inscribes poverty as caused/addressed by individual action and reinforces cultural discourses of a "poor Other" (Watkins-Hayes 2009).

Thinkable poverty politics are also constituted through criminalization of poverty. In the last decade, municipal ordinances against eating, sleeping, standing and urinating in a public space in the US have grown exponentially – in effect criminalizing the bodily existence of homeless people (National Law Center on Homelessness and Poverty 2016). Widespread "legal financial obligations" attach fees to contact with courts or the criminal justice system, including traffic stops, incarceration, arrest, court appearances, "free" legal

representation and mandatory fingerprinting/DNA testing and imprison-
ment of those who cannot pay these debts (Harris 2016). These forms of
criminalization are a thinkable poverty politics, which operates on two levels.
First, they reproduce impoverishment through material sanctions and further political and economic exclusion attached to having a criminal record
(e.g. loss of voting rights, ineligibility for many jobs). Second, they cement
poverty discourses of deviance, irresponsibility and criminality. This double
move prefigures other thinkable poverty politics, such as state-sponsored
violence enacted against impoverished people, including "stop and frisk"
policing, disproportional incarceration of people of colour, intensified detention and deportation of low-income undocumented people and militarization of policing and borders. These violent poverty politics are rendered
"thinkable" – apparently necessary, justifiable and "common sense" – through
the mundane workings of poverty management bureaucracy and cultural
discourses that individualize, stigmatize and criminalize poverty (Cacho
2012; Beckett & Herbert 2009).

Thinkable poverty politics are not only enacted by powerful actors, but
also by impoverished people themselves as they struggle to survive within
systems stacked against them. For instance, Pittman (2018) traces the systematic exclusion of African American custodial grandmothers from safety
net programs designed around white, middle class nuclear family norms, as
well as the strategies grandmothers develop to survive. They broker deals
with birth parents to share cash benefits, omit grandchildren from lease
documents to avoid losing subsidized senior housing, and generate cash
income off the books when lack of childcare makes it impossible for them
to comply with training and work requirements linked to welfare benefits.
These grandmothers enact thinkable poverty politics by making themselves
legible within the rules of age-related benefits (like subsidized housing for
impoverished seniors), while making invisible family arrangements that
would make them ineligible (like custodial care of grandchildren). These
tactics work within, and navigate around, poverty assistance programmes
that are shot through with multiple strands of neoliberal thinkability: White,
middle-class nuclear family ideals, inaccurate assumptions about the availability of living wage work for all and valorization of paid employment over
the work of caring for children. Tracing these tactics shows the limits of
thinkable poverty politics, which in this example allow for survival, but only
barely, and cannot produce transgressions that would directly challenge these
ideological, economic and political systems.

In sum, thinkable poverty politics obscure structural causes of poverty by
doubling down on neoliberal common sense of reforming flawed individuals,
personal responsibility and neutralizing threats from criminal people. These
repeated practices and representations cleanse popular understandings

of poverty so that we no longer speak of collective subjects, public ownership, nor redistribution to address inequality. Breaches of the bounds of speakability are quickly returned to the fold: speaking of income inequality, as President Obama did in his 2014 State of the Union address, is immediately scripted as "class warfare", and calls for race reparations (Coates 2014) are rejected as "racist". And yet Doreen insists that "hegemonies are not totalities". Against all odds, other poverty politics exist. Poverty *is* being contested. It *is* being politicized. We turn next to these refusals and transgressions and the (different) poverty politics they engender.

UNTHINKABLE POVERTY POLITICS

Unthinkable politics refers to struggles that emerge in resistance to existing political economic orders and systems of social valuation (such as white supremacy or middle class normativity) that depend upon the wholesale devaluation of vulnerable subjects, including impoverished people, undocumented immigrants and racialized groups (Cacho 2012). Critical race feminism and black geographies scholarship trace multiple intersecting processes (capitalism, racialization, state violence, bordering, criminalization, settler colonialism, gendering and much more) that render some subjects outside of, and illegible to, existing orders (McKittrick & Woods 2007; Carillo Rowe 2008; Cacho 2012; Nagar & Geiger 2014; McKittrick 2016; Pulido 2016; Borges 2018). Unthinkable politics operate in this terrain, outside, but always in relation to, "common sense" that structures thinkable politics, cracking open hegemonic claims and creating political possibilities in cultural, economic and political realms.

Unthinkable poverty politics are enacted by excluded illegible subjects (and allies) in their refusal to conform to the normative categories undergirding common sense in particular times and places. These politics are unthinkable because they involve alliances among people whose lives, theory and politics have been framed as unimportant and powerless. They are unthinkable because they will arouse the threat (or promise) of state violence and because they employ a language of collectivity forged across difference. Their rebellious political practices create cultural forms and create struggles over values and identities that can reshape "common sense" (McKittrick 2016).

In the face of pervasive economism, individualism, poverty difference and the forms of violence and domination that uphold them, unthinkable poverty politics should not exist, and yet they do! Unthinkable poverty politics take a range of forms, such as subjects disidentifying from common sense framings that they are "deviant" (by challenging racialization, criminalization, homophobia, "illegality" and so on), or intentionally engaging in criminalized

239

actions. Disidentification describes the renegotiation of compromised identities in ways that are not easily legible from within projects of thinkability
(Muñoz 1999; Giles 2018). Whereas the political endgame of legibility is often a
demand for recognition or redress (such as "We Are the Ninety-Nine Percent"),
disidentification facilitates a politics that is often liberatory from within but
opaque from without. For example, queer counterpublics that refuse inclusion in heteronormative publics, or non-market community economies that
spurn inclusion in capitalist markets. Giles's (2018) analysis of Food Not Bombs
(FNB), a global network of anarchist soup kitchens, illustrates unthinkable
poverty politics enacted through disidentification. FNB allies a heterogenous
group of people who reappropriate and consume the "waste" of capitalist food
markets, prepare food in collective community kitchens and redistribute food
to homeless people in public, in violation of city ordinances. These actions
refuse bodily norms of late liberal capitalism with respect to sourcing and
consuming food. FNB activists live outside capitalist markets and openly defy
legal structures that reproduce poverty, such as the criminalization of consuming and sharing food in public spaces. As illustrated through FNB's actions,
unthinkable politics are often enacted in sites typically deemed irrelevant to
challenging poverty, such as the streets, homes, kitchens, dumpsters, bodies
and parks – and yet these grounded politics do indeed challenge normative
framings of poverty. Unthinkability offers an avenue for theorizing how spaces
typically deemed unimportant and actions typically deemed insignificant
(eating, cooking) can constitute transformative poverty politics.

Unthinkable poverty politics are also advanced in actions that incite state-
sponsored violence, making visible the central role of violence in sustaining
capitalist orders and poverty in the current conjuncture. For instance, Black
Lives Matter protestors across the US have occupied streets and public
plazas, often in violation of laws and ordinances governing the sites/forms
of political speech, making racialized state violence highly visible and public.
The movement connects multiple forms of state violence (police shootings,
incarceration, imperialism) to racial capitalism and impoverishment. These
linkages are brought into view through arguments that "Black poverty … is
state violence" (Black Lives Matter nd), as well as through #BLM protests and
free clothing/food giveaways at shopping malls on "Black Friday" (a "holiday"
created by retailers to initiate the profitable December shopping season).
These unthinkable poverty politics push back against existing political economic arrangements and refuse the limits of thinkability. As Ransby (2015
np) has argued, "to seek liberation for black people is also to de-stabilize
inequality in the United States at large and to create new possibilities for all".
Doreen's and Stuart's insistence upon examining the articulation of political,
economic and cultural practices provides a conceptual lens for recognizing
the significance of cultural performances for challenging existing political

arrangements and the contours of (un)speakability that cement them in place. In 2016, as football players from high schools to the National Football League knelt during the national anthem to show solidarity with #BLM, they brought #BLM's critical claims about race, violence and capital to the cultural spaces of sports fields. They called attention to the interconnected workings of nationalism, racism and economic precarity – structures that are typically made unspeakable in football stadiums through repeated performances of patriotism, "colour blindness", and idealized working-class identities.

Unthinkable politics sometimes entail surprising alliances that challenge the links between capital accumulation and the political deployment of difference by linking poverty with race, violence, citizenship, gender, caste and religion. Undocumented activists in the US DREAM movement, for instance, enact unthinkable poverty politics by linking denial of citizenship to poverty, by coming out publically as undocumented (self-consciously risking deportation or detention), and by engaging in practices of allyship between citizens and non-citizens (Swerts 2018). Swerts traces how Chicago DREAMers' everyday practices of organizing, advocacy and protest centre undocumented activists as agenda setters. Simultaneously, citizen allies are prompted to "step back" in these everyday movement practices, to engage in critical learning about their race, class and citizenship privilege, and to deploy their unvulnerability to deportation in material practices in support of the movement (such as driving colleagues whose undocumented status makes it impossible to obtain a driver's licence). Through allyship across differences of citizenship, privileged actors learn about impoverishment and state-sponsored violence as well as understand the lived experiences of unthinkable subjects. These solidarities invite privileged subjects to imagine roles they can play in reworking immigration politics.

It is crucial not to be naïve about histories of suffering and death that shadow unthinkable politics and alliance practices. Not the least because being in relation with privileged subjects runs the risk of reinscribing thinkable modes of hierarchy and the suffering they produce. Yet theorizing the unthinkable politics that may arise from alliances across difference matter because they have the potential to unleash transformative politics. When unthinkable subjects insist on their own (illegible) personhood and engage in practices of allyship with "Others", they articulate experiences of marginalization that have previously pitted people against each other. This can lead to enactments of a politics of collective counter-subjectivity that destabilize systems of social (de)valuation that interpolate some marginalized groups as subjects of (limited) social value through devaluation of Others. Such alliance politics constitute a significant challenge to the forms of thoroughgoing outsiderness that are fundamental to liberal democratic life, and may lead to new bases for, and forms of, solidarities.

SOLIDARITIES AND SPATIAL POLITICS

Repoliticizing poverty requires cracking open the hegemony of neoliberal market fundamentalism and the forms of "common sense" that uphold it; imagining other ways of thinking, doing, relating; and building new political subjects and solidarities. Unthinkable poverty politics are a way of seeking other possible worlds, yet they always also take shape in the shadow of thinkable worlds. This raises vital questions: How do other worlds and other politics come into being out of these alternative imaginings and practices? How might we arrive at political solidarities that can challenge impoverishment from these diverse sites? How are elements of unthinkability already taken up in existing solidarities against poverty?

Thinkable and unthinkable poverty politics are distinct. There is an inherent tension between unthinkable politics and existing solidarities that push for inclusion and recognition but are not necessarily anti-establishment – the latter still articulate with thinkable poverty politics. Yet thinkable and unthinkable poverty politics exist in dynamic dialectical relation with each other, reinscribing dominant orders, refusing to conform to their hegemonic power and revealing alternative subject-framings and actions. Anti-establishment unthinkable politics offer creative lessons for ongoing struggles for redistribution, rights and political voice in our political present.

One critical lesson from unthinkable politics is that refusing the limits of existing "common sense" opens new flights of imagination from which new kinds of poverty politics become imaginable and possible. We see this in ongoing solidarities like those built by Domestic Workers United (DWU), a broad coalition challenging economic exclusion, historical devaluation of care and oppressions on the basis of race and citizenship. DWU organizes low-wage in-home care workers to fight for labour rights and protections, bringing together care workers from the Caribbean, South Asia, Africa, the Philippines and the US, across differences in language, national origin, gender and immigration status. Begun in Brooklyn in the early 2000s, DWU drew on New York City ordinances, federal labour laws and United Nations human rights conventions to articulate a Domestic Workers Bill of Rights that was first passed by the New York State Legislature in 2010, followed by several other states, eventually culminating in Federal action. In 2013, the US Department of Labor extended the protections of the Fair Labor Standards Act to include (some) care workers, giving them access to overtime pay and minimum wage protections (Dean 2013).

These poverty politics, while acting through claims upon rights and law that would seem fall firmly inside the bounds of thinkability, nonetheless encompass creative efforts to get outside the "common sense" spatial politics around which such campaigns are often structured. DWU's strategy for building rights and protections claims rejects fixed spatial imaginaries of jurisdiction that delimit rights/laws as (only) applicable to particular places and scales. Instead, they draw on laws, treaties and regulations protecting human rights at one geographical scale to challenge exclusionary laws at other scales. Whereas local action is often seen as insignificant for shaping national policy change, DWU takes seriously the potential of unruly local action to build geographically extensible support that can be scaled up towards national policy wins (Das Gupta 2006). Further, DWU rejects dominant assumptions about the limited political agency of vulnerable domestic workers or suggestions that these workers are impossible to organize because of differences of language, race, immigration status, gender and national origin. Rather, they engage in unthinkable political work to organize solidarities that *negotiate* difference, rather than *negating* it (organizing far beyond only class/work). DWU brings together diverse care workers to act upon multiple structures of impoverishment that operate in differential ways, making some more vulnerable than others (i.e. the devaluation of care, the criminalization of undocumented workers, lack of living wages). They make rights claims, work towards legislation and advocate with and through laws and ordinances – eminently thinkable forms of political voice and intervention. Yet the solidarities and transformations DWU works towards are materially consequential for a diverse range of impoverished people *and* reveal creative poverty politics that can grow from practices and alliances that push beyond the limits of normative, thinkable poverty politics.

Doreen's and Hall's conjunctural analysis guides our repoliticization of poverty as produced through complex intersections of political, economic and cultural realms working together to produce neoliberal "common sense". Their work offers crucial lessons for our political present. First, calling out the unconscious categories that maintain hegemonic forms of subjectivity is itself an important move in repoliticizing poverty. Struggling against neoliberal common sense about poverty challenges core claims about how a society is organized and its core values. Second, repoliticizing poverty in these ways makes it possible to articulate broadly held attitudes and emotions about injustice that have been silenced but that nonetheless hold the potential to change the terms of debate towards care and collective concern about inequality. A third lesson is about the importance of building political alliances by negotiating difference rather than negating it. The examples traced here demonstrate that working in alliance across difference can challenge categories and norms through which neoliberal governance is achieved, and can open space for solidarities (beyond class) that have been rendered unimaginable. Fourth, these approaches to repoliticizing poverty underscore the necessity and transformative power of creative spatial strategies that exert pressure on existing centralized forms of rule. Spatial strategies starting from local sites (sports fields, dumpsters, parks, cities and homes) and intimate relations (feeding, providing home care, even driving people around) can and do build transgressive politics.

As Doreen insisted, cracking open hegemonies is urgent political work that requires new modes of thought, action and solidarities. She insisted in 2014 that the current neoliberal conjuncture of financialized and globalized capitalism must be re-narrated by the Left as a complex economic, philosophical, moral and political crisis. Repoliticizing poverty, by thinking beyond the social, political, economic, ideological claims that produce and justify current arrangements, is one piece of this work. We offer thinkable and unthinkable poverty politics as a dialectical analytic for bringing back into theoretical and political view what has been obscured in order to sustain hegemonic neoliberal common sense about the causes of impoverishment. Our approach recognizes, and takes seriously, sites and forms of poverty politics that have been unseen or delegitimized. Theorizing thinkable poverty politics brings into view the interplay between structural forces producing impoverishment and the subjectivities, relations and meanings through which the current consensus about poverty is secured. Equally importantly, theorizing unthinkable poverty politics opens alternatives to reductive "common sense" understandings of poverty as well as strategies for disrupting the vast and violent injustices of the current moment of deepening inequality in North Atlantic neoliberal states.[2]

REFERENCES

For works authored and co-authored by Doreen Massey, please see Select Bibliography of Doreen Massey, beginning on p. 371.

Beckett, K. & S. Herbert 2009. *Banished: The New Social Control in Urban America.* New York: Oxford University Press.

Black Lives Matter nd. "The Creation of a Movement". *Herstory.* http://blacklivesmatter.com/herstory/ (accessed 20 April 2017).

Borges, A. 2018. "Ethnographic Alliance: Hope and Knowledge Building through a South African Story". In V. Lawson & S. Elwood (eds) *Relational Poverty Politics.* Athens, GA: University of Georgia Press, 183–200.

---

2. This chapter reworks a lecture in honour of Doreen Massey, delivered at the Royal Geographical Society on 8 October 2016, which appeared in *Soundings* 65: 103–13 (2017). This chapter also draws on a framework developed in the book, Lawson & Elwood (eds), *Relational Poverty Politics* (2018).

Cacho, L. M. 2012. *Social Death: Racialized Rightlessness and the Criminalization of the Unprotected*. New York: NYU Press.

Campbell, B. 2014. "After Neoliberalism: The Need for a Gender Revolution". *Soundings* 56: n.p. www.lwbooks.co.uk/sites/default/files/s56_02campbell.pdf (accessed 8 January 2018).

Carillo Rowe, A. 2008. *Power Lines on the Subject of Feminist Alliances*. Durham, NC: Duke University Press.

Censky, A. 2011. "How the Middle Class Became the Underclass". *CNN Money* 16 February. http://money.cnn.com/2011/02/16/news/economy/middle_class (accessed 20 April 2017).

Coates, T.-N. 2014. "The Case for Reparations". *The Atlantic*, June. www.theatlantic. com/magazine/archive/2014/06/the-case-for-reparations/361631/ (accessed 20 April 2017).

Das Gupta, M. 2006. *Unruly Immigrants: Rights, Activism and Transnational South Asian Politics in the United States*. Durham, NC: Duke University Press.

Dean, A. 2013. "How Domestic Workers Won their Rights: Five Big Lessons". *Yes! Magazine* 9 October. www.yesmagazine.org/people-power/how-domestic-workers-won-their-rights-five-big-lessons (accessed 20 April 2017).

De Hert, T. & J. Bastiaensen 2008. "The Circumstances of Agency: A Relational View of Poverty". *International Development Planning Review* 30 (4): 339–57.

Elwood, S., V. Lawson & E. Sheppard 2016. "Geographical Relational Poverty Studies". *Progress in Human Geography* 41 (6): 745–65.

Giles, D. 2018. "Abject Economies, Illiberal Embodiment, and the Politics of Waste". In V. Lawson & S. Elwood (eds) *Relational Poverty Politics*. Athens, GA: University of Georgia Press, 113–30.

Goode, J. & J. Maskovsky 2001. *The New Poverty Studies*. New York: NYU Press.

Hall, S. 1988. "Toad in the Garden: Thatcherism Among the Theorists". In C. Nelson & L. Grossberg (eds) *Marxism and the Interpretation of Culture*. Urbana, IL: University of Illinois Press, 35–74.

Hall, S. 1994. "The Rediscovery of 'Ideology': Return of the Repressed in Media Studies". In J. Storey (ed.) *Cultural Theory and Popular Culture*. London: Prentice Hall, 111–41.

Hall, S. & D. Massey 2010. "Interpreting the Crisis". *Soundings* 44 (Summer): 57–71.

Harris, A. 2016. *A Pound of Flesh: Monetary Sanctions as a Punishment for the Poor*. New York: Russell Sage.

Hickey, S. & A. Du Toit 2007. "Adverse Incorporation, Social Exclusion and Chronic Poverty". *Working Paper 81*, Chronic Poverty Research Centre, Manchester. www. gsdrc.org/document-library/adverse-incorporation-social-exclusion-and-chronic-poverty/ (accessed 20 April 2017).

Lawson, V. & S. Elwood (eds) 2018. *Relational Poverty Politics*. Athens, GA: University of Georgia Press.

Lawson, V., S. Elwood, S. Canevaro & N. Viotti 2015. "The Poor are Us": Middle Class Poverty Politics in Buenos Aires and Seattle. *Environment and Planning A* 47 (9): 1873–91.

McCartin, J. 2016. "Public Sector Unionism under Assault". *New Labor Forum* 19 January. http://newlaborforum.cuny.edu/2016/01/19/public-sector-unionism-under-assault-how-to-combat-the-scapegoating-of-organized-labor/ (accessed 20 April 2017).

McKittrick, K. 2016. "Rebellion/Invention/Groove". *Small Axe: A Caribbean Platform for Criticism* 49 (March): 79–91.

McKittrick, K. & C. Woods 2007. *Black Geographies and the Politics of Place*. Urbana, IL: Indiana University Press.

Muñoz, J. E. 1999. *Disidentifications: Queers of Color and the Performance of Politics*. Minneapolis, MN: University of Minnesota Press.

Nagar, R. & S. Geiger 2014. "Reflexivity, Positionality and Languages of Collaboration in Fieldwork". In R. Nagar (ed.) *Muddying the Waters*. Urbana, IL: University of Illinois Press, 81–104.

National Law Center on Homelessness and Poverty 2016. *Housing Not Handcuffs: Ending the Criminalization of Homelessness in U.S. Cities*. Washington, DC: National Law Center on Homelessness and Poverty.

Peck, J. 2002. "Labor, Zapped/Growth, Restored? Three Moments of Neoliberal Restructuring in the American Labor Market". *Journal of Economic Geography* 2 (2): 179–220.

Peck, J. & N. Theodore 2012. "Politicizing Contingent Labor: Countering Neoliberal Labor-Market Regulation … From the Bottom Up?" *South Atlantic Quarterly* 111 (4): 741–61.

Pittman, L. 2018. "Safety Net Politics: Economic Survival Among Impoverished Grandmother Caregivers". In V. Lawson & S. Elwood (eds) *Relational Poverty Politics*. Athens, GA: University of Georgia Press, 25–43.

Pulido, L. 2016. "Geographies of Race and Ethnicity II: Environmental Racism, Racial Capitalism and State-Sanctioned Violence". *Progress in Human Geography* 41 (4): 524–33.

Ransby, B. 2015. "The Class Politics of Black Lives Matter". *Dissent,* Fall. www.dissent magazine.org/article/class-politics-black-lives-matter (accessed 20 April 2017).

Rose, N. & P. Miller 1992. "Political Power Beyond the State: Problematics of Government". *The British Journal of Sociology* 43 (2): 173–205.

Roy, A. & E. S. Crane (eds) 2015. *Territories of Poverty: Rethinking North and South*. Athens, GA: University of Georgia Press.

Schram, S. 2000. *After Welfare: The Culture of Postindustrial Social Policy*. New York: NYU Press.

Schram, S. 2015. *The Return of Ordinary Capitalism: Neoliberalism, Precarity, and Occupy*. Oxford: Oxford University Press.

Swerts, T. 2018. "'Check your Privilege': The Micro-Politics of Cross-Status Alliances in the DREAM Movement". In V. Lawson & S. Elwood (eds) *Relational Poverty Politics*. Athens, GA: University of Georgia Press, 166–82.

Watkins-Hayes, C. 2009. *The New Welfare Bureaucrats: Entanglements of Race, Class, and Policy Reform*. Chicago, IL: University of Chicago Press.

Žižek, S. 2000. *The Ticklish Subject: The Absent Centre of Political Ontology*. London: Verso.

Zweig, M. 2000. *America's Working Class Majority: The Best Kept Secret*. Ithaca, NY: Cornell University Press.

PART III

CONNECTIONS

CHAPTER 18

DOREEN MASSEY'S URBAN POLITICAL ECOLOGY

Nik Heynen, Nikki Luke and Caroline Keegan

INTRODUCTION

As the consequences of neoliberal capitalism continue to ravage working people, especially women and communities of colour, there is a need for geographers to keep looking for ways to connect emancipatory theories of change with communities engaged in geographically situated forms of radical praxis. We argue it is equally important to understand and confront the politics of nature as directly experienced through uneven impacts related to energy, waste, toxic exposure and other destructive socio-ecological relations. Within urban political ecology today the question of how to connect models of public engagement and radical praxis to the politics of nature remains open. Better developing these bridging logics between theory and practice is an important place to begin. The question then becomes where to find models for robust and comprehensive geographic, political and ecological thinking in the face of continued narrowing across the discipline? Doreen Massey's contributions are especially important in this context as she exhibited an expansive breadth across her intellectual and political commitments.

In this chapter, we engage some less discussed threads of Massey's radical, feminist praxis and extend her theorization of the politics of nature at the intersections of physical and human geography. Featherstone and Painter (2013: 12) highlight Massey's commitment to nature in arguing, "[T]he bio-physical world is of great importance to her and, as her writings attest, that importance is simultaneously personal, conceptual and political." As such, we draw upon Massey's writings on nature, physical geography, climate change and urban politics to consider their implications for urban political ecology.

We will start the chapter by discussing her first major work, *Capital and Land*, which we argue anticipates urban political ecology in significant ways. Next, we will work through her evolving thoughts around the politics of nature by staging a conversation between her writings on power-geometries, physics, time-space, and nature more generally. Finally, we will consider her coming together with Stuart Hall and Michael Rustin (and others) to publish the *Kilburn Manifesto*, which, we argue, offers encouragement to scholars of urban political ecology to extend the reach of their scholarship. All of these sections, we argue, contain within them connected physical and human geographic logics that provide an integrative response to ongoing fracturing in the discipline of geography. We argue through these cases that her model of prefigurative praxis, embraced and fostered through Massey's originality and intellectual creativity, has much more to offer.

ANTICIPATING URBAN POLITICAL ECOLOGY

A deep reading of Massey's early work foreshadows urban political ecology and discussions about the neoliberalization of nature. Here, we focus specifically on her first book with Alejandrina Catalano, *Capital and Land: Landownership by Capital in Great Britain*, published in 1978.[1] *Capital and Land* anticipates urban political ecology through a robust Marxist analysis of historical property relations. Massey and Catalano trace the evolution of land holdings by the landed gentry/aristocracy, the Church of England and the Crown Estate towards increasingly urban, industrial and financial landownership across the contexts of specific resources like agriculture, forestry and coal mining. Their analysis is avowedly political in its focus on the social justice consequences of resource financialization and privatization. It also prefigures the kinds of questions about natural resources now commonplace in political ecology writ large (see Dempsey 2016; Mansfield 2004; Osborne 2011; Peluso 1992).

Part of the importance of the book is Massey and Catalano's argument that the 1970s British Left misunderstood the ways in which the growth in real-estate investment of the time mattered for wider economic relations. Massey and Catalano (1978: 1) open with a discussion of the political upheaval in response to this

> major "land and property boom" and subsequent collapse, a mushrooming of tenants and community actions against the direct and indirect effects of this boom, and attempts by all three major Parliamentary parties to formulate legislation concerning

---

1. This book was published in the same Edward Arnold series as Castells' (1977) *The Urban Question: A Marxist Approach*, that was edited by David Harvey and Brian Robson.

the economic use of land and the definition of the rights of private landownership.

Massey and Catalano analyse the evolution of changes in the property relations of residential and industrial sectors to explain a broader underlying socioeconomic process.

Massey and Catalano's (1978: 20) framing extends beyond any one resource base and shows how land-based property politics "arise to some extent from differences in the part of the economy which was affected – whether, for instance, it was housing or mining, the city or the rural periphery". They then suggest that particular land uses yield variable problems just as increasingly private land-ownership produced problems. At the same time, they foreshadow the increased role of finance capital across property types by suggesting, "property companies are investing capital so as to generate higher rents from a given piece of land. As long as the new rent can provide an adequate return on the capital invested, property companies will continue to develop" (138). This discussion seems a prophetic insight when we consider that according to Scott (2013: 224) between 1980 and 1989 investment in business and service sector property doubled in London. Ultimately, Massey and Catalano spell out a distinctly political ecological explanation of the geographically situated, financial and regulatory relations that govern distinct systems of land ownership, showing how uneven development is grounded in spatially explicit ways. To this end, they write (1978: 21), "The purpose of this work is, then, to investigate the effect of private landownership on the structure of development of a capitalist society, to understand the significance of the 'land problem' and to account for and analyse the varying forms it takes."

Massey and Catalano (1978: 188) conclude that all instances of private property will proliferate processes of capitalist accumulation and lead to greater displacement and inequality. While this insight forms a basic tenet of Marxist political economy, they prefigured considerable empirical analyses about the political ramifications of these place based dynamics. They do not see the effects of privatization as confined only to issues of distribution of surplus capital, but also internalized within production processes more generally with unanticipated, negative externalities. Specifically, they refer back to "surface ownership's effects on the coal mining industry" and the "succession of forms of landownership in agriculture". Through their focus on the privatization of property, they cover a broad range of resources, all of which have gained prominence as focal research areas in the decades that followed. In the context of their detailed, historical-geographical materialist analysis of interrelated social and economic contradictions that are inherent to the British "land problem", they articulate the immanent contradictions between increased privatization of property and natural resources commodification.

Their analysis of the ways in which private property relations facilitate uneven development within and across particular geographies, resources and industries sets up later work that Massey would develop in *Spatial Divisions of Labour* (see Chapter 8), as well as important sector, resource, and industry-based analyses in political ecology and related fields today (see Bridge 2009; Kay 2017).

Importantly, *Capital and Land*, at the time, contributed to the increasing engagement with Marxist theory in geography, especially with regard to questions of property relations and the spatial politics of resource use. Yet, the book also sketches a spatial understanding of the processes of urbanization as experienced in urbanizing and otherwise rural spaces that echoes Raymond Williams's (1975) *The Country and the City* – published shortly before – and anticipates other interventions like Cronon's (1991) *Nature's Metropolis*. Indeed, the working definitions of the urban central to Massey and Catalano's study could offer a counter-narrative to the narrow framing of cities within much contemporary urban political ecology and cognate discussions, as noted by Angelo and Wachsmuth (2015) in their recent intervention.

Through the 1980s and 1990s, Massey's ideas about land, property and resources continued to inform her evolving analysis. However, there are some sophisticated analytical leaps within her thinking, especially by the early 1990s, that more explicitly work through "power-geometries", physical science and the relationship between dimensions of space-time and nature. Here, we urge a corrective given the near absence of discussion of *Capital and Land* within the urban political ecology literature.

MASSEY'S EXPANDING POLITICS OF NATURE

In an interview with the *Social Science Bites* podcast in 2013, Massey articulated how the politics of nature necessitate broader dialogue between social and natural scientists to develop more integrative logics, goals and solutions to wicked environmental challenges. This too is a key point for thinking about her contributions to urban political ecology. She noted, "I think there is within geography the possibility of bringing together the social and the natural sciences more than we have historically done, and there are vast differences between them, and the process is very hard, but we need to do that." For Massey, this necessity is tied to radical praxis and the pursuit of engaged scholarship, as she explained: "In an age which is faced by environmental problems such as we have, with climate change, with pollution questions, which are utterly social too, then I do think that the natural and the social sciences need to talk to each other more" (Massey & Warburton 2013).

Beginning with her earliest interventions around space-time, Massey worked to establish this dialogue across branches of the discipline through substantive and integrative engagement with research in the natural sciences. In the *New Left Review*, she builds from radical geography, feminist theory and modern physics to argue for the necessity of thinking in terms of "the inseparability of time and space" (1992b: 84). Her engagement with modern physics is a provocation to consider critically the relationships between disciplines and to rethink hardened conceptualizations of space and time (Massey 1999g, 2005). Starting with the axiom of modern physics that space and time are distinct but "inextricably interwoven", Massey opens new ways to engage space and the processes through which it is constituted (Massey 1992b: 77).

But Massey did not see the "hard" science of physics as a model for geographers. In a later paper that continues her efforts to bring space and time together as analytical categories, she also problematizes the blind authority bestowed on "harder" sciences across human and physical geography, noting that the "habit of referring to physics bears witness to an implicit imagination both of a model of science and of a particular relationship between the disciplines. It is an imagination that physical and human geographers share [but that] nonetheless serves to hold us apart" (1999g: 263). Instead, she asserts, these comparisons between the relative "scientific status" of human and physical geography "need to be laid aside. Rather, we should put in a claim for their both being sciences of the complex and the historical" (Massey 1999g: 266). Ultimately, a central objective of Massey's overall project was to challenge central assumptions that both human and physical geographers hold: she pushes us to consider more integrative ways of thinking spatially that overcome limitations imposed through the domination of only thinking historically (Massey 1999g).

During this same period, Massey (1991a) also presented her theory of power-geometries in *Marxism Today* where she outlined how flows of knowledge, power and mobility are experienced differently across different places and social relations. While she did not take on questions of nature explicitly in this text, in line with new discussions in urban political ecology, we can now see the importance of bringing these threads together.

In working through a theory of power-geometries, Massey never loses sight of the connections between theory and praxis. She is explicit in her concern that conceptualizing power-geometries and the role of place opens up "the possibility of developing a politics of mobility and access" (1991a: 26). Here, Massey emphasizes the ongoing importance of place under globalized capital while acknowledging both the troubling delineation of insiders and outsiders and the reactionary nationalism made possible through attempts to recapture a sense of place. She notes, "In this interpretation, what gives a place its specificity is not some long internalised history but the fact that it is constructed out of a particular constellation of social relations, meeting and weaving together at a particular locus" (1991a: 28). By taking seriously the experiences of gendered and racialized bodies, Massey's approach pushes against disembodied studies of capital common in many Marxist-oriented analyses. Massey's take on power-geometries recognizes that only some people are "in charge" of space-time compression, "some initiate flows and movement [while] others don't; some are more on the receiving-end of it than others; some are effectively imprisoned by it" (1991a: 26).

Urban political ecology has much to gain from Massey's concept of power-geometries and her insistence on the lack of separation of space-time, in conversation with her fidelity to greater interaction between physical and human geography. Urban political ecology, especially the early formation of urban political ecology, relies heavily on Smith's (2008) discussions of uneven development and his production of nature thesis. While Smith gestured toward embodying the spatial politics central to both concepts, both lack the explicit theoretical agility required to think though the role of positional politics related to the uneven gendered and racial development of nature and urban nature. Massey's approach, in deeper conversation with Smith, can extend contemporary discussions that put uneven development, global climate crisis, white supremacy, patriarchy and heteronormativity into conversation and study these dynamics relationally (Doshi 2017; Heynen 2016, 2017). While neither of them explicitly engage either Massey or Smith, the work of Mollett and Faria (2013) and Doshi (2017) demonstrates how unevenly interconnected, socio-ecological power relations shape nature and are in turn shaped by nature. We argue that these authors, in addition to writers considering non-human geographies (see Collard & Dempsey 2017), characterize the productive possibilities that await in staging more of these theoretical questions as conversations instead of debates.

CLIMATE CHANGE POLITICS

Returning to Massey's (1999g: 274) distinctive commitment to reading across disciplinary boundaries – engaging substantively with philosophy, physics, economics, geology and geomorphology – offers other important insights with regard to how urban political ecology can move forward. In questioning the theoretical, political and historical commitments that have divided, and continue to divide, human and physical geographers, Massey provides, at least in part, the impetus to consider new opportunities for interdisciplinary research able to grapple with the "large-scale issues such as climate change" (Harrison *et al.* 2004: 435). Massey (2008: 493) reminds us that to think about space in a way that is relatable and useful to a range of people beyond human geography serves the discipline. Yet, more importantly, this project "to bring space alive" is an effort "to bring it alive politically" (Massey & Warburton 2013: 264). Exploring and extending Massey's interrogation of science and space offers numerous, important paths for answering Walker's (2005, 2006) always lingering questions about where the ecology and policy can be found within political ecology.

Throughout her career, Massey worked to trace the relationship between space and power that led her to questions of privatization, the environment and climate change. Weaving together her public intellectual engagements around space and power in South Africa, Nicaragua, the United Kingdom and Venezuela, she observed, "the issue of politics is paramount" (Massey 2008: 494). From the privileged point of the academy, Massey (2008: 495) noted that geographers have numerous ways to engage politically: addressing a wider audience through the popular press and popular education, direct involvement in policy-making and "active participation in struggles and campaigns and through contributing to wider debate in civil society". Each of these, Massey practised throughout her career, yet are forcefully drawn together in the *Kilburn Manifesto*, one of the last texts that Massey wrote as part of a collaboration with Stuart Hall and Michael Rustin. The Manifesto offers a radical rethinking of neoliberalism through several concrete policy issues that demonstrate her deep concern for politics as well as the collective coming together of her relational thinking about nature and public intellectual praxis.

The Manifesto's framing statement lists the serial inaction on the environmental crisis among the failures of Western governments that have embraced neoliberalism (Hall *et al.* 2015: 10). Massey (2015c: 29) refines this into a critique of the UK's supposed greenness, foreshadowed in concrete terms in her discussion about London in *World City* (2007: 194) where she illustrates that "not only must policy be directed to managing the impacts of climate change on London, but it must also work to reduce London's own contribution to the production of that problem". Keenly aware of the consequences of uneven power-geometries, Massey (2015c: 29–30) notes in her contribution to the Manifesto that "environmental destruction and the catastrophes consequent upon climate change will not fall evenly across the world. Probably such ills will fall most quickly, and most heavily, on more impoverished places, which in any case have fewer resources with which to offset such damage." To respond to environmental crisis then requires a new set of shared values (Massey & Rustin 2015: 218).

Importantly, Massey sought to cultivate such public discussion about the environment, space and geography in civil society through her use of a wide range of formats that ensure availability for different audiences. The *Kilburn Manifesto* was initially published in a serialized form in *The Guardian* where she explained that to address climate change "our current notion of wealth, and our commitment to its growth, must be questioned". As if in dialogue with urban political ecology, and harkening back to sentiments in *Capital and Land*, she condemns the use of market-based mechanisms to stem environmental crisis. These critiques were also staged in a film project she carried out with filmmaker Patrick Keiller, in which landscape was used as a provocation to think through environmental politics (see also Massey 2006b; Hinchliffe 2012). In an essay accompanying the film, she elaborates her perspective on the commodification of nature and "challenge[s] the currently hegemonic understanding that markets are natural. They are not. What's more, in their unfettered operation around the planet they destroy what we might actually call 'nature'" (Featherstone *et al*. 2013: 265). Woven throughout these works, Massey sparks public debate about the fundamental values that underlie Western society and the political possibilities for change. Informed by a commitment to address inequality between regions, people, and species, she urges "a wider ideological and political questioning", without which "there can be no conjunctural rupture in the balance of social forces" (Featherstone *et al*. 2013: 266).

Scholars currently working within urban political ecology around neoliberal capitalism and climate change reflect these deep commitments to social justice and critique of neoliberal agendas in urban and environmental policy, ranging across questions of property and infrastructure (Broto & Bulkeley 2013), territorial market calculations of carbon politics (Rice 2010) and international water governance (Kaika 2003) among others. In the process, many seek to cultivate public discussion in coalition with environmental activists (London 1998; Wolford 2010) or by facilitating community dialogue to examine the ways in which climate praxis can be embodied (Rice *et al*. 2015). Massey's insights that cross the boundaries of academia and praxis, as well as the disciplinary boundaries between human and physical geography, physics, economics and philosophy, both anticipated key themes of urban political ecology and provide a framework for future work that takes seriously the material – social, natural and political – impacts of economic change, industrial development and property relations.

CONCLUSIONS

Notions of urban political ecology were developed through explicit engagement across embodied, material politics and robust, relational thinking with a focus on questions of justice, equality, egalitarianism and transformative change. We can use Watts's (2015: 32) insight as a litmus test when he suggests, "Political ecology constructed a theory upon a more-or-less Marxist analysis of political economy in which social relations of production, access to and control over resources, and forms of capital accumulation ... were its central starting points". When revisiting Massey's work in this context, it is striking that she is so rarely discussed within political-ecological scholarship. Like other influential thinkers, a hallmark of Massey's approach was openly working through unstable categories and ideas to move towards revolutionary changes both within geography's thinking about the world and within the world itself. In talking about the motivation behind the *Kilburn Manifesto* she said:

> Instead what we have ... is a world, and an imagination of a world, that is constant change, flow, mobility. Change has become the new stasis; there is no change to this constant change. But it is small change, trivial, within the system. Fashion, technology, a new gadget. New, new, new; but nothing really new at all. And so the very notion of change becomes traduced, and we arrive at the point of what has been called the eradication of the possibility of thinking a radically different future. Big time, historical time, becomes unthinkable.   (Peck *et al.* 2014: 2039)

If indeed the contemporary era is marked by "the eradication of the possibility of thinking a radically different future", what lessons can scholars and activists working within the politics of nature take away from Massey's legacy? Working to emulate Massey's embrace of change, flow and mobility of ideas is a good starting place to avoid intellectual staleness or political paralysis evident on the Left in these times, just as she and Catalano argued was necessary in her first major work in 1978. But perhaps more important is to note the ways she forced intellectually sophisticated discussions about uneven spatial relations to be taken seriously beyond abstract conjecture. To this end, Massey argues, along with Hall and Rustin (2015: 22), that: "The neoliberal order itself needs to be called into question, and radical alternatives to its foundational assumptions put forward for discussion. Our analysis suggests that this is a moment for changing the terms of debate, reformulating positions, taking the longer view, making a leap." What matters just as much as their fierce call for revolutionary change is the forum outside the academy and through the popular press in which they made this argument.

One of the clearest explanations for the model through which she worked came in a response to critiques levelled against her in the early 2000s by Dorling and Shaw (2002). In response to Massey's 2001 essay ("Geography on the Agenda") in which she was critical of the lack of effort geographers take in policy formulation and enactment, Dorling and Shaw (2002: 638) directly responded that

> There is an alternative interpretation as to why geography fails to be on the policy agenda. It is that the discipline of geography is not well suited to stampeding this way and that, following whatever turn is currently fashionable – being relevant when what is seen as relevant can change so quickly.

In her response, Massey (2002c: 645) argued, "It is not a case of sitting at one's desk, having an idea, and rushing out to tell a politician." She described instead a model for a radical praxis in which "working is an endless moving-between" the academy and non-academic spaces. She continued:

> I learned more ... from working in "Ken Livingstone's" GLC, with Labour Party regional policy-makers, with the ANC, and with a whole range of community groups, all the while mulling things over, relating them to wider arguments, writing, talking, thinking, arguing, than I ever could have done in an academic arena alone.
>
> (Massey 2002c: 645)

With notable exceptions, political ecologists concerned with current issues, and professing their dedication to transformative scholarship, have rarely moved so fluidly between the realms of academia, formal politics and public engagement.

Doreen Massey was aware about who she was writing for, considerate of her own positionality, grounded in a feminist praxis and motivated by questions of physical geography and nature in ways largely unrecognized. She provides a model of how to be reflexive and engaged. She urges us as scholars to ask of ourselves why we do the work we do, what service it does, for whom we work, and what are the wider political projects in which we are engaged. These qualities allowed her to produce transformative ideas throughout her life's work, which generated concepts that in turn travel far beyond the academy in support of projects of emancipatory change. It is for all of these reasons that scholars working in urban political ecology should revisit her many generative ideas.

REFERENCES

For works authored and co-authored by Doreen Massey, please see Select Bibliography of Doreen Massey, beginning on p. 371.

Angelo, H. & D. Wachsmuth 2015. "Urbanizing Urban Political Ecology: A Critique of Methodological Cityism". *International Journal of Urban and Regional Research* 39 (1): 16–27.

Bridge, G. 2009. "Material Worlds: Natural Resources, Resource Geography and the Material Economy". *Geography Compass* 3 (3): 1217–44.

Broto, V. C. & H. Bulkeley 2013. "Maintaining Climate Change Experiments: Urban Political Ecology and the Everyday Reconfiguration of Urban Infrastructure". *International Journal of Urban and Regional Research* 37 (6): 1934–48.

Castells, M. 1977. *The Urban Question*. London: Edward Arnold.

Collard, R.-C. & J. Dempsey 2017. "Capitalist Natures in Five Orientations". *Capitalism Nature Socialism* 28 (1): 78–97.

Cronon, W. 1991. *Nature's Metropolis: Chicago and the Great West*. New York: WW Norton.

Dempsey, J. 2016. *Enterprising Nature: Economics, Markets, and Finance in Global Biodiversity Politics*. Malden, MA: Wiley-Blackwell.

Dorling, D. & M. Shaw 2002. "Geographies of the Agenda: Public Policy, the Discipline and its (re)'turns'". *Progress in Human Geography* 26 (5): 629–46.

Doshi, S. 2017. "Embodied Urban Political Ecology: Five Propositions". *Area* 49 (1): 125–8.

Featherstone, D. & J. Painter 2013. "There is No Point of Departure: The Many Trajectories of Doreen Massey". In D. Featherstone & J. Painter (eds) *Spatial Politics: Essays for Doreen Massey*. Hoboken, NJ: John Wiley & Sons, 1–18.

Featherstone, D., S. Bond & J. Painter 2013. "Stories So Far". In D. Featherstone & J. Painter (eds) *Spatial Politics: Essays for Doreen Massey*. Hoboken, NJ: John Wiley & Sons, 253–66.

Hall, S., D. Massey & M. Rustin (eds) 2015. *After Neoliberalism? The Kilburn Manifesto*. London: Lawrence and Wishart.

Harrison, S., D. Massey, K. Richards, *et al.* 2004. "Thinking Across the Divide: Perspectives on the Conversations Between Physical and Human Geography". *Area* 36 (4): 435–42.

Heynen, N. 2016. "Urban Political Ecology II: The Abolitionist Century". *Progress in Human Geography* 40 (6): 839–45.

Heynen, N. 2017. "Urban Political Ecology III: The Feminist and Queer Century". *Progress in Human Geography* 42 (3): 446–52.

Hinchliffe, S. 2012. "A Physical Sense of World". In D. Featherstone & J. Painter (eds) *Spatial Politics: Essays for Doreen Massey*. Hoboken, NJ: John Wiley & Sons, 178–88.

Kaika, M. 2003. "The Water Framework Directive: A New Directive for a Changing Social, Political and Economic European Framework". *European Planning Studies* 11 (3): 299–316.

Kay, K. 2017. "Rural Rentierism and the Financial Enclosure of Maine's Open Lands Tradition". *Annals of the American Association of Geographers* 107 (6): 1407–23.

London, J. K. 1998. "Common Roots and Entangled Limbs: Earth First! and the Growth of Post-Wilderness Environmentalism on California's North Coast". *Antipode* 30 (2): 155–76.

Mansfield, B. 2004. "Neoliberalism in the Oceans: 'Rationalization,' Property Rights, and the Commons Question". *Geoforum* 35 (3): 313–26.

Mollett, S. & C. Faria 2013. "Messing with Gender in Feminist Political Ecology". *Geoforum* 45: 116–25.

Osborne, T. 2011. "Carbon Forestry and Agrarian Change: Access and Land Control in a Mexican Forest". *Journal of Peasant Studies* 38 (4): 859–83

Peck, J., D. Massey, K. Gibson & V. Lawson 2014. "Symposium: The Kilburn Manifesto: after neoliberalism?" *Environment and Planning A* 46 (9): 2033–49.

Peluso, N. L. 1992. *Rich Forests, Poor People: Resource Control and Resistance in Java.* Berkeley, CA: University of California Press.

Rice, J. L. 2010. "Climate, Carbon, and Territory: Greenhouse Gas Mitigation in Seattle, Washington". *Annals of the Association of American Geographers* 100 (4): 929–37.

Rice, J. L., B. J. Burke & N. Heynen 2015. "Knowing Climate Change, Embodying Climate Praxis: Experiential Knowledge in Southern Appalachia". *Annals of the Association of American Geographers* 105 (2): 253–62.

Scott, P. 2013. *The Property Masters: A History of the British Commercial Property Sector.* London: Taylor & Francis.

Smith, N. 2008. *Uneven Development: Nature, Capital, and the Production of Space.* Athens, GA: University of Georgia Press.

Walker, P. 2005. "Political Ecology: Where is the Ecology?" *Progress in Human Geography* 29 (1): 73–82.

Walker, P. 2006. "Political Ecology: Where is the Policy?" *Progress in Human Geography* 30 (3): 382–95.

Watts, M. 2015. "The Origins of Political Ecology and the Rebirth of Adaptation as a Form of Thought". In T. Perreault, G. Bridge & J. McCarthy (eds) *The Routledge Handbook of Political Ecology.* London and New York: Routledge, 19–50.

Williams, R. 1975. *The Country and the City.* New York: Oxford University Press.

Wolford, W. 2010. *This Land Is Ours Now: Social Mobilization and the Meanings of Land in Brazil.* Durham, NC: Duke University Press.

CHAPTER 19

THE SOCIOGEOMORPHOLOGY OF RIVER RESTORATION: DAM REMOVAL AND THE POLITICS OF PLACE

Frank Magilligan, Christopher Sneddon and Coleen Fox

INTRODUCTION

Although known for her extensive work on labour, gender and the socio-economic production of place, Doreen Massey also engaged with the links between physical and human geography (Harrison *et al.* 2004) and the pursuit of common epistemological ground between these subdisciplines (Massey 1999g). In particular, she suggested that these subfields share important engagements with questions regarding how space, time and "space-time" are conceptualized, arguing that a historical perspective has been a crucial shared focal point for these seemingly disparate subdisciplines. Her vision for greater engagement between physical and human geography was not merely a plea for encouraging greater interdisciplinary cooperation; rather, she articulated it as a way for each to integrate some of the associated ontological and epistemological scaffolding of the other in order to begin to address larger societal questions and to develop a fuller understanding of the social and physical production of space. She further encouraged physical geographers, in particular, to shed some of their "physics envy" and embrace some of the broader theoretical and normative approaches in the social sciences. In many ways, Massey anticipated and inspired contemporary efforts to rethink and transcend the human–physical divide (e.g., Lane 2014; Lave *et al.* 2014; Ashmore 2015). Using our ongoing work on river restoration in New England, we take up Massey's challenge to integrate physical and human geography. Our analysis of dam removals reveals the benefits of a geographically holistic approach, taking into account contestations over interpretations of history, scientific expertise, knowledge production, identity and micropolitics (Fox *et al.* 2016). The result is a more robust understanding of the relationship between river restoration and the production of place.

River restoration is a big business; within the United States (US), it is a $1 billion per year enterprise (Bernhardt *et al.* 2005) with many scientists claiming that the practice is rapidly outpacing the science of river restoration (Wohl *et al.* 2015). River restoration is an all-encompassing term that includes bank stabilization, fish ladders, flow modification, channel design, land acquisition, riparian/wetland management and dam removal – approaches that vary in price and engineering/geomorphic design and intensity. Moreover, river restoration is done for various reasons: in some instances it might be regulated as part of a mitigation strategy, as in the case of stream construction/management (Lave *et al.* 2010; Doyle *et al.* 2015); it might be promoted by a local community for aesthetics (Kondolf 2006; Kondolf *et al.* 2007); or, as exemplified by dam removal, it might be required to reduce risk from a potential hazard (Doyle, Harbor & Stanley 2003; O'Connor, Duda & Grant 2015; Tullos *et al.* 2016). The efficacy of these restoration approaches vary widely, in part because there is minimal post-project assessment (Bash & Ryan 2002; Downs & Kondolf 2002; Kibler *et al.* 2011), or in most instances, the restoration goals were never explicitly stated thus there is a significant institutional and scientific disconnect between project design and outcomes (Palmer *et al.* 2005; Bernhardt *et al.* 2007).

This expansion of restoration activities coincides with the emergence of a nascent epistemological turn in physical geography that is gradually confronting the social dimensions and broader impacts of scientific research, and is thus beginning to incorporate epistemological and ontological perspectives from the social sciences into environmental scientific practices (Wilcock *et al.* 2013; Tadaki *et al.* 2014; Lane 2014; Ashmore 2015; Mould *et al.* 2017). There has been a strong theoretical challenge from social scientists and physical geographers alike to embrace a "critical physical geography" that asks physical geographers to reflect more on the politics and practice of scientific endeavours and more fundamentally on the nature of knowledge (Lave *et al.* 2014; Lane 2016) although the contours of this engagement remain undefined and poorly articulated (Blue & Brierley 2016). In addition, the realization that rivers and watersheds exist in human dominated landscapes – whether invoking the trope of the Anthropocene or not – indicates that the range of restoration efforts may be ecologically and geomorphically limited, thus requiring what some scholars call intervention ecologies (Hobbs *et al.* 2006, 2009, 2011; Morse *et al.* 2014; Truitt *et al.* 2015; Collier 2015) or uncharted socio-ecological futures (Robbins 2014; Mansfield *et al.* 2015).

Although numerous restoration efforts exist, we focus on dam removal as a thought-provoking entry point into the power relations within and between state and non-state actors, the efficacy and evolution of institutional structures for environmental governance, and the *pas de deux* of social and physical scientists. Dam removal, although trying to achieve something biophysical, is at its core a social process where realization requires a constellation of historically contingent institutional, political and economic forces to align within a particular location. It is thus, in Massey's (1999g: 274) terms, a complex phenomenon that forces us to come to grips with "time-bound processes" that remain open and indeterminate because they operate within physical and cultural landscapes where space is "the sphere of the existence of multiplicity". Dam removal is also, crucially, a value-laden assessment of what the dammed/undammed landscape – in effect, "nature" – should look like. More than 1400 dams have been removed in the US (O'Connor *et al.* 2015; Foley *et al.* 2017) to achieve an array of environmental and economic goals (Pohl 2002), with many more slated for removal. Yet despite the seemingly strong momentum to remove dams, many of these removals have been fraught with conflict (Lejon *et al.* 2009; Fox *et al.* 2016; Magilligan *et al.* 2017).

As part of our effort to understand the social dimensions of dam removal and the longstanding and often heated conflict over their removal, we have amassed academic and grey literature information on over 125 removed dams and over 50 ongoing planned removals in New England, implemented detailed semi-structured interviews with more than 40 stakeholders – including dam removal opponents and proponents – and attended numerous public meetings. In the remainder of this chapter, we develop Massey's discussion of the role of, and need for, deeper conversations between physical and human geography (Massey 1992b, 1999g, 2001c) in the context of river restoration by dam removal. Moreover, we attempt to show how dam removal is situated politically and ideologically, how the dam removal process reveals the theoretical elements of the nascent fields of critical physical geography and sociogeomorphology, and how its implementation is fraught with place-based conflicts over identity and claims to historical legitimacy.

TOWARDS A SOCIOGEOMORPHOLOGY OF RIVER RESTORATION

Although Massey pushed physical geographers to leave a comfort zone characterized by prediction, physical laws and "physics envy", she never articulated a precise template for that radical ontological and epistemological turn. Recently, however, physical geographers are attempting to craft that blueprint. Ashmore (2015), for example, contends that geomorphologists need to recognize the sociogeomorphic, hybrid components of landscapes and to see riverscapes as the co-production of social and natural processes. This socionature perspective, he contends, would enable geomorphologists to better comprehend the ways in which contemporary fluvial landforms and landscapes evolved to their present condition (cf. Urban 2002), thereby providing insights into river management as well as helping to identify which fluvial processes currently drive or will drive the system in the future. Geomorphologists have had a longstanding tradition of documenting human impacts on fluvial form (Hooke 2000; Gregory 2006), but Ashmore (2015) pushes geomorphologists to recognize that social norms and processes need to be evaluated on par with fluvial processes in understanding how the river came to be and how it might evolve (see also Blue & Brierley 2016). He suggests four possible ways forward for physical geographers to engage with a sociogeomorphological viewshed: (1) to develop a more place-based and historical conceptualization of rivers; (2) to move beyond predictive models; (3) to develop new paradigms; and (4) to embrace a greater realization that geomorphologists are active actors in the quest for understanding rivers (Ashmore 2015). In addition to echoing Massey's (2001c) defence of an "open historicity" when it comes to approaching biophysical and social change, his themes, especially the first three, map especially well onto the science and management of dam removals, yet their application lacks a physical groundedness that most geomorphologists can appreciate. In the next several sections, we present how these potential new directions and ontological turns might manifest in river restoration and, specifically, how the debates and outcomes of dam removal in New England specifically incorporate and reveal historical contingencies and contemporary environmental values; engage with and push the limits of prediction; potentially help the development of new paradigms for both physical and social scientists; and perhaps illustrate the opportunities for physical geographers to engage with a sociogeomorphic perspective at a more than cursory level.

"The past isn't dead, it isn't even past"

In New England dams are – and will continue to be – an integral part of its industrial history (Steinberg 1991), and with over 14,000 dams scattered across the New England landscape (Magilligan *et al.* 2016), there are few tributaries or mainstem systems unaffected by a dam. Because dams, no matter their size, have ideologically and materially become part of the lived landscape for hundreds of communities, their removal – and associated conflicts – often represent struggles over whose sense of place, history and nature predominate. These dams, in short, have taken on a symbolic materiality greater than their mere physical presence (Fox *et al.* 2016). This embodiment of historical identity can greatly complicate the removal process and can, in many instances, rebut even the most ecologically reasoned rationale for removal. As one regional coordinator of dam removals said

We call it the first "fill-in the-blank" argument. At almost every public meeting about a dam removal, someone gets up objecting to the removal by saying "that pond is where I first kissed my wife" or "that's where I caught my first fish". The most important lesson we have learned over the years is that the hardest part of removing a dam isn't removing the dam, it's convincing people to remove a dam.

That entrenched historicized view becomes further manifest institutionally. Because many of these dams have been in place for decades or even a century, the removal process, by law, must engage with Section 106 of the National Historic Preservation Act of 1966 – operationalized by each state's State Historical Preservation Office (SHPO). When asked about their involvement with dam removal, one SHPO coordinator indicated that she would like to save every dam in the state. And when queried about what part of the dam they wanted to save, they said "we look at a resource holistically: the mill, the dam, and impoundment ... the dam might not be historical itself, but it might be part of an historical element" to the point that they would authorize, if need be, the retrofitting of the dam for micro-hydropower even if it meant "altering the dam to save the impoundment". Even as removal advocates strive to generate a future of free flowing rivers, they must continuously negotiate with history.

But the historical perspective recently advocated by Lane (2014) and Ashmore (2015) is more about encouraging physical geographers to be aware of and sensitive to the contingencies of place. This focus on contingencies is probably the most fertile and overlapping ground linking physical and human geography, yet there are subtle distinctions in the ways contingencies get envisioned and incorporated by these two subdisciplines. Physical geographers have been engaging with contingencies for decades (Phillips 2004, 2007), but primarily as a fixed, generally time invariant boundary condition of the physical system (e.g. regional geology, network structure etc.) that dictates or limits the potential array of morphological outcomes. The place-based contingencies of human geography provide a different yet important rubric concerning conflicts over removal as these contingencies reveal whose vision of the physical and social landscape gets actualized, and predicting the outcomes of an intervention like dam removal demands an appreciation of both physical and sociohistorical processes (e.g., a region's industrial past, local political dynamics, changing environmental norms). Perhaps this is best illustrated by the removal of the Homestead Woolen Mill Dam on the Ashuelot River in New Hampshire in 2010. The fluvial response to removal was unexpectedly muted: there was minimal erosion of the reservoir sediment and knickpoint migration was spatially limited both immediately post-removal and in subsequent years including the regionally catastrophic flooding associated with Hurricane Irene (Gartner *et al.* 2015). Several social and physical forces coalesced to generate the limited geomorphic outcome. The physical context conditioned the limited headcutting as the initial phase of knickpoint erosion exhumed a buried Pleistocene coarse gravel riffle that blocked further upstream knickpoint migration. But the physical boundary condition only partially explains the lack of headcutting as social processes also coalesced to constrain unabated erosion. For fear of losing a prized iconic historical covered bridge immediately upstream of the dam, in a hotly contested encounter over the removal process (Magilligan *et al.* 2017), local residents demanded that the engineering firm construct a series of channel-armouring rock vanes at the location of the former dam to limit channel erosion. Moreover, as Gartner *et al.* (2015) show, the presence of two upstream flood control dams has limited extreme flows that would normally evacuate sediment and engender sustained knickpoint headcutting. Therefore, understanding the scientific process and outcomes requires a consideration of the physical constraints imposed on the system in concert with the contemporary and historical social realities.

"You want me to predict what?"

Prediction is at the core of the scientific process, and is perhaps even more salient with resource management as the outcomes may have important social impacts. For river restoration, prediction can be highly elusive as the goals of the project may not be well defined; the requisite background data may be lacking or inadequate; or the scale of the restoration outcome may exceed the technical expertise. As the example from the Homestead Woolen Mills Dam reveals above, predicting the magnitude and timing of geomorphic outcomes is predicated on the extant social setting in combination with the geomorphic boundary conditions. This confluence represents the conundrum – or opportunity – that physical geographers confront: how to predict outcomes of an intervention when not all of the social and physical parameters are established or properly parameterized? This ambiguity often befalls any field-based scientific endeavour where few of the dials of the experiment are properly calibrated, but this uncertainty may be more profound for river restoration where the scale of the outcome may extend well beyond the outcome space of the combined social and physical intervention. Dam removal occurs in a fixed location, but unlike bank stabilization or reach scale restoration efforts, the geomorphic effects may extend hundreds of metres to tens of kilometres – both upstream and downstream of the dam – thus there is a scale shift where the technoscientific predictive space extends from the scale of the individual grain to the landscape. Because of the political and regulatory environment of the intervention, state actors typically demand specified outcomes, yet the scientific knowledge may not be sufficiently robust enough at this coarse scale to meet institutional and regulatory requirements. At the same time, however, the state is not necessarily demanding a grain-by-grain modelling exercise, but it is, at a minimum, requiring justification that remediation efforts, especially for sediment characteristics (volume and potential contamination), will not damage existing infrastructure, degrade existing ecosystems (e.g. loss of wetlands) or diminish water quality (Tullos *et al.* 2016).

Achieving or predicting successful outcomes of a restoration intervention, however, is more than meeting regulatory requirements, yet defining a singular metric of success can often be quixotic and context dependent (see Blue & Brierley 2016). The frequent ambiguity about restoration initiatives or goals can complicate prediction or even question its necessity as a scientific metric of success thus blurring the line between scientific objectives and social goals. There are many reasons to remove a dam, and for environmental agencies, watershed councils, and NGOs like American Rivers and The Nature Conservancy, support for dam removal is based loosely on a somewhat abstract ideological goal of "improving watersheds", without a clear scientific definition of what that might mean. The specific or spatially explicit response matters less because the actual removal of the dam is often the only de facto desired outcome.

For example, the fundamental cosmology of American Rivers, the leading NGO supporting and funding dam removals in the United States, is that dam removal creates a free-flowing river, and that a free-flowing river is in a "more natural state". American Rivers' support of restoring rivers is predicated on this more natural state – thus their predicted outcome (a free-flowing river) is always achieved by removal. In other words, the prediction (restoring a free-flowing river) is simply achieved by the action (removing the dam). Similarly, ambiguously defined restoration outcomes call the question of whether a scientific or social measuring stick is needed. For example, if the goal of the removal is to restore fish passage, the removal of the dam, by default, serves that purpose. What bothers many ecologists with this calculus is that there is no well-defined consensus as to what parameter best represents the correct "metric of success" for restoring fish passage: is it merely the presence or absence of key species; is it fish abundance; or is it a sustained viable fish community following the removal? As one scientist working on the Elwha Dam removal remarked, partially in jest, "considering all the money that was spent on the removal, if we do not have a picture of a 10-year-old kid reeling in a 50 lb. salmon above the former Elwha Dam by next year, we're in trouble". In contrast to prediction from a modelling or statistical analysis, from a sociogeomorphic perspective, these value-based metrics of success may simply reflect a tautological goal.

Ashmore (2015) queries whether physical geographers can move beyond prediction; perhaps one extension of that plea would be to ask how well physical scientists and decision-makers can contend with, and be content with, what exists in the penumbra of prediction – i.e. uncertainty? Debates about uncertainty plague the discussion of climate change, but that does not suggest that there is no evidence or scientific consensus (Oreskes 2004; Cook *et al.* 2016), nor does it imply that nothing can be done politically, socially or institutionally because the scientific predictions aren't 100 per cent infallible (Winkler 2016). The problem of presenting and dealing with uncertainty is inherently social and political, and it can undermine river restoration in the same ways that climate mitigation efforts fail to launch, or, more recently in the US, fail to get the necessary political support. Dam removal is an intervention clouded by uncertainty, and it thus lies at the intersection of science, policy and environmental conflict.

Thus as Ashmore (2015) questions whether physical geographers, or scientists in general, can move beyond prediction, the other side of that same coin is whether watershed managers/management can move beyond uncertainty? The predictive capacity of watershed managers to make informed science decisions is supported by basic laws of physics, cumulative knowledge production by geomorphologists, and physically based models – all associated with error bands of uncertainty – but that should not preclude making informed decisions (Clark 2002; Lane 2014). Conveying that information and uncertainty to other stakeholders is in many ways what foments conflict, especially for dam removals where removal decisions can occur in town meetings and other local venues where the court of public opinion can reign supreme often to the dismay of environmental agencies and NGOs (Magilligan *et al.* 2017).

Debates in these localized settings often orbit around a general mistrust of state or federal agency personnel or NGOs who are perceived as outsiders, and a greater trust in local, historicized knowledge (Sneddon *et al.* 2017). Ironically, although scientific information presented by "scientists/ technicians" often has to pass the litmus test of "uncertainty", our ethnographic data shows that the veracity of local knowledge concerning, for example, what is likely to happen to flow rates once a dam is removed, is rarely if ever vocally questioned or doubted. That is not to suggest that local knowledge rules the decision outcome. During an interview with the Penobscot Trust Foundation (the NGO responsible for removing two of the largest dams in New England) several interviewees said it didn't deter their mission. As one scientist from a Maine NGO said, "we don't try to contradict these erroneous scientific statements in these local settings as we know that the Courts believe we have the expertise and training to present the correct and appropriate scientific evidence". However, that disavowing of local voices and local knowledge may not operate effectively where the decision tree is more localized, as compared to the discussions about removing two large privately owned power plants that required a suite of local, state and federal agencies along with the Tribes and a watershed-scale NGO/Foundation. The importance of these anecdotally based, but empirically limited, local claims cannot be ignored by agency scientists or removal advocates even when they may seem to violate fundamental first principles (e.g., conservation of mass). As the State Director of one of the offices of TNC said, reflecting on a town vote to save the dam, "never underestimate the power of an irrational argument". Although as Lane (2014) has pointed out, despite the differences in vernacular/jargon, there may be some instances where local knowledge claims align well with sound hydrologic reasoning. We caution, however, that where the sociohistorical stakes are high and the local public forum is powerful, there may be an important mismatch between "lived science" and "pure science" and to neglect this context – no matter how "good" the pure science – can have significant impacts on the restoration trajectory. Relatedly, this is where human geographers' attention to politics and identity can complement such debates by linking local knowledge claims against removal to an implicit political strategy seeking to complicate and delay the removal process.

New paradigms and novel ecosystems

Unlike other restoration efforts such as bank stabilization or channel (re)design that generate localized and generally steady-state "predictable" responses, dam removal can generate far-reaching effects and potentially unknown outcomes. Moreover, since these dams may have been in place for centuries, the boundary conditions of sediment supply, rainfall–runoff relationships, and lateral connectivity may differ radically from the reach scale and watershed scale conditions that existed at the time of dam emplacement – which for New England in the early nineteenth century differed enormously from prehistorical "natural" conditions. Therefore, post-removal riparian recovery occurs in an uncharted social and physical domain. Besides the uncertainty of geomorphic response, the vegetation community structure and species composition of a particular reach and its immediate riparian zone differ radically from when the dam was initially constructed and from the initial prehistorical vegetation. Therefore, what's to be anticipated, predicted or appreciated ecologically from a dam removal in terms of the resulting landscape where few of the initial parameters remain the same or where invasive exotics may now dominate the successional trajectory of this now liberated landscape? Do we have the scientific rigour to predict the potential ecological outcomes or, more importantly, the social capacity to appreciate and/or tolerate the new post-removal novel ecosystem (Truitt *et al.* 2015)?

This uncertainty of landscape response to any of these ecological interventions suggests that we may be entering into a new era of resource and watershed management that may require a new epistemological and onto-logical framing of landscape(s). Do we have – or need – as Ashmore (2015) ponders, new paradigms in a sociogeomorphological context to meld the physical and social within this new parameter space-time?

In essence, dam removal is an intervention in an already altered land-scape. According to Hobbs *et al.* (2014), for these large-scale interventions the historically and socially produced landscape changes may represent an entirely new ecosystem or new set of boundary conditions, and it may not be possible to put the biophysical genie back in its bottle once the intervention has happened. In some instances, the responses to removal may generate new hydrologic, sedimentological and geomorphic systems unlike the pre-removal present and very different from the pre-dam construction past. In addition to the new localized boundary conditions at and/or near the dam, global scale changes in climate and sea-level generate unknown impacts on biotic and abiotic processes. Again, Massey's (1999g, 2001c) emphasis on open historicity and critique of "timeless" geomorphologic laws resonates with these observations; there is no timeless referent for river systems unen-cumbered by dams that restoration advocates can adhere to.

Besides developing new paradigms for detecting and predicting what might happen following dam removal, we may also need to develop new paradigms and new types of science and politics, which account for the new landscapes associated with existing dams and its novel ecosystems. Most dams, depending on size and management operation, essentially shift the hydrologic regime from a lotic (free-flowing river) to a lentic (e.g. a pond) system and may generate a vastly different ecosystem that local residents believe is the "natural state" of the system and therefore worth protecting. Some of the fiercest conflicts over dam removal in New England have pivoted around this struggle between local residents defending the existing novel ecosystem against environmental agencies and NGOs who perceive the dammed landscape as an unnatural condition that needs to be "fixed" and returned to a natural state (Fox *et al.* 2016). Moreover, the appreciation of the dammed novel ecosystem can further manifest institutionally. In an interview with staff at *Save the Bay*, a regional NGO located in Rhode Island (RI), members noted that some of the intense opposition to dam removal comes from ecologists in the RI Department of Environmental Management (DEM). Because of urbanization and development in this coastal setting, the only freshwater wetlands are those that occur in ponds associated with a dam, which now house rare species that DEM wants to protect, and they see dam removal as a threat to these threatened species.

PHYSICAL AND HUMAN GEOGRAPHY AS A SHARED SOCIOGEOMORPHIC PROJECT

River restoration provides a timely opportunity for encouraging greater dialogue between physical and social scientists, but more importantly, river restoration, to be evaluated holistically or grasped theoretically, requires that engagement. And as Demeritt (2009) reminds us, this dialogue need not be confined to geography's long disciplinary division, but must occur across the social and physical sciences. River restoration can be envisioned both as a biophysical intervention in a social system *and* a social intervention in a biophysical setting. As our research in New England has shown, the decision to remove a dam constitutes a complex set of social and physical interactions, and the outcome of that intervention may be difficult to predict scientifically. It is also the culmination of material processes occurring across several timescales without predetermined end points, processes that occur and interact in specific spaces (Massey 1999g). For some removal advocates, that inability to predict specific outcomes is immaterial to the decision-making process as the act itself is the necessary outcome. While for the state and local stakeholders, identifying the specific outcomes is a legal and even ethical obligation and a fundamental part of the process. This institutional requirement demands some modicum of predictability that often tests the limits of the scientific community. Although physical models exist to predict some of the outcomes of removal (cf. Downs *et al.* 2009), the lack of post-removal monitoring across a range of landscape settings limits total prediction. That should not suggest that dam removals be halted until perfect predictive capacity exists; nearly every human intervention into non-human realms (the *construction* of dams being a case in point) is characterized by varying degrees of uncertainty as to how that intervention might play out. So where might we go from here?

What the multiple examples of dam removal from New England demonstrate is that river restoration has never been and will never be amenable to objective and pure scientific assessment, and is always already subject to multiple political and material rationalities. This forces the epistemological hand of both physical and human geographers, since neither can easily approach river restoration via a conceptual framework that privileges the biophysical/social or the natural/cultural, a position that Massey would appreciate. The fact that many dam removal efforts generate controversy only heightens the inseparability of the physical and the human within a geographical perspective, a view that accords well with Latour's "matters of concern" (Latour 2005, 2007), an emerging paradigm that strives to resituate the role of scientific knowledge within emergent public controversies. Latour (2005) argues that many of the controversies that science is asked to address and provide supportable responses to – in theory offering a rational path for decision-makers hoping to resolve said controversies – are insoluble in this sense. The most seemingly intractable environmental controversies – whether revolving around climate change, genetically modified organisms, endocrine disruptors, or endangered species conservation – often gather together scientific knowledge assertions that are either at odds with other expert knowledge claims or, as in the case of climate change, integrated within antagonistic political disputes. There is an alternative: highlight instead how science and scientific knowledge production turn around issues, objects, things or whatever label we choose that emerge as "problems" once defined so by diverse public actors (Latour 2007). Massey (2001c) – whose thoughts on these matters certainly overlapped with Latour's (see Massey 1999g) – talks about issues, objects and phenomena in different terms, arguing we must understand any phenomenon as a result of its relative location within a specific set of historical processes, processes that in turn co-produce the space of the phenomenon. Focusing on the issue of concern brings these space-time processes into relief.

The starting point for scientists should thus not be matters of "fact", where answers are already understood because scientists know, for example, how rivers will function once a dam is removed, but matters of concern. Within this rubric, the crucial question becomes how to improve the ecological integrity of a physical landscape when that landscape is simultaneously loved or valued in its altered state. People *care* about such landscapes, and their arguments for opposing removal under these terms are as rational as the scientific claims for removal (Fox *et al.* 2016). As a matter of concern, as opposed to a matter of facts yet to be determined, the restoration of dammed rivers is more amenable to productive inputs from a wider variety of actors, including scientists, and ultimately more open to democratic processes. This perspective echoes that of John Dewey and the pragmatic tradition, where ontological and epistemological questions are secondary to the social and ecological issues and matters – specific to their own times and places – which should draw on natural and social sciences as crucial resources for the overall betterment of society (Marres 2007). While somewhat idealized given the nature of political-economic forces at multiple scales that are often implicated in environmental controversies, conflicts over dam removal and river restoration lend themselves to a conceptual framing that perceives the things in question – i.e. rivers, dams, landscapes – as rife with socio-ecological meaning contingent, à la

Massey, on multiple interpretations of time and space. Approached from this angle, the combined questions and methodologies of physical and human geographers are imperative to any workable, democratic process that leads to settlement of the conflict in question and, in theory, greater understanding of similar controversies in the future.

CONCLUSION

Doreen Massey implored physical geographers to reflect on their "physics envy" and engage with epistemological and ontological positions from across the social sciences, ones that did not fall back on "timeless", eternal laws of nature or conceptions of space as dead, static zones where physical and social processes play out. As a crucial part of this plea, she tasked geographers of all stripes to live up to their epistemological potential: "We must be spatial, as well as historical, sciences" (Massey 1999g: 273). As we have tried to show herein, discussions about how river restoration – especially by dam removal – occurs and its associated mechanics and goals provide an important vehicle for that engagement. With its assembling of and dependence on ecological and geomorphic knowledge, river restoration is predicated on some scientific foundation, yet at the same time the process is loaded with social and cultural elements that demand an integrated radical "intradisciplinary" perspective (cf. Domosh 2017) to simultaneously actualize and unpack. Perhaps what gives river restoration by dam removal a special saliency for geographers is that it is a symbolic and material space occurring at the landscape scale, and geographers – both physical and human – have found that space to be an effective intellectual comfort zone even while recognizing that "landscapes" are constructed historically and biophysically. We contend that river restoration, because it activates and embodies numerous social and physical processes that co-produce space over time, might prove a fertile intellectual nexus for physical and human geographers to jointly explore common questions surrounding their epistemological and ontological foundations.[1]

1. This research was funded in part by the US National Science Foundation (BCS-1263519) and from a Seed Grant from the Dartmouth College Rockefeller Center for Public Policy. We would like to thank the numerous individuals who agreed to be interviewed by us, and also all a big thanks to all of the undergraduate students who have helped on this project over the years, especially Anna Wearn, Chloe Gettinger, Soohyung Hur and Brendan Schuetze. Finally thanks to Doreen Massey for championing so many intellectual and social justice causes and for being an inspiration to many generations of geographers.

REFERENCES

For works authored and co-authored by Doreen Massey, please see Select Bibliography of Doreen Massey, beginning on p. 371.

Ashmore, P. 2015. "Towards a Sociogeomorphology of Rivers". *Geomorphology* 251 (December): 149–56.

Bash, J. S. & C. M. Ryan 2002. "Stream Restoration and Enhancement Projects: Is Anyone Monitoring?" *Environmental Management* 29 (6): 877–85.

Bernhardt, E., M. Palmer, J. D. Allan, *et al.* 2005. "Synthesizing U. S. River Restoration Efforts". *Science* 29 April: 636–7.

Bernhardt, E., E. Sudduth, M. Palmer, *et al.* 2007. "Restoring Rivers One Reach at a Time: Results from a Survey of US River Restoration Practitioners". *Restoration Ecology* 15 (3): 482–93.

Blue, B. & G. Brierley 2016. "'But What Do You Measure?' Prospects for a Constructive Critical Physical Geography". *Area* 48 (2): 190–7.

Clark, M. J. 2002. "Dealing with Uncertainty: Adaptive Approaches to Sustainable River Management". *Aquatic Conservation: Marine and Freshwater Ecosystems* 12 (4): 347–63.

Collier, M. 2015. "Novel Ecosystems and Social-Ecological Resilience". *Landscape Ecology* 30 (8): 1363–9.

Cook, J., N. Oreskes, P. Doran, *et al.* 2016. "Consensus on Consensus: A Synthesis of Consensus Estimates on Human-Caused Global Warming". *Environmental Research Letters* 11 (4): 048002.

Demeritt, D. 2009. "Geography and the Promise of Integrative Environmental Research". *Geoforum* 40 (2): 127–9.

Domosh, M. 2017. "Radical Intradisciplinarity: An Introduction". *Annals of the American Association of Geographers* 107 (1): 1–3.

Downs, P. & G. M. Kondolf 2002. "Post-Project Appraisals in Adaptive Management of River Channel Restoration". *Environmental Management* 29 (4): 477–96.

Downs, P. W., Y. Cui, J. K. Wooster, *et al.* 2009. "Managing Reservoir Sediment Release in Dam Removal Projects: An Approach Informed by Physical and Numerical Modelling of Non-Cohesive Sediment". *International Journal of River Basin Management* 7 (4): 433–52.

Doyle, M., J. M. Harbor & E. H. Stanley 2003. "Toward Policies and Decision-Making for Dam Removal". *Environmental Management* 31 (4): 453–65.

Doyle, M., J. Singh, R. Lave & M. Robertson 2015. "The Morphology of Streams Restored for Market and Nonmarket Purposes: Insights from a Mixed Natural-Social Science Approach". *Water Resources Research* 51 (7): 5603–22.

Foley, M. M., J. R. Bellmore, J. E. O'Connor, *et al.* 2017. "Dam Removal: Listening in". *Water Resources Research* 53 (7): 5229–46.

Fox, C. A., F. J. Magilligan & C. S. Sneddon 2016. "'You Kill the Dam, You Are Killing a Part of Me': Dam Removal and the Environmental Politics of River Restoration". *Geoforum* 70 (March): 93–104.

Gartner, J. D., F. J. Magilligan & C. E. Renshaw 2015. "Predicting the Type, Location and Magnitude of Geomorphic Responses to Dam Removal: Role of Hydrologic and Geomorphic Constraints". *Geomorphology* 251 (December): 20–30.

Gregory, K. 2006. "The Human Role in Changing River Channels". *Geomorphology* 79 (3–4): 172–91.

Harrison, S., D. Massey, K. Richards, *et al.* 2004. "Thinking Across the Divide: Perspectives on the Conversations between Physical and Human Geography". *Area* 36 (4): 435–42.

Hobbs, R. J., S. Arico, J. Aronson, *et al.* 2006. "Novel Ecosystems: Theoretical and Management Aspects of the New Ecological World Order". *Global Ecology and Biogeography* 15 (1): 1–7.

Hobbs, R. J., E. Higgs & J. A. Harris 2009. "Novel Ecosystems: Implications for Conservation and Restoration". *Trends in Ecology and Evolution* 24 (11): 599–605.

Hobbs, R. J., L. M. Hallett, P. R. Ehrlich & H. A. Mooney 2011. "Intervention Ecology: Applying Ecological Science in the Twenty-First Century". *BioScience* 61 (6): 442–50.

Hobbs, R. J., E. Higgs, C. M. Hall, *et al.* 2014. "Managing the Whole Landscape: Historical, Hybrid, and Novel Ecosystems". *Frontiers in Ecology and the Environment* 12 (10): 557–64.

Hooke, R. L. 2000. "On the History of Humans as Geomorphic Agents". *Geology* 28 (9): 843–6.

Kibler, K. M., D. D. Tullos & G. M. Kondolf 2011. "Learning from Dam Removal Monitoring: Challenges to Selecting Experimental Design and Establishing Significance of Outcomes". *River Research and Applications* 27 (8): 967–75.

Kondolf, G. M. 2006. "River Restoration and Meanders". *Ecology and Society* 11 (2): 42.

Kondolf, G. M., S. Anderson, R. Lave, *et al.* 2007. "Two Decades of River Restoration in California: What Can We Learn?" *Restoration Ecology* 15 (3): 516–23.

Lane, S. N. 2014. "Acting, Predicting and Intervening in a Socio-Hydrological World". *Hydrology and Earth System Sciences* 18 (3): 927–52.

Lane, S. N. 2016. "Slow Science, the Geographical Expedition, and Critical Physical Geography: Slow Science and Critical Physical Geography". *The Canadian Geographer/ Le Géographe Canadien* 61 (1): 84–101.

Latour, B. 2005. "From Realpolitik to Dingpolitik or How to Make Things Public". In B. Latour & P. Weibel (eds) *Making Things Public: Atmospheres of Democracy*. Cambridge, MA: MIT Press, 14–43.

Latour, B. 2007. "Turning around Politics: A Note on Gerard de Vries' Paper". *Social Studies of Science* 37 (5): 811–20.

Lave, R., M. Doyle & M. Robertson 2010. "Privatizing Stream Restoration in the US". *Social Studies of Science* 40 (5): 677–703.

Lave, R., M. W. Wilson, E. S. Barron, *et al.* 2014. "Intervention: Critical Physical Geography". *The Canadian Geographer / Le Géographe Canadien* 58 (1): 1–10.

Lejon, A. G. C., B. M. Renöfält & C. Nilsson 2009. "Conflicts Associated with Dam Removal in Sweden". *Ecology and Society* 14 (2): 4.

Magilligan, F. J., B. E. Graber, K. H. Nislow, *et al.* 2016. "River Restoration by Dam Removal: Enhancing Connectivity at Watershed Scales". *Elementa: Science of the Anthropocene* 4 (May): 108.

Magilligan, F. J., C. S. Sneddon & C. A. Fox 2017. "The Social, Historical, and Institutional Contingencies of Dam Removal". *Environmental Management* 59 (6): 982–94.

Mansfield, B., C. Biermann, K. McSweeney, *et al.* 2015. "Environmental Politics After Nature: Conflicting Socioecological Futures". *Annals of the Association of American Geographers* 105 (2): 284–93.

Marres, N. 2007. "The Issues Deserve More Credit: Pragmatist Contributions to the Study of Public Involvement in Controversy". *Social Studies of Science* 37 (5): 759–80.

Morse, N. B., P. A. Pellissier, E. N. Cianciola, *et al.* 2014. "Novel Ecosystems in the Anthropocene: A Revision of the Novel Ecosystem Concept for Pragmatic Applications". *Ecology and Society* 19 (2): 12.

Mould, S. A., K. Fryirs & R. Howitt 2017. "Practicing Sociogeomorphology: Relationships and Dialog in River Research and Management". *Society and Natural Resources* 31 (1): 106–20.

O'Connor, J. E., J. J. Duda & G. E. Grant 2015. "1000 Dams Down and Counting". *Science* 348 (6234): 496–7.

Oreskes, N. 2004. "Beyond the Ivory Tower: The Scientific Consensus on Climate Change". *Science* 306 (5702): 1686.

Palmer, M. A., E. S. Bernhardt, J. D. Allan, *et al.* 2005. "Standards for Ecologically Successful River Restoration: Ecological Success in River Restoration". *Journal of Applied Ecology* 42 (2): 208–17.

Phillips, J. D. 2004. "Laws, Contingencies, Irreversible Divergence, and Physical Geography". *The Professional Geographer* 56 (1): 37–43.

Phillips, J. D. 2007. "The Perfect Landscape". *Geomorphology* 84 (3): 159–69.

Pohl, M. 2002. "Bringing Down Our Dams: Trends in American Dam Removal Rationales". *Journal of the American Water Resources Association* 38 (6): 1511–19.

Robbins, P. 2014. "No Going Back: The Political Ethics of Ecological Novelty". In K. Okamoto & Y. Ishikawa (eds) *Traditional Wisdom and Modern Knowledge for the Earth's Future.* Tokyo: Springer, 103–18.

Sneddon, C. S., F. J. Magilligan & C. A. Fox 2017. "Science of the Dammed: Expertise and Knowledge Claims in Contested Dam Removals". *Water Alternatives* 10 (3): 677–96.

Steinberg, T. 1991. *Nature Incorporated: Industrialization and the Waters of New England.* Cambridge and New York: Cambridge University Press.

Tadaki, M., G. Brierley & C. Cullum 2014. "River Classification: Theory, Practice, Politics". *Wiley Interdisciplinary Reviews: Water* 1 (4): 349–67.

Truitt, A. M., E. F. Granek, M. J. Duveneck, *et al.* 2015. "What Is Novel about Novel Ecosystems: Managing Change in an Ever-Changing World". *Environmental Management* 55 (6): 1217–26.

Tullos, D. D., M. J. Collins, J. R. Bellmore, *et al.* 2016. "Synthesis of Common Management Concerns Associated with Dam Removal". *Journal of the American Water Resources Association* 52 (5): 1179–1206.

Urban, M. A. 2002. "Conceptualizing Anthropogenic Change in Fluvial Systems: Drainage Development on the Upper Embarras River, Illinois". *The Professional Geographer* 54 (2): 204–17.

Wilcock, D., G. Brierley & R. Howitt 2013. "Ethnogeomorphology". *Progress in Physical Geography* 37 (5): 573–600.

Winkler, J. A. 2016. "Embracing Complexity and Uncertainty". *Annals of the American Association of Geographers* 106 (6): 1418–33.

Wohl, E., S. N. Lane & A. C. Wilcox 2015. "The Science and Practice of River Restoration". *Water Resources Research* 51 (8): 5974–97.

CHAPTER 20

FILM AND THINKING SPACE

Geraldine Pratt with Jessica Jacobs

Perched on what appears to be an uncomfortable grey cube, Doreen Massey is shown on grainy CCTV footage being interviewed by curator Hans Ulrich Obrist (H.U.O.) and architect Rem Koolhaas, as part of the Serpentine Gallery's first 24-hour interview marathon in 2006 (Obrist & Koolhaas 2006). It is an art celebrity event and among those being interviewed are Damien Hirst, Doris Lessing, Brian Eno and Zaha Hadid. H.U.O. notes that Doreen's text, "London Inside-Out", was key to their preparations of "this whole marathon. It's a text that we always came back to." As the interview progresses, H.U.O. returns to her influence: "You are, as a professor of geography, in terms of influence, venturing into all kinds of other fields. Many of my artist friends are unusually inspired by your books." And then he asks, "I was wondering what are your inspirations. What are your … I mean heroes is kind of a big word." She ventures that she does not have heroes. Rather, there are lives that she finds fascinating:

> Tina Modotti, a photographer who mixed art, politics: a weird life
> of continual commitment and travelling everywhere. And died in
> a taxi on the way home from a party at a rather early age [age 45].
> As lives go, that seemed to me to be pretty impressive. She was in
> Spain for the civil war. She was in Mexico for the revolution. And
> so on and so forth. But I don't have heroes.

Asked more explicitly about her disciplinary influences, she turns instead to her geographical preoccupations, foremost that of reconceptualizing space. When Rem Koolhaas asks her "Is this space you are talking about the same space that architects work in?", she replies, "I've wondered about that for donkey's years because I have worked with architects a lot." She concedes that her preoccupation is more abstract. Perhaps

> the way it relates to architectural practice is thinking of space as the intersection of stories, as socially produced, as always moving. And one of the essential contradictions with architecture is that you are essentially enclosing it, pinning it down, carving off bits of it. And that's just a tension.

It is unclear at what hour in the 24-hour marathon Massey was interviewed – for his exertions, H.U.O. was admitted to hospital directly after (Max 2014).

Massey's appetite for thinking through her ideas about space in conversation appears to have been immense and places her – as a geographer – at the centre of key conversations across the arts. Her 2003 essay on Danish–Icelandic artist Olafur Eriasson's art installation, *The Weather Project* (commissioned for the Turbine Hall of the Tate Modern) is most readily accessed on the artist's website, where he acknowledges his debts to her thinking (see Massey 2003b). Massey's commentary on Eriasson's installation opens a conversation about materiality, art and space: "Origin and destination have lives of their own. This project of undoing that traditional counterposition of active subject and passive object is an element in Olafur Eliasson's practice. He challenges the static, given, implacable 'objecthood' of art." Even more extensive are her engagements with film and television – and it is to these that we turn. If architecture proved too enclosing and territorializing a practice for thinking concretely with her ideas around space, film appeared to offer far richer opportunities to think through the limits and with the possibilities of representation, politics and space. That Doreen Massey had most to say about film and television is not surprising given her close involvement in the production of film and video teaching modules as an instructor at the Open University (Weinbren 2015). She knew these media from the inside out. Using film and television to communicate about geography and space, her observations about film as a spatial medium have also been far reaching.

IN *SCREEN*

As part of an emerging subfield focused on cinema and the city, in 1999 the journal *Screen* published a special issue on the theme Space/Place/City. In lieu of an editorial introduction, Karen Lury thought it would be interesting to stage a conversation between herself and a prominent scholar in the field of geography. She "was very pleased" when Doreen Massey agreed to participate. And so a series of exchanges occurred, mediated by an electronic technology that had, Lury notes, "radically altered the experience of time and space in the modern world" (Lury & Massey 1999: 229): the fax machine.

Film theorist, Malini Guha, describes the effect of reading this fax-conversation as a graduate student; she does this by quoting Gayatri Spivak (who is describing the influence of Derrida on her own thinking): "It becomes internalized. You are changed in your thinking and that shows in your work" (Spivak 2016 in Guha 2016: n.p.). Massey made three precise interventions in this short exchange with Karen Lury. First, she insisted on particularity: "for reasons both of politics and better analysis". "I start feeling itchy when arguments about the relation city/cinema evoke such generalizations – apparently ahistorical and non-geographical – as 'the city' and 'the urban sensibility'" (Lury & Massey 1999: 230). Second, she unbounded "the city" and its relations to cinema by taking aim at the starring role given to the urban stroller, the *flâneur*, in much of the scholarship on modernity, film and the city. She drew attention to "the massive mobilities of imperialism and colonialism which were underway – beyond, way beyond, the little worlds of *flânerie* – at the same period of history" (Lury & Massey 1999: 231). Guha notes: "in bringing this critique to bear on the founding narrative of the cinematic city, she unveils the way in which the movements, displacements, and stories of dispossession that transpired on a global scale during the age of imperialism are often diminished" (Guha 2016: n.p.). That is, Massey unveiled the underlying eurocentricism of much cultural analysis of modernity. Massey's comment reoriented Guha's thinking about film and the city "towards its structuring absences" and redirected her attention to the transits and mobilities associated with imperialism, as integral to and not extraneous from an understanding of cinema and cities. And third, Doreen Massey laid out her central claims about spatiality: space as process and becoming, a bringing together of different journeys and trajectories, a sphere of possibility, and a medium of encountering difference, unexpected juxtapositions and coexisting multiplicities and temporalities.

Film, Massey thought, "is fantastic" (Lury & Massey, 1999: 232) for representing aspects of this spatiality. It can portray intense and unexpected juxtapositions and the simultaneity of multiple temporalities in and across space. It can make visible our relationships with those to whom we are connected in other parts of the world: "Precisely because of its mobility, its ability to travel, to make new juxtapositions, new cartographies [...], film has the potential powerfully to present this other aspect of our spatial world as well" (Lury & Massey 1999: 232). "Then comes film", wrote Walter Benjamin so long ago, "and burst this prison-world asunder by the dynamite of the tenth of a second, so that now, in the midst of its far-flung ruins and debris, we calmly and adventurously go traveling" (Benjamin 1968 [1935]: 236). Like Benjamin, Massey was enthusiastic about the political potential of film: to engage and reorder our geographical imaginations of the world.

And still she wondered about the limits of representation. In her longstanding quarrel with representations of space as static, Massey wondered whether it was possible to represent space otherwise. Guha notes that so many of those working on spatiality in film, televisual and other media studies have been drawn to Massey, and then she questions what it was that drew Massey to "the terrain of audio-visual culture as conducive to further exploration" of space (2016: n.p.). In the *Screen* fax-conversation with Lury, Massey gave an indication of what facet of film theory held her attention. She asked of film theory: "does representation necessarily 'freeze the flow of experience'?". This question, she noted, is typically asked of the written word and photography and "often gets tortured when it comes to film" (Lury & Massey 1999: 233). Lury attempted an answer, calling attention to the indeterminacy of audience reception and interpretation but, for Massey, this did not address the crux of her question: "This is the issue which I find really difficult" (Lury & Massey 1999: 234). Working with Patrick Keiller some years later seemed to offer possibilities for puzzling towards a more satisfying answer.

CONVERSATIONS OFF SCREEN

Asked how they came to work together on *The Future of Landscape and the Moving Image* project, the collaborators – Doreen Massey, filmmaker Patrick Keiller, historian Patrick Wright – describe a series of previous encounters, marking Massey's more abstract point that: "Happenstance juxtaposition … in the formation of the spatial new narratives are born" (Massey 1997b: 224). From Massey: "Our [hers and Patrick Keiller's] paths crossed on a number of occasions – at the Serpentine Gallery Marathon, for instance. It seemed right that we should have a more sustained conversation" (Stevens 2010a). From Keiller:

> By the end of 2005, I was beginning a second three-year research fellowship at the Royal College of Art, when the Arts and Humanities Research Council announced an interdisciplinary programme, *Landscape and Environment* […] I first met Doreen after asking if I could interview her for *The Dilapidated Dwelling* [a previous film], and we kept in touch. It turned out, also, that Doreen, Patrick and I had all contributed to *The Unknown City* [Borden *et al.* 2002]. For the current project I was interested in following up a conversation we [he and Massey] had had about "nature".   (Stevens 2010b).

This series of encounters amount to "a constellation of social relations held in a precarious and open specificity", what Massey calls spatiotemporal events (1997b: 224). First asked to join the project as advisors, the roles of Doreen Massey and Patrick Wright soon expanded and the project became more collaborative, with each committed to producing an item informed by their discussions: Massey an essay ("Landscape/Space/Politics"; see Massey 2011d); Keiller a film (*Robinson in Ruins*) and Wright a book (in process). In line with her theorization of space, Massey makes clear that they were not engaged so much in an interdisciplinary conversation but rather had "a collective focus coming out of different trajectories" (Stevens 2010a).

The collaboration was immensely productive. They brought together shared political preoccupations that began from a concern about how nostalgia for a lost, settled, English agricultural past (that never existed) works in relation to a pervasive and politically dangerous sense of displacement. As the conversations continued over time, the focus of the questioning moved. Patrick Keiller recalls that: "Doreen Massey suggested that, rather than debating this [question of displacement], or whether, if real, such common experience offers any basis for a commonality, perhaps the project could attempt to dispose of this idea of 'displacement'" (Keiller 2008: n.p.). The question of dwelling, belonging and mobility was rethought: "Rather than that dwelling-saturated question of our belonging to a place, we should be asking the question of to whom the place belongs. Who owns it?" (Massey 2011d: n.p.). This shifts the frame and opens political possibility: "Materially, in terms of power, the 'national' working class (of whatever ethnic origin) has no more ownership than does the recent migrant. There is common cause here" (Massey 2011d: n.p.).

Their method of working seems to epitomize what Massey had been writing about space for years: open to possibility and creative disruption, with the purpose of uncovering multiple stories in the landscape: "At one point in the project, Patrick K and I decided it would be good to spend more time together in parts of the landscape he had engaged with. And so we set off" (2011d: n.p.). Or: "one of my instinctive practices when thinking about a particular place is to get out the geology map. [For more on Massey and geology, see Sayer, this volume.] It was thus that I found myself tracing the intermittent outcrops in our area of a sometimes rubbly limestone called Cornbrash" (Massey 2012a: 93).

> What interests me now is why it seemed to matter so much. Partly it was the simple fun of the chase. Partly, maybe, there was an element of wonder at the serendipity or perhaps in more theoretical terms the psychogeographical coincidence – that our method had thrown up. (2011d: n.p.)

Patrick Keiller's method as a filmmaker is to film without a script and discover the story in the images he has created. The filming was one of the "research methods" that they used to find their narrative. On archival research, Patrick Wright has this to say:

> I use archives a lot, but I don't go to them as a source of evidence: sometimes, indeed, they seem to work best as surreal alienation devices – a way of thickening things up a bit and of reintroducing confusion and complexity to an oversimplified if not entirely depleted and amnesiac public culture.
>
> [W]e never conducted anything resembling a collective site visit while the film was being made. What we did do, in the end, is explore a number of common preoccupations. We had a lot of discussions, many of them prompted by viewing sequences of film. (Stevens 2010c)

The collaboration on *The Future of Landscape* project afforded an opportunity to think more deeply about the question that Massey had posed to film theorists a decade earlier: "Does filmic representation necessarily 'freeze the flow of experience'?" Her essay for the project is a wide-ranging reflection on the themes taken up in the film; it is also a deep analysis of the form of the film as a mode of representing her preoccupations with space. The formal structure of Keiller's cine-essay, *Robinison in Ruins,* is a journey: it takes the form of a narration (by Vanessa Redgrave) of an unplanned circumambulation that a fictional character, Robinson, takes on foot in 2008 through the southern England countryside, mostly in Oxfordshire and Berkshire. The film visits overlooked, historically resonant locations, including the Pelican Inn, where in 1795 the Speenhamland system was devised – a system of wage supplement aimed to mitigate the extremes of rural poverty that came alongside the "freeing up" from feudal social bonds and the development of a market system in labour and land. Karl Polanyi, writing in response to the Great Depression, locates the Speenhamland system at the origins of "the twentieth-century catastrophe", being played out once again as the financial crisis in 2008 when the film was being created. Robinson reads and quotes from Polanyi (2001 [1944]) on his travels. The film is a journey but not a linear one; it is a process of discovering stories in the landscape, a jumble of stories from different times. Keiller has said that *Robinson in Ruins* is an attempt to create a film that disrupts the "problem" of linearity and an effort to incorporate some of the "spatial possibilities" developed in his preceding work (which was a 30-screen installation, a moving-image reconstruction of Chhatrapati Shivaji Terminus in Mumbai staged in an exhibition space at Le Fresnoy near Lille France). *Robinson in Ruins* attempts in a single film "an awareness of what is continually traversing the story-line laterally" (Pattison & Keiller 2011), that is, it attempts to work against the temporal linearity and fixed spatiality of film.

Keiller then was an exceptionally congenial companion for Massey's philosophical ruminations on space, and she reflects deeply on how the form of *Robinson in Ruins* represents space (as she has theorized) without freezing the flow of experience. She makes a number of striking observations in approaching an answer to the question that she had posed to Karen Lury. "There is about [*Robinson in Ruins*]", she notes, "a stillness" (Massey 2011d: n.p.). Noting the recurrent use of long takes – the camera stays on butterflies working the teasel, for instance, for 4 minutes and 15 seconds – she notes: "These long takes give us, in the midst of the rush and flow of globalization, a certain stillness. But they are not stills. They are about duration. They tell us of 'becoming', in place" (Massey 2012a: 90). Of the butterflies with the teasel, "the moments spent with the teasel tell of what it takes to survive, just to go on, from season to season" (2012a: 90).

She notices that the film is a series of takes at particular locations: "The camera comes to rest, and films" (Massey 2011d: n.p.). The movement or journey between locations is not filmed. Travelling across a landscape – indeed the very concept of landscape, Massey had earlier noted (2006b), induces the effect of smoothing; its trick of visual continuity "implies a kind of present reconciliation" (2011d: n.p.). As a journey, *Robinson in Ruins* might have performed this trick: "But the form of the film itself tells us that this is in no sense a simple surface. The camera does not film while moving. It films when it stops, and at each point when it does, we dwell on a story" (Massey 2011d: n.p.). The stories do not settle into each other: "the stories shoot out of the soil, speaking to today" (Massey 2012a: 91). There are few wide-angle orienting shots and the camera work disallows closure or reconciliation. As film critic Michael Pattison observes: "There's a gorgeous shot of a combine harvester [...] and just as the harvester is about to leave the frame for its final time, Keiller cuts so as to deny us visual and therefore narrative closure: this is serious cinema at its most teasing" (Pattison & Keiller 2011). This lack of closure also matters politically: "those loose ends give space its openness to the future" (Massey 2011d: n.p.).

Massey argues that *Robinson in Ruins* also opens the viewer to the agency of the non-human. In an interview in *New Statesman* Keiller claims of his long takes:

> It's not so much whether one wants to make a long take, it's "can you bear to stop?" But it also had something to do with the way that the subjects moved. For instance the [shot of a] foxglove,

which goes on for a very long time, seemed to be ... I mean it's obviously completely oblivious to the camera, but there seemed to be a performance going on here [...] so I couldn't stop.

(Trilling 2010)

Of the long take of the foxglove, Massey writes of being forced to look *without ownership*: "Quite the opposite. They [the images] force me to submit, to the foxglove's implacable otherness, to its vulnerability to human actions, yes, but also its utter indifference to us – to the camera, to the person behind the camera, to us watching the film" (2012a: 93).

ON SCREEN IN *SIGNIFICANT GEOGRAPHIES*

Another conversation with Doreen Massey – this one video-recorded by Jessica Jacobs on her phone in Massey's office at Open University in February 2016, where Massey sits, leaning casually on an adjacent desk, surrounded by boxes of files, tellingly labelled "alt docs" (see Jacobs 2016) (Photo 20.1). This is a short (7 minute) test video that was intended to raise funds for a larger project that she and Jacobs were working on together: a documentary film that Massey wanted to call *Significant Geographies: Reflections from a Paper Age*. Echoing her response to H.U.O.'s earlier question about her heroes, "Doreen was clear from the outset [of conceptualizing the film project] that she didn't want it to be a telling of her life but an argument and demonstration of how geography has influenced the political world at some crucial junctures" (Massey & Jacobs 2016).

The boxes of files, Massey tells us, are a

non random collection of things that I couldn't bear to throw away. When I pull them out and look at them I remember not just the thing itself – the materiality of them (it's lovely), but I remember a bunch of people and sets of activities. And I think putting them all together, one of the things it really gives me a feel for is a kind of ecology of the left that used to exist, and maybe still does in a different form [...] And so one reason that they're all here, I couldn't bear [...] It's like saying goodbye to old friends if I got rid of them. So I kept them.

The film was intended to take shape around a series of interviews that Massey would have as she revisited people with whom she was involved in various activities in the 1970s through 1990s: the birth of neoliberalism (the Greater London Council and the miners' strike of 1984–85); in debates about unpaid labour with the ANC in South Africa in 1993 during the transition from apartheid; and new spatialities of democracy in Latin America. These interviews were to be augmented with visual storytelling inspired by the tangible documents that surrounded her in her office. Through these specific entry points, the film would document the "ecology of the left" in the 1970s through to the 1990s and, crucially, the academy's relationship to it. In the short test video, Massey explains:

> It reminds me of what may have been a different kind of relationship between the academy and political activity. There was a totally porous boundary in lots of ways. And a lot of us – obviously not everybody, but a lot of us were getting our most penetrating theoretical insights through something we were doing outside: on the streets, in feminism, in some organization around unemployment say. And you'd bring that back in. So your questions came out of that stuff. It was almost like the outside world had more impact on your intellectual work than the common notion of impact now, where you are supposed to sit in a library and dream something up and take it out.

The title of the planned video references the folders of paper in Massey's office, but also pre-digital political activity – the spatiotemporality and sheer physicality of typing, photocopying, and sending materials by post – and the academy's relationship to it. It references how activism using paper and all the associated machinery was so labour intensive and how this labour in turn shaped the activism for which it was produced. Jacobs recalls Massey's relationship to paper and pen:

> everything she did was handwritten [...] she had this amazing love affair with paper and the written word. [...] She understood the digital but her way of thinking chimed with the way she wrote and spoke. There was a wonderful spatiotemporal relationship between her thought processes, her intellectualism, and the way that she physically transformed that into the written word. It synced with her. It helped her clarity.[1]

Ironically perhaps, Massey and Jacobs wanted to use film to connect paper activism with film, to start a conversation with younger generations of "digital

---

1. Conversation with Jessica Jacobs, 11 July 2017. Unless otherwise indicated, all further quotes are Jessica Jacobs' words from this conversation.

natives" by documenting – through the use of digital tools – an earlier analogue relationship between politics and the academy.

> There was this idea that paper thinking was less caught up with neoliberal culture and it would be useful for students today – who have never known anything different – to reflect on that.
>
> Doreen loved the visual and was a great fan of cinema, so she immediately understood how filming "Paper Geographies" could bring the textual quality of her archive material to life for a generation raised on digital media. Film would allow us to show the immediate contrast between this paper world and the one that followed; it would allow the project to explore how the way we collect information affects the way we understand the world.
>
> (Jacobs 2016)

But "it was not an archival project in a traditional sense. Doreen was not nostalgic, by any means. It's about finding something new" through the project: about reconnecting politics and the academy in new/old ways and understanding how some of the work of and knowledge produced in previous campaigns might be reconfigured and reused in political struggles in the present and future.

Jacobs and Massey were using digital technology as their research methodology, as a process of doing research, and Jacobs anticipates that much of their thinking about and analysis of the relationship between paper and filmic geographies would have come through this process. If their collaborative project is no longer possible, their aspirations live on for digital technology as a research methodology and as a lively mode of representing space. Jacobs, a former PhD student supervised by Massey, has been instrumental in pushing geography as a discipline to further explore this methodology, for example, by co-organizing the newly established Film specialty group, as well as AAG Shorts, a short film competition, at the annual meetings of the American Association of Geographers. Of the over 40 submissions to AAG Shorts in 2017 she says:

> I felt there was a real sense of what a geographical film is. So many people working independently not knowing each other coming up with the same issues and themes: about marginal communities, about places under threat, this idea of being able to bring those places visually into the center by using film, showing everyone what's going on because you don't know it or can't go there. [...] And it wasn't anthropological [...] There was this sense that there is a geography of film that emphasises the relationship of people

Photo 20.1 Massey in her office with her papers. Still from *Significant Geographies: Reflections from a Paper Age*

to place, where landscape is given agency, and becomes more than a passive background for human interaction."

Film is, as Massey noted in the *Screen* conversation with Karen Lury, "fantastic" for its ability to travel and create new imaginative mappings of the world. As a discipline, geography seems poised to take up and take further Massey's longstanding curiosity about the potential of film to represent space in more dynamic, theoretically sophisticated and politically engaging ways.

And film of course is itself an extraordinary archive, with an ethereal materiality of its own. As Jessica Jacob notes, "Though there is quite a bit of her online, I wish there was more. That's the other sad thing. When you are gone, film becomes important again. Because you can read and read her. But to listen to and watch her is something else."

REFERENCES

For works authored and co-authored by Doreen Massey, please see Select Bibliography of Doreen Massey, beginning on p. 371.

Benjamin, W. 1968 [1935]. "The Work of Art in the Age of Mechanical Reproduction". In H. Arendt (ed.) *Illuminations*. New York: Schocken Books, 217–51.

Borden, I., J. Kerr, J. Rendell, with A. Pivaro 2002. *The Unknown City: Contesting Architecture and Social Space*. Cambridge, MA: MIT Press.

Guha, M. 2016. "The Practice of the Interdisciplinary: A Tribute to Doreen Massey" *Mediapolis* 31 August. www.mediapolisjournal.com/2016/08/practice-interdisciplinary-tribute-doreen-massey/ (accessed 3 July 2017).

Jacobs, J. 2016. "Doreen Massey: Reflections from a Paper Age". *Coordinates* 28 July. www.coordinatessociety.org/single-post/2016/07/28/Doreen-Massey-Reflections-from-a-Paper-Age (accessed 9 January 2018).

Jacobs, J. 2017. Personal Communication. 11 July.

Keiller, P. 2008. Blog Entry. 18 June. https://thefutureoflandscape.wordpress.com (accessed 17 January 2018).

Lury, K. & D. Massey 1999. "Making Connections". *Screen* 40 (3): 229–38.

Max, D. T. 2014. "The Art of Conversation: The Curator Who Talked his Way to the Top". *The New Yorker* 8 December. www.newyorker.com/magazine/2014/12/08/art-conversation (accessed 3 July 2017).

Obrist, H. U. & R. Koolhaas 2006. "Interview Marathon 2006: Doreen Massey". www.youtube.com/watch?v=BVl6AgmAW4k&t=254s (accessed 17 January 2018).

Pattison, M. & P. Keiller 2011. "Q&A: Five questions for Patrick Keiller". *idFilm* 12 January. www.idfilm.net/2011/01/q-five-questions-for-patrick-keiller.html (accessed 3 July 2017).

Polanyi, K. 2001 [1944]. *The Great Transformation: The Political and Economic Origins of Our Time*. Boston, MA: Beacon Press.

Spivak, G. with S. Paulson 2016. "Critical Intimacy: An Interview with Gayatri Chakravorty Spivak". *Los Angeles Review of Books* 29 July. https://lareviewofbooks.org/article/critical-intimacy-interview-gayatri-chakravorty-spivak/ (last assessed July 3, 2017).

Stevens, A. 2010a. "The Future of Landscape: Doreen Massey". *3:AM Magazine* 29 September. www.3ammagazine.com/3am/the-future-of-landscape-doreen-massey/ (accessed 3 July 2017).

Stevens, A. 2010b. "The Future of Landscape: Patrick Keiller". *3:AM Magazine* 14 July. www.3ammagazine.com/3am/the-future-of-landscape-patrick-wright/ (accessed 3 July 2017).

Stevens, A. 2010c. "The Future of Landscape: Patrick Wright". *3:AM Magazine* 9 September. www.3ammagazine.com/3am/the-future-of-landscape-patrick-keiller/ (accessed 3 July 2017).

Trilling, D. 2010. "Politics and the English Countryside. The Film Interview: Patrick Keiller on 'Robinson in Ruins'". *New Statesman* 24 November. www.newstatesman.com/blogs/cultural-capital/2010/11/robinson-film-landscape-idea (accessed 3 July 2017).

Weinbren, D. 2015. *The Open University: A History*. Manchester: University of Manchester Press.

CHAPTER 21

GEOGRAPHICAL IMAGINATIONS OF PENSION DIVESTMENT CAMPAIGNS

Kendra Strauss

INTRODUCTION

Doreen Massey's early and ongoing dedication to understanding the inherent interconnection of the "social" and the "economic" in the production of space, place and scale made her an economic geographer ahead of her time. Her interests and commitments were heterodox and wide-ranging; she refused to be confined by subdisciplinary conventions that still channel economic, labour, urban and feminist geographers along particular topical, conceptual and theoretical lines, and she was deeply engaged with the ongoing work of probing foundational assumptions (what Castree (2004), at once critical and celebratory, called critical human geography's "shibboleths"). Her formative work on spatial divisions of labour; her insistence on the relational nature of place, space and identity; her concept of "power-geometries"; and her elaboration of the notion of geographical imaginations ranged across thematic concerns. These concerns included economic ("industrial") restructuring, urbanization and the world city, the concept of space in social theory and political philosophy, and neoliberalism and conjunctural politics. Her work is also grounded, unafraid to speak to regional- and national-level processes, globalizing trends and "the urban" from a *localist* standpoint, whether in London, Manchester, a Cambridge science park, or on a train to Keswick in England's Lake District.

Given this breadth, it is perhaps unsurprising that economic and labour geographers (myself included!) – most notably in the last decade or so, after building on the insights in *Spatial Divisions of Labour* (1984b) – often acknowledge, but seldom fully engage, Massey's work. In this chapter I want to explore how Massey's articulation of geographical imaginations relates to her concept of power-geometries, and the utility of these ideas for exploring the pension divestment movement (that is, the movement to pressure pension funds to divest from the fossil fuel industry because of the need to combat climate change). I examine a subset of arguments for and against divestment grounded in legal debates about fiduciary duty, and explore how Massey's approach can be used to highlight the dynamics of institutional power that these debates reveal – and the political strategies for interrogating and reshaping those dynamics. As legal geographers have amply demonstrated, law's power relies on spatial logics, both overt and implicit (Blomley 1994; Delaney 2015), as well as on the social construction of legal space (Strauss 2016). Legal arguments for the compatibility of fiduciary duty with divestment hinge, I argue, on geographical imaginations that envisage political-legal responsibility (of the investor subject) in relation to *both* a politics of place *and* the "smooth" spaces of portfolio management, diversification and global capital flows.[1] These imaginations evoke power-geometries that pension fund trustees, managers and plan members are both positioned within, and have the capacity to (re)shape. In this sense they can be seen to enable a politics of place beyond place (Massey 2007). I conclude by arguing, however, that the concept of power-geometries needs to retain and foreground Massey's emphasis on relationality, because it can otherwise have the effect of eliding the very multiplicity of relations in which pension funds are enmeshed.

The rest of the chapter is structured in the following way. I first offer a brief overview of the relationship in Massey's work between geographical imaginations and power-geometries. I then turn to a description of the mainstream divestment movement in North America, especially as it relates to the organization 350.org started by the environmentalist Bill McKibben. I examine the nascent legal literature on divestment and its arguments about fiduciary duty, and the geographical imaginations that imbue the normative positions taken in this work; arguments about responsibility, morality and ethics that evoke (or perhaps interpellate) a particular kind of investor subject. I conclude with thoughts on the promise – and some limitations – of a more in-depth engagement with Massey's work in economic geographies of institutional investment and the politics of socially responsible investing (SRI).

GEOGRAPHICAL IMAGINATIONS AND POWER-GEOMETRIES

Massey was not the first geographer to explore the idea of geographical imaginations (see, among others, Gregory 1998; Pred 2000), but in her

---

1. As Langley (2017: 70) argues in Christophers *et al.*, *Money and Finance after the Crisis*, there are a variety of metaphors and similes used to describe the flows of financial capital in a globalized world of networks, including Clark's (2005) discussion of their mercurial characteristics. In financial discourse and the financial media, these flows and the markets they sustain are naturalized as frictionless, or smooth.

broader body of work it took on special salience in the context of her insistence on the relational nature of space, and on the idea that particular hegemonic understandings of space and place underpin particular hegemonic politics. She identified, for example, "general geographical imaginations and implicit conceptualisations of space ... that are utterly essential, even if most often implicit, to our more general 'political cosmologies'" (Massey 2007: 20). To illustrate, she contrasted a spatial imaginary of autonomous, preformed, already territorialized regions, associated with modernity and the era of the nation-state, with one that saw that "the character of a region, or the economy of a place, is a product not only of internal interactions but also of relations with elsewhere" (Massey 2007: 20). The contrast led Massey into a discussion of globalization that refused both the understanding of the "local" as the eternal and inevitable victim of global forces, and also the simple inversion of this dynamic that posits the local as a privileged site of autonomy, resistance and authenticity (see also Nagar *et al.* 2002).

It is not only space but also the nature of relational connections that Massey sought to unpack in her analysis, which in *World City* was part of her extended critical engagement with London. "[I]n some imaginations of the geography of the world", she wrote, "'space' is a surface, across which investments/migrants/connections flow and forces march" (2007: 22). London is discursively constructed in this way, Massey pointed out, with material consequences for the political economy of the UK and beyond. While paying lip service to connectivity, multiplicity and difference, such imaginations underpin political cosmologies that deny coevality and multiple trajectories by eliding other understandings and other actors and narrowing the range of political questions that can be asked. Massey argued that these elisions are naturalized: "Geographical imaginations are most often implicit. When deftly mobilised they pass us by as self-evident, barely recognised as being the framing assumptions that they are. And in part precisely as a result, they are powerful elements in the armory of legitimation of political strategies" (23–4). What is legitimized is often a sociospatial imaginary that reinforces inequalities within and between places and ascribes positive agency to certain individuals, groups and sectors while denigrating or discounting the contributions and capacities of others (as Massey showed in her discussions of finance versus manufacturing, and London versus "the North").

Geographical imaginations, then, are not only ways that we (individually and collectively) orient ourselves in space and make sense of spatial relationships through processes of mental ordering and projection. They are also the narratives of places, their histories and meanings, that are part of processes of identity formation and the exercise of social power. There can be, in fact always are, competing geographical imaginations of places and of the spatial processes involved in place-making. Conversely, seemingly neutral or objective entities – institutions, markets, infrastructures – are always underpinned by geographical imaginations that are integral to their legitimation.

In Massey's work, hegemonic discourses of globalization that privilege and give agency to the "global" over the "local", that celebrate London and its financial sector as the UK's "golden goose" while minimizing, erasing or pathologizing other regions, places and domains of socioeconomic life, are examples of dominant – though contested – geographical imaginations. Other geographers have used the concept in similar ways to explore narratives of economic development. Mark Graham, Casper Andersen and Laura Mann, for example, compare historical and contemporary technologies of connectivity in East Africa to illustrate how "links between non-proximate places are created and imagined" in and through popular imaginations of spatio-temporal connectivity. "Not only", they write, "is the organisation of connectivity and spatial relations crucial to the enaction of specific modes of colonial or contemporary capitalist production (Lefebvre 1991 [1974]), but so too are the ways that non-proximate connectivity is imagined and thus ultimately brought into being" (2015: 335).

It thus follows that Massey's insistence on space as a "discrete multiplicity" (Massey 2005) can be linked to a set of contingencies involving the spatial and political imaginaries from which forms of relationality (of which connectivity is one) emerge. Cumbers and Routledge (2004: 819) argue that competing geographical imaginations, a "plurality of alternatives" to the current political-economic order, are most likely to emerge – in ways that reflect different contexts and contingencies – in "localized spaces" where the contradictions between the competing demands of capital accumulation and social reproduction are manifest. Thus political-geographical imaginations (Harvey 2000; Strauss 2015) both shape, and are the product of, grounded or spatially embedded practices associated with "the local", *and* of the kinds of connections and relations that we associate with spaces of globalization. But there is no teleology to this process. Certain kinds of localities, or articulations of "localness", do not automatically produce progressive political imaginations or forms of resistance. Nor do Massey's concepts of relationality and multiplicity deny power relations and assume a flat, networked horizontality among places or people.

Graham *et al.* thus explicitly link geographical imaginations with power-geometries, drawing on Massey's (1993b) discussion of the concept in relation to her critique of understandings of space-time compression that do not account for social differentiation in discourses of mobility and technological connectivity. Power-geometries are given brief treatment in Massey's early piece, given that concept later "travels" in ways its originator could hardly have anticipated at the time (Massey 2008). "[T]he *power-geometry* of it all" (emphasis in the original) refers to the way "different social groups and different individuals are placed in very distinct ways in relation to ... flows and interconnections" (Massey 1993b: 62). It is, again, helpful in thinking through not only personal or group mobilities, but also the technological mobilities discussed above and the flows of commodities, capital, labour, waste and non-human life through which what we think of as globalization is enacted.

Massey, elaborating the concept of power-geometries further in relation to "ways of 'imagining globalization'" (Massey 1999b: 27), was explicit that her understanding encompasses two distinct dimensions: "And this is meant both in the sense of the power relations in the social spheres we are examining and in the sense of power-relations embedded in the power-knowledge system which our conceptualizations are constructing." In other words, they are produced through processes that are at once material and representational (Barney 2009). Thus the concept of power-geometries is broadened out beyond differentially positioned individuals and groups (power relations in the social sphere) to include the shifting, contingent configurations of spatial relations themselves and the representations that undergird or seek to undermine them (power relations in the power-knowledge system).

Put another way, power-geometries are the structured but dynamic uneven sociospatial relations that both shape, and emerge from, geographical imaginations. They are, in this sense, the reason London is fêted as the UK's economic engine, and the forms of power that flow from that discourse to further reinforce it (in government decision-making, for example, or the locational decisions of firms). What, then, are the geographical imaginations at play in the fossil fuel divestment movement, and how do they interact with the power-geometries of institutional investors and pension funds?

THE RELATIONAL GEOGRAPHY OF FOSSIL FUEL DIVESTMENT

Pension funds and other institutional investment vehicles, such as university endowments, have long been targets for political campaigns seeking to enact social change. The divestment of such vehicles, during the apartheid era, from the stocks of South African corporations or those corporations with significant business in South Africa is the exemplar for many contemporary campaigns (for discussions see Massie 1997; cf. Gosiger 1986). There are interesting and significant differences, however. Unlike campaigns targeting South Africa, and more recently Israel and Sudan, fossil fuel divestment campaigns do not focus on a specific political regime (although they are critical of governments that do not enact policies to combat global warming). Instead they focus on harmful products, more in keeping with socially responsible investment (SRI) principles that shun tobacco, firearms or pornography.

But fossil fuels are not (or not only) consumed or used by individuals and social groups in ways that cause harm to themselves and proximate others. They are essential to the very fabric of contemporary economies, woven into transport systems, consumer products, packaging, everyday technologies and systems of agriculture. In that sense, what is being advocated and sought is (at least implicitly) not just a change in socially damaging patterns of production and consumption, but a fundamental transformation of the socio-economic system of fossil capital (Huber 2013; Malm 2016). The goal of this transformation is to halt, or at least slow and restrain, anthropogenic climate change caused by greenhouse gas emissions.

Given the scope and nature of this goal, divestment is generally portrayed and pursued as one part of a multipronged strategy to address climate change. This chapter focuses on 350.org, a "global climate movement" started by students and the American environmentalist Bill McKibben in 2008 (350. org n.d.). Many divestment campaigns are local and grassroots in nature, and focus on a range of issues that encompass explicit "environmental" goals, but may also articulate broader issues of uneven development and dispossession, ongoing colonialism and indigenous rights, gendered dimensions of socio-ecological change, and corporate attempts to manipulate and muzzle climate science research.[2] In the context of this diversity, 350.org can be seen as a significant mainstream face of the movement. It is therefore by no means the only organization advocating and facilitating collective mobilizations against climate change, but significantly for this chapter, it is the movement on which the emerging legal literature on divestment focuses.

McKibben's (2012) article in *Rolling Stone* magazine, "Global Warming's Terrifying New Math", was significant in advocating for divestment as a key strategy. McKibben forcefully argued that individual and collective forms of behavioural change have failed, including legislative-political change at the national and international levels, and that the fossil fuel industry should be the target of climate politics. He linked this argument explicitly to the divestment campaign against the government of South Africa under apartheid.

> Once, in recent corporate history, anger forced an industry to make basic changes. That was the campaign in the 1980s demanding divestment from companies doing business in South Africa … "Given the severity of the climate crisis, a comparable demand that our institutions dump stock from companies that are destroying the planet would not only be appropriate but effective", says Bob

---

2. Recent commentaries have highlighted tensions between divestment campaigns and political direct action, including by indigenous activists, although some student-led movements embrace both. See, for example, "Tarsands Divestment and its Discontents" (McSorley 2014).

Massie, a former anti-apartheid activist who helped found the Investor Network on Climate Risk. The message is simple: We have had enough. We must sever the ties with those who profit from climate change – now.   (McKibben 2012)

McKibben went on to highlight university endowments and public pension funds as key targets. These institutions are future-orientated because of their investment goals, and they have clearly defined constituencies – students, workers, retirees. As members of these constituencies, individuals potentially acquire concrete political subjectivities shaped by direct links with a threatened future. In other words, more than the amorphous mass of consumers and citizens confounded by the seeming insignificance of their actions, students and pension plan members have a route to political action through their relationship with distinct capitals that act in the world through the medium of financial flows.

And those flows are far from insignificant. As the OECD (2016: 22) has recently highlighted, pension systems are increasingly backed by accumulated assets (as opposed to being funded on a pay-as-you-go basis by current contributions) (see also Clark 2000): in 2015, assets invested through pension vehicles in 35 OECD countries reached a cumulative USD 39.6 trillion (OECD 2017).[3] Most of those assets are held in investment vehicles of various types. Direct investments in corporate stocks, mutual fund holdings and investments in "alternative" asset classes (real estate, hedge funds, private equity) all potentially include companies involved in the fossil fuel industry. While individual holdings might seem relatively small, institutional investment in the energy sector is extremely significant. When the Deepwater Horizon oil spill caused BP's share price to nosedive, the British press fretted about its effect on the value of assets in UK pension funds, most of which held (directly or indirectly) BP stocks (PA 2010).

Pensions have long been both targets, and instigators, of SRI campaigns (Hebb 2008; Quarter *et al.* 2008). For just as long, the question of the desirability and efficacy of pensions as vehicles for political activism has been the subject of sometimes acrimonious debate. More recent accounts of the apartheid boycott sometimes gloss over the degree of disagreement about the proper role of university endowments and pension funds – clearly on

---

3. Compare with US university endowments – Harvard's endowment was over US$36 billion in 2015. The combined value of the five largest endowments (Harvard, Yale, the University of Texas system, Princeton and Stanford) in 2015 was over US$131 billion (NACUBO 2016), about equivalent to a medium-sized pension public system (the Florida Retirement System Pension Plan had US$141.9 billion assets in 2014) (OECD 2015: 8). It is clear why university endowments are such a significant target of the divestment movement.

display in the pages of *Minerva* in the mid-1980s (Bok *et al.* 1986). While similar debates rage about the legitimacy (is social change a legitimate goal of pension fund governance?) and efficacy (do the investment choices of pension funds have a real impact on share price or corporate decision-making?) of such campaigns, an important terrain of debate is whether factors other than the maximization of investment returns can be legally integrated into pension fund decision-making and governance. This pivots on the question of fiduciary duty, which is one of the key subjects of the emerging academic literatures on divestment, and where legal scholars look again to the experiences of the apartheid divestment movement (see, for example, Troyer *et al.* 1985).

FIDUCIARY DUTY AND THE PRUDENT INVESTOR

Although the academic literature on divestment is relatively small, the debates so far largely mirror debates about the apartheid investment campaign, in the sense that they tend to focus on two main types of claims: whether the strategy is likely to be effective and is therefore justified despite potential negative impacts on returns (see, for example, Ritchie and Dowlatabadi (2015)'s pessimistic assessment of the divestment scenario for the University of British Columbia's endowment), and whether it is permissible under the rules that govern the management of trusts (pension funds and endowments) (see Richardson (2011) on fiduciary responsibilities in relation to SRI goals). Interestingly, two recent additions to the legal literature on divestment address both questions but focus on definitions of fiduciary duty as an overarching issue that potentially speaks to both set of claims. In other words, they posit a redefined fiduciary role as *accommodating* – or even impelling – a duty to act on the basis of moral or ethical norms. Moreover, they articulate a legal-geographical imagination in which lawyers and judges create the legal space for divestment to flourish, based on a relational sense of responsibility to distant others (distant in space *and* time). In seeking to recast power-geometries of legal and corporate responsibility, however, other sociospatial power dynamics remain unacknowledged.

What is fiduciary duty? A fiduciary is a person holding the character of trustee, and thus required to act primarily for another's benefit in a relation involving the management of property or affairs (Richardson 2011). Trustees of pension funds are governed by fiduciary duties, which in the United States are set by trust and corporate law and guided by principles of loyalty, prudence and impartiality (Sarang 2015; Schneider 2015). These qualities are related, implicitly and explicitly, to norms of utility maximization and modern port-folio theory. The debates about the interpretation of fiduciary duty thus hinge on whether "there are other instances when non-financial factors can and should play a role in institutional investment decisions" (Sarang 2015: 311).

As geographers have noted, economic models of behaviour abstract decision making from its sociospatial context and deny the contingent and relational nature of market-making processes (Bathelt & Gluckler 2003; Ettlinger 2003; Strauss 2009). The legal arguments in favour of divestment, such as those of Schneider and Sarang, do not challenge this elision of place, context or spatial relations; the idea that diversification can achieve a perfect calibration of risk and return, for example, through the channelling of capital flows into optimal investment vehicles involving optimal sectoral and geographical allocations goes unquestioned. Yet, in both analyses, herd mentality among money managers (Schneider 2015: 612) and the practice of looking to peers to guide the "prudent man" [*sic*] standard are acknowledged. The redefinition of fiduciary duty must thus involve the *re-inscription* of rationality, in which a global environmental consciousness and the costs of climate change enter the calculus of prudent, loyal and impartial decision-making. This rationality is buttressed by the concept of stranded assets (in which fossil fuel reserves represent a future financial risk), and by the conclusion that because fossil-fuel free portfolios perform similarly to other portfolios, fiduciary duties are not breached (Schneider 2015). Thus Sarang (2015: 340) argues that "Fiduciary duties are entirely compatible with making the decision to divest, especially given the risk involved in holding fossil fuel stocks and their stranded assets."

The geographical imaginations at play here evoke a decision-landscape in which trustees, responsible to their "local" constituents (students, plan members, sponsoring institutions), are nevertheless bound to recognize the relational time-spaces of investing and their impacts on distant others and on "the planet". These impacts are described in three ways. First is the global consequences of climate change, "the wide-ranging physical, social and economic chaos climate change will cause" (Sarang 2015: 309) that is also uneven in its impacts: "people who are already socially, economically, politically, culturally and institutionally, or otherwise marginalized are at a heightened risk for experiencing the effects of climate change" (Schneider 2015: 601). Second, however, are localized impacts *directly* affecting trusts' constituencies. Schneider (602) notes, in documenting the growing divestment movement among public pension plans, the actions of Providence, Rhode Island, "a coastal city uniquely situated to feel the effects of climate change", which "requested" divestment in solidarity with the movement. In addition, in the analysis of a suit brought against the university by Harvard students seeking divestment of its endowment, Sarang notes that the plaintiffs alleged the university was in violation of its charter for investing in fossil fuel companies that contribute to sea level rise and increased storm activity, which will harm the Harvard campus and impact the education and welfare of students. Third, the same lawsuit reflects a claim on behalf of future generations, widely made in the divestment literature, that the duty to combat the activities of fossil fuel companies is a duty to both future selves (current students in adulthood, current workers in retirement) and to a global "constituency" not yet born. This claim challenges the notion of the prudent investor, pointing out that a standard stricter than the "morals of the marketplace" (Judge Cardozo, quoted in Schneider 2015: 304) must take account of the future implications of current investment decisions.

These claims mobilize a broadened sense of fiduciary responsibility grounded on spatiotemporal relations and forms of connectivity that deny the methodological individualism of economistic definitions of maximization and optimization. Massey's concepts thus offer an expanded vocabulary – potentially enriching, rather than replacing, critical approaches to decision-making, context, behaviour or affect – for seeking to theorize the political dimensions of "economic" decision-making and their sociospatial implications. As Massey wrote in *For Space,* her analysis

> build[s] upon the vision of space as constituted through the practices of engagement and the power-geometries of relations, of the structuring of space (both through enclosure and through flow) *through* such relations, and through an understanding of those relations as differentially (and unequally) empowering in their effects
> (2005: 99)

This vision of space offers a starting point for identifying and exploring the specific power-geometries both invoked and challenged by legal analyses of divestment, for highlighting the geographical imaginations that shape and are shaped by those power-geometries, and for identifying the different foundations of reworked notions of responsibility. The latter includes a politics of place beyond place (Massey 2007, see Chapter 10): "an outward-looking politics of place ... presenting a challenge to the hegemonic imagination, whether that be of globalization, or of local place, *or indeed simply of what is possible*" (188, emphasis added). It is from this relational position that the litigation by Harvard students can be understood as a place-based politics that makes explicit its relationality, to other places and people distant in both space and time.

CONCLUSION

The mainstream divestment movement is also contentious for what some claim it fails to challenge: capitalism, colonialism (McSorley 2014) and potentially the very nature of massive private (treated as charitable, with tax exempt status) endowments of elite universities. If economic geographers are to use Massey's work to further enrich the project of relational thinking and analysis, power-geometries themselves must be treated as multiplicities. This is not so much a critique of Massey's own work, as of its potential interpretation or usage: that power-geometries are treated as "maps" of delineated power relations that fix those relations in time and space, undermining their relationality and their contingent geographical imaginations.

This is not to deny the existence of structural dependencies, embedded institutions and durable relations of power; rather, it is to point out that sociospatial power relations are grounded in *multiple* power-geometries. I will finish with one short illustration of what I mean. The divestment movement, when targeting pension funds, tends to focus on the power relations between plan members, trustees and managers, and those (current and future) impacted by today's investment decisions. The wider power-geometries in which public pension funds are themselves enmeshed are seldom acknowledged. Public pensions face challenging times. Treated both as cash cows for hedge funds and money managers – the egregious case of Rhode Island's pension fund being used to funnel money into a hedge fund directly associated with its now governor is only one example (Siedle 2016) – and as targets of benefits retrenchment campaigns against public sector workers, they are also being redefined by social and environmental justice goals that stretch their mandate and capacity. That is not to argue that divestment strategies are wrong, or should not target pension funds. But the power-geometries of pension funds also involve a broader pension politics (with their own geographical imaginations), including the widespread conversion of defined benefit plans into defined contribution and hybrid plans – which reinstate the methodological individualism of the individual investor and shift risk from the sponsor to the plan member. Analyses that take seriously Massey's commitment to relational analyses of space and politics must do the work of grappling with such multiplicities. This is both an opportunity, and a challenge, for engaging more fulsomely with Massey's thought in economic geography.

REFERENCES

For works authored and co-authored by Doreen Massey, please see Select Bibliography of Doreen Massey, beginning on p. 371.

350.org n.d. "About 350: History". https://350.org/about (accessed 16 January 2018).
Barney, K. 2009. "Laos and the Making of a 'Relational' Resource Frontier". *Geographical Journal* 175 (2): 146–59.

Bathelt, H. & J. Gluckler 2003. "Toward a Relational Economic Geography". *Journal of Economic Geography* 3 (2): 117–44.

Blomley, N. 1994. *Law, Space, and the Geographies of Power*. New York: Guilford Press.

Bok, D., H. Calkins, R. Macdougall, *et al.* 1986. "The Policy of American Universities Towards Divestment in South Africa". *Minerva* 24 (2–3): 246–343.

Castree, N. 2004. "Differential Geographies: Place, Indigenous Rights and 'Local' Resources". *Political Geography* 23 (2): 133–67.

Clark, G. 2000. *Pension Fund Capitalism*. Oxford: Oxford University Press.

Clark, G. 2005. "Money Flows Like Mercury: The Geography of Global Finance" *Geografiska Annaler: Series B Human Geography* 87 (2): 99–112.

Cumbers, A. & P. Routledge 2004. "Alternative Geographical Imaginations – Introduction". *Antipode* 36 (5): 818–28.

Delaney, D. 2015. "Legal Geography I: Constitutivities, Complexities, and Contingencies". *Progress in Human Geography* 39 (1): 96–102.

Ettlinger, N. 2003. "Cultural Economic Geography and a Relational and Microspace Approach to Trusts, Rationalities, Networks, and Change in Collaborative Workplaces". *Journal of Economic Geography* 3 (2): 145–71.

Gosiger, M. C. 1986. "Strategies for Divestment from United States Companies and Financial Institutions Doing Business with or in South Africa". *Human Rights Quarterly* 8 (3): 517–39.

Graham, M., C. Andersen & L. Mann 2015. "Geographical Imagination and Technological Connectivity in East Africa". *Transactions of the Institute of British Geographers* 40 (3): 334–49.

Gregory, D. 1998. *Geographical Imaginations*. London: Blackwell.

Harvey, D. 2000. *Spaces of Hope*. Edinburgh: University of Edinburgh Press.

Hebb, T. 2008. *No Small Change*. Ithaca, NY and London: ILR Press/Cornell University Press.

Huber, M. T. 2013. *Lifeblood*. Minneapolis, MN: University of Minnesota Press.

Langley, P. 2017. "Financial Flows: Spatial Imaginaries of Speculative Circulations". In B. Christophers, G. Mann & A. Leyshon (eds) *Money and Finance After the Crisis: Critical Thinking for Uncertain Times*. Chichester: Wiley-Blackwell, 69–90.

Lefebvre, H. 1991 [1974]. *The Production of Space*. Oxford: Blackwell.

Malm, A. 2016. *Fossil Capital*. London: Verso.

Massie, R. 1997. *Loosing the Bonds*. New York: Nan A. Talese/Doubleday.

McKibben, B. 2012. "Global Warming's Terrifying New Math". *Rolling Stone* 19 July. www.rollingstone.com/politics/news/global-warmings-terrifying-new-math-20120719 (accessed 9 January 2018).

McSorley, T. 2014. "Tarsands Divestment and Its Discontents". *Briarpatch* 6 January. https://briarpatchmagazine.com/articles/view/tarsands-divestment-and-its-discontents (accessed 9 January 2018).

Nagar, R., V. Lawson, L. McDowell & S. Hanson 2002. "Locating Globalization: Feminist (Re)readings of the Subjects and Spaces of Globalization*". *Economic Geography* 78 (3): 257–84.

NACUBO 2016. "U.S. and Canadian Institutions Listed by Fiscal Year (FY) 2015 Endowment Market Value and Change* in Endowment Market Value from FY 2014 to FY 2015". *National Association of College and University Business Officers and Commonfund Institute*. www.nacubo.org/Documents/EndowmentFiles/2015_NCSE_ Endowment_Market_Values.pdf (accessed 9 January 2018).

OECD 2015. *Annual Survey of Large Pension Funds and Public Pension Reverse Funds*. Paris: Organisation for Economic Co-operation and Development.

OECD 2016. *OECD Pensions Outlook 2016*. Paris: Organisation for Economic Co-operation and Development.

OECD 2017. "Global Pension Statistics – OECD". *Organisation for Economic Co-operation and Development*. www.oecd.org/finance/private-pensions/globalpensionstatistics.htm (accessed 16 May 2017).

PA 2010. "BP Share Price Slide Hits UK Pension Funds". *Independent* 2 June. www. independent.co.uk/money/pensions/bp-share-price-slide-hits-uk-pension-funds-1989503.html (accessed 9 January 2018).

Pred, A. 2000. *Even in Sweden*. Berkeley, CA: University of California Press.

Quarter, J., I. Carmichael & S. Ryan 2008. *Pensions at Work*. Toronto: University of Toronto Press.

Richardson, B. 2011. "Fiduciary Relationships for Socially Responsible Investing: A Multinational Perspective". *American Business Law Journal* 48 (3): 597–640.

Ritchie, J. & H. Dowlatabadi 2015. *Fossil Fuel Divestment: Reviewing Arguments, Implications and Policy Opportunities*. Victoria: Pacific Institute for Climate Solutions.

Sarang, S. 2015. "Combating Climate Change Through a Duty to Divest". *Columbia Journal of Law and Social Problems* 49 (2): 295–341.

Schneider, N. 2015. "Revisiting Divestment". *Hastings Law Journal* 66 (2): 589–615.

Siedle, E. 2016. "Rhode Island Politicians' Billion-Dollar Pension Hedge Fund Gamble Loses $500 Million". *Forbes* 29 September. www.forbes.com/sites/edwardsiedle/2016/ 09/29/rhode-island-politicians-billion-dollar-pension-hedge-fund-gamble-loses-500-million/ (accessed 9 January 2018).

Strauss, K. 2009. "Cognition, Context, and Multimethod Approaches to Economic Decision Making". *Environment and Planning A* 41 (2): 302–17.

Strauss, K. 2015. "These Overheating Worlds". *Annals of the Association of American Geographers* 105 (2): 342–50.

Strauss, K. 2016. "Sorting Victims from Workers Forced Labour, Trafficking, and the Process of Jurisdiction". *Progress in Human Geography* 41 (2): 140–58.

Troyer, T., W. Slocombe & R. Boisture 1985. "Divestment of South Africa Investments – the Legal Implications for Foundations, Other Charitable Institutions, and Pension Funds". *Georgetown Law Journal* 74 (1): 127–61.

CHAPTER 22

DOREEN MASSEY AND LATIN AMERICA

Perla Zusman

INTRODUCTION

"Learning from Latin America". This is the title chosen by Doreen Massey for her article published in *Soundings* in 2012. In the context of the neoliberal crises in Europe and North America, when competition and subordination among countries was deepening, Massey noticed the construction of other kinds of relations in Latin America during the first decade of the twenty-first century. These relations challenged the geometries of power within states and in the world-system as well. Massey thus invited European countries to look at and learn from the Latin American experience.

From her perspective, governments in the region were striving to expand democratic practices. Some examples are, first, the advancement of social self-organization mechanisms that channelled social demands towards the state; second, the passing of rules aimed at ending the monopolistic control of mass media and opening spaces to a diversity of expressions, mainly the voice of communities; and third, the establishment of cooperation and solidarity bonds through organizations such as UNASUR (Union of South American Nations) or CELAC (Community of Latin American and Caribbean American states including Cuba but excluding Canada and the USA) (Massey 2012b).

It is not common for European intellectuals to show interest in the Latin American experience and to reflect upon it as a model for developing a different kind of relation between European countries themselves. Even critical perspectives have labelled Latin American governments populist, and disqualified them on the grounds that they did not conform to European modes of governance (regarded as the model of civilization and rationality).

The purpose of this chapter is to understand the significance of Latin America in Doreen Massey's professional trajectory, while outlining some ties between her view of space and discussions and practices in countries where subaltern sectors became protagonists after their conception of space and territorial claims were voiced and heard.

> In Latin America, new forms of thinking are being born all the time.　　　　　　(Román Velásquez & García Vargas 2008: 340)

Doreen Massey developed a peculiar sensitivity to the third world and, particularly, Latin America. She understood that conquest and colonization processes across the region in the sixteenth century showed the dark side of modernity (Mignolo 1995). In general, European historiography and geography have ignored these processes, which positioned Europe at the heart of the world-system and the remaining continents in the periphery. The American peoples' concepts of space-time became invisible when the continent was integrated into European concepts of space and time, as Doreen Massey noted in her introduction to *For Space*, recognizing that the space travelled by Spanish conquerors on their way to Tenochtitlan, the Aztec city (the current city of Mexico) is depicted as an uninterrupted surface, as something already given, devoid of historical density (Massey 2008).

Doreen Massey also understood that the scars left by this manner of incorporating the Americas into the world-system can be seen today in the naturalization of structural inequalities and the hierarchies established in terms of class, gender and race. In her view, Latin America thus needed to tread its own paths to find emancipation. This would entail calling into question all the development policies promoted at different times from outside the continent, with the complicity of some sectors such as (white) local elites, which only widened the existing social, cultural and regional gaps (Massey 1999b).

For Doreen Massey, travelling to Latin America was more "stimulating" than visiting her US colleagues because political discussions and practices made her "think" (Román Velásquez & García Vargas 2008: 340). From her perspective, both state and popular experiences were interwoven, designing alternative forms of political organization.

Three conjunctures have marked the relationship of Doreen Massey with the region.

1. Her integration into the planning policy of Nicaragua within the framework of the Sandinista Revolution;
2. her participation in the making of documentary films about Mexico City and Brazil for the BBC during the 1990s; and
3. her association with governments in Latin America, which, after taking over, sought to expand the rights of the population in terms of distribution of wealth and cultural recognition.

In this context, the introduction of her concept of power-geometries stands out, particularly in relation to the Bolivarian revolution led by Hugo Chávez in 2007 and the invitation she received subsequently from the government of Venezuela to take part in discussions about this idea.

During the 1980s, when Margaret Thatcher started to promote neoliberal policies in England, some Latin American countries, such as Argentina, Uruguay and Chile, struggled to reinstate democracy. Others, such as Cuba and Nicaragua, were striving to protect themselves from the United States' attacks against revolutionary actions and to reorganize their economy, space and society from a socialist perspective.

In Nicaragua, after defeating the regime led by Anastasio Somoza (July 1979), the Sandinista movement started a process of political and economic reorganization that prioritized health, education and land distribution. In 1985, Doreen Massey travelled to Nicaragua. With her participation in INIES (the Nicaraguan Institute for Economic and Social Research), she endeavoured to contribute to the understanding of why informal settlements continued to expand in urban areas even though the revolution had advanced policies intended to stop rural–urban migration, such as agrarian reform. Massey saw that Managua (the capital city of Nicaragua) offered more secure conditions than rural areas, which remained battle grounds in the war against "the *contras*". Urban areas, in turn, received the benefit of programmes ensuring access to more affordable housing and oriented to taking care of the social needs of neighbourhoods. However, war and the US embargo created budgetary expenses that jeopardized the continuity of these new services for populations migrating from the rural areas (Massey 1986, 1987a).

During the 1990s, Massey visited Mexico City and Brazil,[1] which were under neoliberal governments after the Washington Consensus (1989),[2] to make documentary films for the BBC. The leading question throughout her journey across Mexico City was "whose city is it?" She interviewed different city inhabitants, from rural migrants living in deteriorated areas in historic neighbourhoods to advisors to transnational companies. Their answers led her to think that Mexico City was developing within the framework of a dispute between its different groups, some of which were more powerful (bankers, state, companies) than others. This process, however, could not be interpreted only in light of the local context because all actors were involved

---

1. We could only access the documentary film about Mexico City. See BBC (1999).
2. The term Washington Consensus refers to the economic reforms required from Latin America and African countries by entities such as the IMF, World Bank and the US Department of the Treasury during the 1990s. These bodies asked countries to liberalize their economy and streamline state intervention if loans were to be granted to help them overcome their economic crisis. This paved the way for the privatization of sectors that were under the dominance of the state (utilities, telephony, transport means, among others).

in global networks. Against the background of neoliberal policies, Doreen Massey noticed that the gaps between the rich and the poor were becoming wider, at the same time that new forms of political and economic organization emerged that challenged the established order and privileges, opening possibilities of change for the future.

In 2008, Doreen Massey was invited to Venezuela to participate in activities, provide advice on and disseminate the concept of power-geometries. This idea had been incorporated into the five motors of socialism[3] under the Hugo Chávez administration in Venezuela. The invitation renewed her energies. She delivered lectures at universities, participated in popular fora and even appeared on television. In parallel, she wrote a non-academic text that was included in materials for free distribution as part of the socialist readings programme (*Programa de Lectura Socialista*) (Albet & Benach 2012). The use of the power-geometry concept in the articulation of strategies for including formerly excluded sectors in politics raised a set of challenges, reflected in the following questions:

> What responsibilities does one have in a situation like this? Whose concept is it [power-geometries] now? The best, minimalist, way may be not to insist on anything, but to aim to enrich discussion through elaboration ("what I was trying to get at when I came up with this idea…;" "the way I've tried to use it…;" "the kind of thing it can be used for in the UK…"). Yet inexorably one becomes part of the politics of the place ("well I think it needs to be used in the economic as well as the political sphere…;" "you have to look here at the balance between centralisation and popular power…").
>
> (Massey 2008: 496)

In this conjuncture, Doreen Massey noticed that Latin American politicians were searching for different theories that could support them in developing a form of democracy that would consider the limits of the liberal project and Latin America's special characteristics as well.[4] Participating in this search

---

3. With the aim to achieve a set of political, economic, social, military and territorial transformations, Hugo Chávez proposed the development of five strategic processes that he called the "motors to socialism". The motors were the following: the enabling law to Socialism, constitutional reform, popular education, communal power and the new power-geometry. The basis of this new power-geometry was the creation of communal councils that would act as a mechanism of self-organization at the local level to attain the socialist reorganization of the nation's geopolitics. Through this structure, Chávez wished to give voice not only to the poorest rural regions but to city inhabitants excluded from society as well.

4. Within this framework, the television program hosted by Canal Encuentro (an educational and cultural television channel in Argentina) called *Diálogos con Laclau* (Conversations

led her to accept an invitation from her friend Ernesto Laclau in 2012 to present her viewpoints in a seminar in Buenos Aires organized by *Debates y Combates* [Debates and Combats] (a journal launched by Laclau himself in 2011 to create a forum of ideas equivalent to the *New Left Review* in the Ibero-American world).[5]

At the *Debates y Combates* seminar, Doreen Massey reinforced some of the ideas about Latin America she had already presented in her article in *Soundings*. She pointed out that Latin America was sketching alternatives to the neoliberal economic model, particularly targeted towards raising its voice, constructing its identity on the basis of differences, reducing poverty and offering responses that would not promote austerity to overcome the crisis. However, at the same time, she suggested that this outlook posed new challenges. Thus, she wondered whether it would be possible to go beyond a private economic model, purely based on redistribution, to initiate a new era that would give rise to a much less unequal economy. Along these lines, she speculated about the actual potential of the new forms of popular power, and the multiple modes of articulation between the state and social movements that were striving to change societies.[6]

These ideas reflect Doreen Massey's commitment to the project of radical democracy. Outlined by Chantal Mouffe and Ernesto Laclau, radical democracy put forward the possibility of setting up political articulation between distinct groups (in terms of class, sexuality and race) to unite their proposals, negotiate and transform the state (Schuliaquer 2015).

In the three conjunctures just described, Doreen Massey reflected on the design of political alternatives seeking to challenge existing power-geometries. She had always been interested in identifying both the limitations and the potential of these projects. She emphasized that the radicalization of proposals could only be ensured by dealing successfully with the profound economic inequalities prevailing in Latin America.

---

with Laclau) can be mentioned. The program was aired in 2011 and showed a series of conversations that Ernesto Laclau held with European intellectuals such as Antonio Negri, Chantal Mouffe, Jacques Rancière, Gianni Vattimo, Stuart Hall, Etienne Balibar, and other Argentine personalities. Doreen Massey was also invited to speak with Laclau in this television series.

5. Her trip to Buenos Aires was followed by a visit to Santiago de Chile, where she participated at the meeting of Government Experts on the Development of Territorial Policies in Latin America and the Caribbean organized by the Economic Commission for Latin America and the Caribbean (CEPAL). Her presentation can be found at CEPAL (2015).

6. Doreen Massey's (2012c) presentation at the *Debates y Combates* seminar can be found at: www.youtube.com/watch?v=9wn4QphR4Yw

THE OPENNESS OF SPACE AND LATIN AMERICAN VIEWS

Doreen Massey's conception of space demonstrates her commitment to the radical democracy project in more depth. This is reflected by the connections that she established between "the imagination of the spatial and the imagination of the political" (Massey 2005: 10). First, for Massey, space is as a sphere for the coexistence of multiplicity, that is to say, the simultaneous coexistence of others "with their own trajectories and their own story to tell" (Massey 2005: 11). Second, she understood that space is a product of interrelations, meaning that "it does not exist prior to identities/entities and their relations" (Massey 2005: 11). Third, she conceived space as being under ongoing construction, open to the future "with connections yet to be made, juxtapositions yet to flower into interaction, (...) relations which may or may not be accomplished" (Massey 2005: 11).

Conceptualizing space as a sphere for the coexistence of multiplicity helped her understand that societies experienced dissimilar space-time trajectories and that they were not dominated by the dictates of Eurocentric readings and practices. This opened the way for a dialogue between her perspective about space and time and the approaches developed under decolonial perspectives, which render visible subaltern views of space and time. The approach that identities/entities are redefined by connections interacts with the evidence that territorial claims redefine their stances based on emerging conflicts. Finally, the openness of space to the unexpected is linked to views that put forward the difficulty of foreseeing the future trajectories of social movements anchored on a territorial basis.

In the early 2000s, decolonial interpretations systematized and deepened the criticisms of linear and teleological views that rendered invisible the stories and geographies of the peoples in Latin America (Mignolo 1995; Lander 2000; Castro Gómez 2004; Porto-Gonçalves 2009; Rivera Cusicanqui 2010; Escobar 2014). This plural field of study[7] attempted not only to give visibility to historical and current violence and exclusion processes (on the basis of class, gender, sex and race), but also to subaltern world conceptions in Latin America by recognizing their "status of majority" (Rivera Cusicanqui 2010: 60).

Within this framework, sociologist Silvia Rivera Cusicanqui has examined Andean cosmologies (specifically the Quechua and Aymara ones). Rivera Cusicanqui identified images such as the "world-upside-down" in the drawings in *Primera Nueva Crónica y Buen Gobierno [First New Chronicle*

---

7. Within this field it is possible to find different postures. Some are critical of postcolonial perspectives, especially those of a feminist character. Among them, it is also possible to identify different degrees of commitment to social movements.

and Good Government] (1615) by Waman Puma de Ayala as a testimonial of his journey across the Alto Peru Viceroyalty. This image seems to be a counterpoint to the meeting between Moctezuma and Cortés described by Doreen Massey, where the Aztecs were identified as a people without history. With the idea of "a world-upside-down", Waman Puma de Ayala accounts for an Andean world experiencing Hispanic colonization as "a space-time dump", a "hecatomb", and "a cycle of catastrophes and renovation" (Rivera Cusicanqui 2010: 21–2).

Based on the idea of a "cycle of catastrophes and renovation", Cusicanqui introduced an additional aspect of the Andean cosmovision of space and time. This cosmovision "is neither linear nor teleological (...) it moves by cycles and spirals, (...) it charts a course without failing to return to the same point". Along these lines, "the past-future are embedded in the present: regression or progression, repeating or leaving behind the past are at stake in each conjuncture" (Rivera Cusicanqui 2010: 55, our translation).

These types of space-time conceptions are present in the current territorial claims by native and rural populations, shepherds and Afro-descendants in Latin America. Within this context, the term "territory" as both an academic and a political practice category reflects the relations between space and power, not in terms of the political technologies of government (Elden 2013), but, rather, in terms of the material and symbolic appropriation of these areas by subaltern sectors (Haesbaert 2011). From this perspective, a territory is formed on the basis of practices and types of knowledge that interrelate human, animal and plant elements, along with ancestors and divinities (Nates Cruz 2011; Escobar 2014).

Even though territorial claims often embody disputes about distribution of wealth and cultural recognition rights, the characteristics of territories and the identities of their populations are redefined within the framework of each individual conflict. In Latin America, the struggle waged by subaltern sectors since the 1990s has been geared towards the recognition of their rights of occupation of ancestral territories. The dispute led to constitutional reforms under which, in some cases, states recognized the territoriality of communities inside their own borders (Argentina, Brazil, Colombia, Venezuela). In other cases, territorial recognition was reaffirmed by a manifestation of the plurinational character of the state (Ecuador, Bolivia).[8] In this context, land-grant programmes were developed. Although controversies about the defence of territoriality, self-governance and self-management

---

8. Both Constitutions recognize the culturally heterogeneous character of both countries. This involves a juridical recognition of those populations (particularly indigenous) that had been invisibilized during the political and economic social history of these states and a rupture with the mono-cultural Westphalian state.

rights continue, from the 2010s onward these populations focused primarily on defence of their entitlement to spaces subsequently appropriated by transnational companies (with the complicity of states) for extractive practices (e.g. forestry, mining, soybeans, real estate or construction of dams) (Cruz 2007; Porto-Gonçalves 2009; Agnew & Oslender 2010; López Sandoval, *et al.* 2016). Communities not only defend their right to remain in and move around the territories to which they claim ownership, but also act as the guardians of nature (Ulloa 2004; Palladino 2017). In some cases, partnerships have been established among environmental groups, indigenous peoples and university intellectuals. This encounter of dissimilar space-time trajectories, in some situations, has pivoted around the idea of environmental justice (Svampa & Viale 2014; Göbel & Ulloa 2014).

These territorial disputes challenge existing power-geometries. Territories are defined through relations established within the framework of conflicts. These conflicts both reflect connections materialized through the juxtaposition of different territorialities (belonging to states, transnational companies and subaltern sectors), and lead to the emergence of new connections, hence, new territorialities (sometimes resembling a network, such as the links that are woven when subaltern sectors join the organizations that fight for environmental justice). In summary, the territorialities of the subaltern sectors are neither fixed nor closed, but remain open and under an ongoing process of reconfiguration.

For Rogerio Haesbaert, the term *territory* in the Latin American context has a similar significance to how Doreen Massey conceptualizes *place*. Territory is open, originates from a mixture of relations, and its ongoing redefinition stems from conflicts. While through her conceptualization of place, Doreen Massey attempted to defend the right of migrants to live next door to European inhabitants, the idea of territory seeks to capture the strategies of subaltern groups that have been deterritorialized (deprived of their territories) by the dynamics of capital to reterritorialize (regain these territories) and transform the access to their territories into a human right (Haesbaert 2013). In line with this, both the idea of place and the idea of territory appear as concepts that intend to capture the open character of becoming.

FINAL THOUGHTS

Latin American political experiences, especially those that have put state and social movements in conversation, enriched Doreen Massey's theoretical and political thinking. These experiences showed her that existing power-geometries may be challenged and that political alternatives may be constructed. They also helped her to recognize the challenges and limitation in dealing successfully with the structural social inequalities of the region.

Her sensitivity to the problems in Latin America, in addition to her dialogue with Chantal Mouffe and Ernesto Laclau, were critical in creating a spatial dimension for the proposal of radical democracy. In this sense, the conception of space as a sphere for the coexistence of multiplicity opened the way to introduce into the Latin American political scene the space-time notions of social sectors that had historically been rendered invisible in the narrative of European modernity.

The dialogue we seek to open between Doreen Massey's conception of space/place and the idea of territory upheld by scholars and by the political practice of subaltern sectors in Latin America suggests ways to turn the academic arena into a meeting point for concepts that, although different, may be driven by comparable political projects. At the same time, this dialogue can enhance the importance of imbuing the ideas of space, place and territory with political content if they are to be conceived as concepts that remain open to the future, to the unexpected.

REFERENCES

For works authored and co-authored by Doreen Massey, please see Select Bibliography of Doreen Massey, beginning on p. 371.

Agnew, J. & U. Oslender 2010. "Territorialidades Superpuestas, Soberanía en Disputa: Lecciones Empíricas desde América Latina". *Tabula Rasa* 13: 191–213.

Albet, A. & N. Benach 2012. *Doreen Massey. Un Sentido Global del Lugar*. Barcelona: Icaria.

BBC 1999. "México City". www.youtube.com/watch?v=ONUxYewyDdo (accessed 9 January 2018).

Castro Gómez, S. 2004. *La Hybris del Punto Cero. Ciencia, Raza e Ilustración en la Nueva Granada (1750–1816)*. Bogotá: Pontificia Universidad Javeriana/Instituto Pensar.

CEPAL 2015. *Memoria del Primer Encuentro de Expertos Gubernamentales en Políticas de Desarrollo Territorial en América Latina y el Caribe*. Santiago: Naciones Unidos.

Cruz, V. C. 2007. "Territórios, Identidades e Lutas Sociais na Amazônia". In F. G. B. Araújo & R. Haesbaert (eds) *Identidades e Territórios: Questões e Olhares Contemporâneos*. Rio de Janeiro: Access, 93–122.

Escobar, A. 2014. *Sentipensar con la Tierra: Nuevas Lecturas Sobre el Desarrollo, Territorio y Diferencia*. Medellín: Ediciones UNAULA.

Elden, S. 2013. *The Birth of Territory*. Chicago, IL: University of Chicago Press.

Göbel, B. & A. Ulloa (eds) 2014. *Extractivismo Minero en Colombia y América Latina*. Bogotá: Universidad Nacional de Colombia.

Haesbaert, R. 2011. *El Mito de la Desterritorialización: del "Fin de los Territorios" a Multiterritorialidad*. México: Siglo XXI.

Haesbaert, R. 2013. "A Global Sense of Place and Multi-Territoriality: Notes for Dialogue form a "Peripheral" Point of View". In D. Featherstone & J. Painter (eds) *Spatial Politics: Essays for Doreen Massey*. Oxford: Wiley-Blackwell, 146–57.

Lander, E. 2000. *La Colonialidad del Saber: Eurocentrismo y Ciencias Sociales. Perspectivas Latinoamericanas*. Buenos Aires: Clacso.

López Sandoval, M. F., A. Robertsdotter & M. Paredes 2016. "Space, Power and Locality: The Contemporary Use of Territorio in Latin American Geography". *Journal of Latin American Geography* 16 (1): 43–67.

Mignolo, W. 1995. *The Darker Side of the Renaissance*. Ann Arbor, MI: University of Michigan Press.

Nates Cruz, B. 2011. "Soportes Teóricos y Etnográficos Sobre los Conceptos de Territorio". *Revista Co-herencia* 8 (14): 209–29.

Palladino, L. (2017) "Cuidar el Monte, Devenir Indígena. Re-territorialización y Comunalización Ticas a partir del Conflicto Territorial (2015–2016)". *Revista Cardinalis* (8): 6–31.

Porto-Gonçalves, C. W. 2009. "De Saberes y de Territorios: Diversidad y Emancipación a partir de la Experiencia Latino-Americana". *Polis Revista de la Universidad Bolivariana* 8 (22): 121–36.

Rivera Cusicanqui, S. 2010. *Ch'ixinakaxutxiwa: una reflexión sobre prácticas y discursos descolonizadores*. Buenos Aires: Tinta Limón.

Román Velásquez, P. & A. G. Vargas 2008. "Hay que Traer el Espacio a la Vida". *Signo y Pensamiento* 53: 327–43.

Schuliaquer, I. 2015. "Laclau, Sin Fin de Ciclo". *Anfibia* 11 October. www.revistaanfibia. com/ensayo/laclau-sin-fin-de-ciclo/ (accessed 9 January 2018).

Svampa, M. & E. Viale 2014. *Maldesarrollo*. Buenos Aires: Katz Editores.

Ulloa, A. 2004. *La Construcción del Nativo Ecológico*. Bogotá: Instituto Colombiano de Antropología e Historia ICANH-COLCIENCIAS.

CHAPTER 23

GRASSROOTS STRUGGLES FOR THE CITY OF THE MANY: FROM THE POLITICS OF SPATIALITY TO THE SPATIALITIES OF POLITICS

Helga Leitner and Eric Sheppard

INTRODUCTION

Particularly during the last two decades of her career, Doreen Massey concerned herself with the politics and ethics in and of cities. In some ways this begins in her classic essay "A Global Sense of Place" (Massey 1991a), set in London's Kilburn neighbourhood where she lived. Yet some prime interventions are the little-known critique (with Ash Amin and Nigel Thrift) of Blairite urban policy in the United Kingdom, *Cities for the Many Not the Few* (Amin *et al.* 2000), and her final monograph on London: *World City* (Massey 2007). These writings emanated from her own active engagement in London's politics and policy-making, as a close advisor to "red" Ken Livingston, both as Leader of the Greater London Council until Margaret Thatcher abolished it (1981–6), and as the newly created Mayor of London (2000–8).[1]

Yet, for all her interest in urban politics, Doreen paid relatively little attention to its raggedy edges of urban activism and social movements, such as more-than-capitalist urban commoning. From her writings, her interest seems to have been in empowering the rights to the city for the many by primarily engaging with formal politics. In this brief chapter, we critically interrogate the implications of her relational conceptualization of place and spatiotemporality, and of the politics of spatiality, for urban activism. Endorsing her argument that "you can't [just] take a theory off the shelf and use it" (Hoyler 1999: 73), we write back to her theoretical reflections from our own studies of urban

---

1. It is perhaps no coincidence, paralleling former Venezuelan president Hugo Chávez' interest in Doreen's concept of power-geometries, that the reinvigorated progressive Labour coalition, under the leadership of London politician Jeremy Corbyn, titled their 2017 election manifesto "For the Many not the Few".

commoning in Los Angeles and Jakarta. From this perspective, we argue that spatialities and politics are dialectically interrelated, implying that her politics of spatiality should be extended to embrace the spatialities of real world politics (realpolitik). The chapter is organized as follows. We summarize those aspects of her rich body of sociospatial theory of particular relevance here (section 2), introduce the two urban commoning case studies (section 3), and critically reflect on and extend her politics of spatiality (section 4).

TOWARDS A POLITICS OF SPATIALITY

Doreen's explicitly relational theorization of spatiotemporality, distinct from, yet resonating with David Harvey's more capitalocentric dialectical approach to relational space and Ed Soja's sociospatial dialectic (Sheppard 2006, 2018), has issued a profound challenge to the kind of place-based thinking that seems natural from the perspective of social theory. We wish to stress two aspects that clearly inform her thinking on urban politics and policy, one flowing from the other: The relationality of place, and power-geometries. Using Kilburn High Road as her case study, Massey (1991a) advocates for a progressive, outward-looking approach to place as a space of heterogeneity – against the kind of enclosed, conservative, defensive and homogenizing thinking mobilized, for example, in claims about "making America great again". Yet she also seeks to critically interrogate that relationally produced heterogeneity, arguing:

> Different social groups have distinct relationships to this anyway differentiated mobility: some people are more in charge of it than others; some initiate flows and movement, others don't; some are more on the receiving-end of it than others; some are effectively imprisoned by it. (Massey 1991a: 26)

She mobilizes *power*-geometry (Massey 1991a: 25) to draw out a much larger argument; against those seeking to reduce globalization to space-time compression and the annihilation of space by time, she shows how flows and movement can work to empower select bodies – and places – at the expense of others.

Power-geometry, filled out in subsequent essays (Massey 1993b, 1999b), amounts to a fundamental spatial critique of those promoting capitalist globalization as the means to development for all – as a rising tide that lifts all boats. Without elaborating in detail here (see contributions by Roberts and Heynen *et al.* in this volume), the nub of Doreen's argument is that the asymmetric connectivities linking places under capitalist globalization, whereby some grow wealthy at the expense of impoverishment elsewhere, are a primary vector of uneven geographical development. This undermines stageist conceptions of development (Rostow 1960), based on place-based thinking (Sheppard 2016), which aver that the differences that matter between countries can be reduced to temporal position, measured by how far they have advanced along the only possible development trajectory. Thus, from political economy, via feminist and poststructural theory, Doreen arrived at a postcolonial sensibility:

> this different imagining of globalization – in my terms, a truly spatialized understanding ... – would refuse to convene spatial differences under the sign of temporality. It would reject the tales of inevitability that necessarily accompany such singular narratives ... It would in other words hold open the possibility of the existence of alternative narratives.   (Massey 1999b: 43)

In an essay published that same year, written for a series of lectures at the University of Heidelberg, Doreen turns to the political implications of her approach to spatiality. She begins with three propositions about space (Massey 1999e: 28):

1. Space is a product of interrelations. It is constituted through interactions, from the immensity of the global to the intimately tiny ...
2. Space is the sphere of the possibility of the existence of multiplicity; it is the sphere in which distinct trajectories coexist; it is the sphere of the possibility of the existence of more than one voice. Without space, no multiplicity; without multiplicity, no space. If space is indeed the product of interrelations, then it must be predicated upon the existence of plurality. Multiplicity and space are co-constitutive.
3. Finally, and precisely because space is the product of relation-between, relations which are necessarily embedded material practices *which have to be carried out*, it is always in a process of becoming; it is always being made. It is never finished; never closed.

The political implications are likewise threefold. She argues that proposition one aligns with a politics committed to anti-essentialism and the constructed nature of identities and things. Proposition two underwrites political themes of difference, multiplicity and the importance of granting space to diverse, differently positioned voices. Proposition three, with its focus on uncertainty and becoming, stresses the importance of political discourses that challenge notions of inevitability and progress, approaching the future as genuinely open.

The implications of this for urban politics are taken up in *Cities for the Many not the Few* and *World City* (Amin *et al.* 2000; Massey 2007). Noting that neoliberal politics prioritizes the wealthy few, while rendering invisible the many who make urban life possible, "for the many" runs as a leitmotif through both books. The former, co-authored with Ash Amin and Nigel Thrift, is written as a progressive policy response to Blairite policies targeting cities and communities, in particular the Rogers Report (Urban Task Force 1999). Their recommendations advocate for advancing rights to the city (Lefebvre 1968) for those made invisible, focusing on "recovery of the commons in our cities" (Amin *et al.* 2000: 36). These proposals are framed in terms of advancing capabilities through lifelong education and learning, socialization (a knowledgeable and discursive political community, fostering sociability, inculcating civic values), transversal city politics (forging solidarities across heterogeneous urban communities, a level playing field for agonistic, direct democracy), and creating breathing and stopping places in cities for the many, enabling them to escape the drudgery of everyday work lives. In this chapter we show how the urban majority on the ground is attempting to advance their rights to the city, and the struggles, trials and tribulations this entails.

World City lays out Doreen's views on urban politics through the case study of her adopted hometown, drawing on her almost fifty years of engaging with it. We focus here on her ethico-political conclusion, framed around the question of what it means to be a Londoner and the politics of place beyond place. Drawing on her relational thinking about place, she works through the notion that London, as a global/world city (Massey 2007), profoundly shapes what happens across the world, advancing the notion of a politics of place beyond place. In her view, what it means to be a Londoner is that living in this place is connected with life elsewhere: "[t]he brilliance of today's London … is … dependent for its ordinary, daily, social reproduction on an array of workers from the rest of the world" (100). She advocates for making Londoners conscious of their responsibilities to the people and places affected by and affecting London:

> the identity of this place must take account not only of the outside within, the internal hybridity, but also, as it were, of the inside without; … the question "where does London (or any city) end?" must at least address the issue of those recruited into the dynamics of the urban economy and society by the long lines of connections of all sorts that stretch out to the rest of the country and on around the planet … And this in turn raises questions of unequal interdependence, mutual constitution, and the possibility of thinking of placed identity not as a claim *to* place but as

the acknowledgement of the responsibilities that inhere in *being placed.*   (Massey 2007: 216, italics in original)

These two books thus stress different issues, one asserting rights within the city of the many and not just the view, the other asserting a cosmopolitanism that highlights responsibility towards distant others, whereby residents of a world city take account of the impact of their actions on those people and places connected to that city.

URBAN COMMONING IN LOS ANGELES AND JAKARTA

Doreen's thinking on urban politics, as articulated in *Cities for the Many Not the Few*, focuses on nudging the UK New Labour government in order to facilitate and promote various forms of solidarity: political community, sociability and civic values. This explores the question of how citizens can be socialized into forms of citizenship based on solidarity. From our perspective, however, this focus on engineering conditions of possibility for solidarity overlooks the significance of grassroots actions creating auto-constructed spaces beyond the state, where solidarity is constructed/enacted through commoning, and from which claims can be launched for rights to the city. Thus in this section, we clear ground for reflecting critically on her approach through case studies of such initiatives, taken from our fieldwork in Los Angeles and Jakarta. Focusing on the marginalized outside within, that is, those in-migrants who are yet to be acknowledged as full and proper urban residents, we examine both strategies of commoning and struggles for rights to the city.

Los Angeles is frequently invoked alongside London as a global or world city (Abu-Lughod 1999; Knox & Taylor 1995), but Jakarta is larger (32 million in the greater metropolitan area vs. 19 million) and just as globally connected – albeit in a much less empowering way (Robinson 2006). Population growth in both cities (much faster in Jakarta, which still receives 200,000 annually) has been shaped to a large degree by longstanding vectors of in-migration, including substantial marginalized populations with delimited rights – undocumented international migrants in the case of Los Angeles, and low-income domestic migrants moving into Jakarta from across Indonesia. Both groups face restricted rights due to their political status, placing in question their status as someone with the right to inhabit and work in the city. For undocumented immigrants in Los Angeles the challenge is obtaining a driver's licence that serves as an ID card; in the case of Jakarta, migrants face difficulties in qualifying for a Jakarta ID – the Indonesian identity card (Kartu Tanda Penduduk). Beyond political marginality, their status as urban residents is complicated by other factors. They face economic precarity and denigration from wealthier residents (many of whom also stem from migrant families). They also actively maintain connections with the households and extra-metropolitan places they left behind.

In both places, albeit in variegated ways, commoning has enabled these residents to claim rights to the city. In the case of Jakarta, auto-constructed kampungs (informal settlements) have long functioned as spaces of refuge for new in-migrants, where they can set foot in the city in order to secure livelihoods and make broader citizenship claims. In Los Angeles, auto-constructed spaces such as worker centres and other immigrant organizations have provided immigrants with much needed services, representation, care and solidarity, from where they have launched important social justice campaigns asserting the right to live and work in the city.

Los Angeles' worker centres

In Los Angeles, the needs of low wage, mostly immigrant workers, who were unwelcome in unions, were the immediate impetus for the self-organized creation of worker centres. Organized around workers in a particular industry (e.g. The CLEAN Car Wash Campaign), of a particular ethnicity (e.g. The Pilipino Workers Center) or place/community based (the Korean Immigrant Worker Alliance), centres focus on four spheres of action: service (e.g. legal assistance), education (e.g. English as a Second Language classes, vocational training courses, popular education and leadership training and development), organizing low-wage workers, and general advocacy for immigrants' and workers' rights. More abstractly, worker centres can be seen as attempts to construct solidarity and engage in commoning practices to enact rights to the city. They are characterized by a flat, inclusive governance structure that encourages and allows active participation and the equal voice of members in defining agendas and service needs, as well as in organizing, skill and leadership development.

Beyond practising flat governance, worker centres are spaces of care. Female directors, who in 2016 made up the majority of centre directors in Los Angeles, have made care and care-work, usually associated with the private sphere of the family, an integral element in the running and everyday operation of worker centres. This also includes designing the physical space of centres, bringing elements of domestic space, such as a kitchen, into the public space of the worker centre. The care-work in the centres helps foster an affirmative atmosphere for negotiations across differences. These are facilitated through learning about the Other, e.g. their life histories, their experiences of hardship, discrimination – thus allowing connections to be made across difference.

Yet worker centres are also sites where difficult conversations about racism, sexism, homophobia and religious intolerance within the group are discussed and addressed head-on through sustained conversations, rather than being avoided. As one centre director highlighted, it is not enough to talk about racism, sexism, homophobia and religious intolerance, and to learn about the Other. Rules of conduct with the Other – such as the banning of racist and sexist talk, religious intolerance, and expressions of homophobia – are important in guiding conduct with the Other and facilitating negotiations across differences.

The inward space of the worker centre – the "withdrawal" dimension that attempts to construct solidarity through commoning practices – is complemented by their outward, "agitational" dimension: spaces of publicity through which campaigns for social justice are launched often in collaboration with other worker centres. It is in these spaces of publicity that claims to the right to the city are most visible. In Los Angeles, worker centre campaigns have ranged widely from single centre and issue campaigns to coalitions of several worker centres and other social justice and civil rights organizations, including legal aid clinics, making demands on the local and national state. Many of these campaigns have pushed the city of Los Angeles to pass new ordinances. Innovative campaigns include the Forever 21 Campaign (an effort to organize garment workers), and the $15 minimum wage campaign.

The most recent success story is the multiracial campaign to pass a $15 minimum wage ordinance, which was approved by the Los Angeles City Council in March 2016. Key to this success was the Raise the Wage Campaign, spearheaded by a coalition of labour groups, faith-based organizations, worker centres, day labourer centres, labour unions, and other advocacy and community organizations. Within the campaign, worker centres were instrumental for the wage theft and enforcement provisions. Wage theft, the practice of employers not paying workers their full wage by violating minimum wage laws, stealing tips, withholding overtime pay, and forcing workers to clock out and continue working, is a pervasive problem amongst the low-wage workers who participate in the worker centres, and Los Angeles has particularly high rates of wage theft (Milkman *et al.* 2010). Yet the success of this campaign, like its origins, was far from a simple matter. Paralleling how the space of worker centres is often one of encounter and negotiation across social difference, the solidarity required for a successful campaign could not be taken for granted (see Leitner & Nowak 2018). Indeed there were intense debates amongst worker centres over the meaning of wage theft and the precise nature of the demands to be made on the state. From the perspective of the Black Worker Center, discrimination of black workers on the job market and their unemployment was conceived as the ultimate wage theft, whereas other worker centres were arguing that the ultimate wage theft is unpaid work (Leitner & Nowak 2018).

The campaign against wage theft became not only a moment for political action, but also a claim for equality. Through extensive discussions, demands were amended and commitments to common objectives were enshrined in a Memorandum of Understanding (MOU) among the leaders of the different organizations before the start of the campaign. The MOU, which made each organization on the steering committee accountable to work in good faith towards passing the full package agreed upon, was itself, of course, the outcome of intense negotiations across differences (for further details see Leitner & Nowak 2018).

Jakarta's kampungs

If commoning is a rare exception to private property and possessive individualist norms in Los Angeles, the opposite is the case in Jakarta. Sharing the experience of cities across the postcolony encountering (rather than promulgating) development (Escobar 1995), Indonesia's and Jakarta's peripheral positionality within globalizing capitalism has meant that the pace of in-migration into the capital city has long overwhelmed its formal (capitalist) labour and housing markets. Kampungs as urban residential spaces (the original Bahasa Indonesia meaning is village) date back to Dutch colonialism: spaces tolerated by the Dutch, between their European thoroughfares, where the "natives" could live in their own way. After colonialism, kampungs flourished as auto-constructed spaces where migrants could live, raise a family, and from which they could find and undertake informal employment. We adopt Abdoumaliq Simone's term "the urban majority" (Simone 2014b) to describe kampung residents. Estimates of the proportion of greater Jakarta's residents currently living in kampungs range between 30 and 60 per cent: 9–18 million (Leitner & Sheppard 2018). Jakarta's urban majority occupies kampung spaces dominated by informal land tenure arrangements, with the bulk of the population employed in the informal sector. Individual kampung districts often house migrant populations from a particular place in the Indonesian archipelago who reproduce here ways of life, culture and politics they carried with them, which over the years become hybridized with Indonesia's, Java's and Jakarta's cultural, economic and political norms. They are quiet, and can be green spaces by Jakartan standards: The majority of the alleyways are too narrow for cars (although the explosion of motorbikes has created some traffic), and families work to green the alleys and grow food in whatever open spaces are available. They are also spaces of sociability, where families are expected to help one another in difficult times and where much time is spent socializing outside. Indeed, the better-off kampungs are attractive spaces of alternative urban living – of urban commoning.

Yet the auto-constructed spaces of kampungs are not pure commons, "off-limits to the logic of [capitalist] market exchange and market valuation" (Harvey 2012: 73). Rather, as Simone has analysed at length (Simone 2014a), they "embody both commoning and competition, collective and self-interest, with the balance between these continually in motion" (Leitner & Sheppard 2018: 441). Individual residents are quick to create and take advantage of opportunities to acquire money and power, social relations are shot through with gender, ethnic and political power relations, and kampungs actively seek to take advantage of market opportunities they can provide to those working in nearby factories, shopping centres and high-rise office blocks, such as cheap rents and food.

Kampungs also range widely in terms of quality of life and political status. Some, dating back to the 1950s, are occupied by families who have secured their urban place and livelihoods, some of whom are middle class, with formal employment and freehold home ownership (Hak Milik). The Jakartan government tolerates them, has invested in upgrading infrastructure in some through a series of kampung improvement programmes, and provided political legitimation by designating them as "legal" kampungs. At the other end, are desperate and unhealthy kampungs occupied by recent in-migrants trying to set foot in the city and with little access to the all-important Kartu Tanda Penduduk for Jakarta. The Jakartan government withdraws legitimacy for these kampungs by designating them "illegal", on the grounds that they occupy locations deemed off-limits to settlement, such as within 10 metres of waterways, along railway lines and under highway overpasses (based on regulations passed often long after they were established).

There are different ways in which these auto-constructed urban commons can help secure and advance their residents' rights to the city, albeit involving continuous struggle. First, kampungs are not simply settlements providing shelter, but support everyday practices necessary for securing material means of livelihood. Designating a kampung as legal reinforces these possibilities for its residents, even if some, such as renters, possess no land rights or cannot establish legal residence in the city. Yet legal kampungs are disappearing, particularly in locations of high land value where residents are selling their land rights to Indonesia's large and influential land developers, who then replace kampungs with modern high-rise commercial and residential developments. Rejecting offers by some developers to relocate into the new projects, residents by and large move to other, cheaper and more peripheral kampungs – an indication of the importance of such spaces for securing rights to the city for the urban majority (Leitner & Sheppard 2018).

The picture is very different for residents of those kampungs designated as illegal: accelerating again in the last three years, a number of these kampungs have been razed at very limited notice, with residents who are not fleet of foot losing all they possess. The putative rationale for eviction is that these kampungs stand in the way of flood mitigation and urban greening projects, and violate public order laws. Compensation is uneven (only available for those with a Kartu Tanda Penduduk for Jakarta), and limited to relocation into public housing blocks that are by and large inimical to the informal livelihood practices and social networks of those evicted (Leitner & Sheppard 2018). In the face of these challenges, social movements and NGOs are working closely with residents of illegal kampungs to push back against the erosion of whatever informal rights to the city they have been able to secure (Padawangi 2014). For example, Rujak (www.rujak.org) has helped residents propose alternative rehousing designs that meet their needs (to little avail to date), the Jakarta Legal Aid Institution (Lembaga Bantuan Hukum Jakarta) has documented and publicized evictions and helped residents take their concerns to court, Ciliwung Merdeka (https://ciliwungmerdeka.org/) has deployed community mapping and surveys to make the case for de facto rights of occupancy, and thus legal status, of the "illegal" Kampung Pulo and Bukit Duri, and the Urban Poor Consortium has helped the urban poor support themselves and protest eviction. Success stories may be few and far between, but contestations continue.

REFLECTIONS

Concluding, we reflect on the politics of spatiality enacted through the urban majority in LA and Jakarta, arguing that Doreen's politics of spatiality (Massey 1999e) can and should be extended to consider the spatialities of politics. Examining these struggles we have shown how the outside within – undocumented immigrants of colour in LA and in-migrants to Jakarta in informal settlements – lays claim to rights to the city while maintaining connections with their places of origin. In part, these struggles confirm the relevance of Doreen's three principles for a politics of spatiality. First, commoning is crucial for the construction of solidarity and equality, granting space to diverse, differently positioned voices. Second, commoning is riven with uncertainties as it rubs up against larger political-economic structures; it is always in the process of becoming – or better, always needs to be struggled over and reaffirmed in the urban everyday.

While Doreen derived her political principles from her theorization of space, strangely she did not embed spatiality within these principles, thereby overlooking the spatialities of politics – although this does make an appearance as politics beyond place in *World City*. As our case studies also show, while urban commoning brings out aspects of her politics of spatiality, there are also distinct spatialities to the politics shaping these struggles. In short, our claim is that spatialities and politics are dialectically interrelated. These spatialities are explored in greater detail elsewhere (Leitner & Nowak 2018; Leitner & Sheppard 2018), but we provide two examples here. First, in order to create space for the differently positioned voices of the urban majority, it is vital to create, maintain and defend material spaces (worker centres, kampungs) where alternative modes of politics can be developed and enacted, and from which broader claims to the right to the city can be advanced. Second, the urban majority does not restrict itself to place, but also engages in a politics beyond place. Thus immigrants in Los Angeles draw on organizing experiences in their home country, but also engage with their places of origin through Hometown Associations. Similarly, Indonesian NGOs actively connect kampung initiatives within and across cities, but also internationally. Even if world cities might bear a particular responsibility towards distant others (Massey 2007), questions of a politics beyond place are relevant across the globally interdependent world of ordinary cities (Robinson 2006).[2]

REFERENCES

For works authored and co-authored by Doreen Massey, please see Select Bibliography of Doreen Massey, beginning on p. 371.

Abu-Lughod, J. L. 1999. *New York, Chicago, Los Angeles: America's Global Cities.* Minneapolis, MN: University of Minnesota Press.

Amin, A., D. Massey & N. Thrift 2000. *Cities for the Many Not the Few.* London: The Policy Press.

Escobar, A. 1995. *Encountering Development.* Princeton, NJ: Princeton University Press.

Harvey, D. 2012. *Rebel Cities: From the Right to the City to the Urban Revolution.* London: Verso.

Hoyler, M. 1999. "Interpreting Identities: Doreen Massey on Politics, Gender, and Space-Time". In M. Hoyler (ed.) *Power-Geometries and the Politics of Space-time.* Heidelberg: Department of Geography, University of Heidelberg, 47–82.

Knox, P. & P. Taylor (eds) 1995. *World Cities in a World-System.* Cambridge: Cambridge University Press.

Lefebvre, H. 1968. *Le Droit a la Ville.* Paris: Editions Athropos.

Leitner, H. & E. Sheppard 2018. "From Kampungs to Condos? Contested Accumulations through Displacement in Jakarta". *Environment and Planning A* 50 (2): 437–56.

---

2. We acknowledge support from the US National Science Foundation (grant number BCS-1636437) for the research in Jakarta.

Leitner, H. & S. Nowak 2018. "Making Multi-Racial Counter-Publics: Toward Egalitarian Spaces in Urban Politics". In K. Ward, A. Jonas, B. Miller & D. Wilson (eds) *The Routledge Handbook on Spaces of Urban Politics*. New York: Routledge, 451–64.

Milkman, R., A. L. González, V. Narro, *et al.* 2010. *Wage Theft and Workplace Violations in Los Angeles: The Failure of Employment and Labor Law for Low-Wage Workers*. Los Angeles, CA: UCLA Institute for Research on Labor and Employment.

Padawangi, R. 2014. "Reform, Resistance and Empowerment: Constructing the Public City from the Grassroots in Jakarta, Indonesia". *International Development Planning Review* 36 (1): 33–50.

Robinson, J. 2006. *Ordinary Cities: Between Modernity and Development*. London: Routledge.

Rostow, W. W. 1960. *The Stages of Economic Growth: A Non-Communist Manifesto*. Cambridge: Cambridge University Press.

Sheppard, E. 2006. "David Harvey and Dialectical Space-Time". In N. Castree & D. Gregory (eds) *David Harvey: A Critical Reader*. Oxford: Blackwell, 121–41.

Sheppard, E. 2016. *Limits to Globalization: Disruptive Geographies of Capitalist Development*. Oxford: Oxford University Press.

Sheppard, E. 2018. "Socio-Spatial Dialectic". In A. M. Orum (ed.) *The Wiley-Blackwell Encyclopedia of Urban and Regional Studies*. Hoboken, NJ: Wiley-Blackwell, forthcoming.

Simone, A. 2014a. *Jakarta: Drawing the City Near*. Minneapolis, MN: University of Minnesota Press.

Simone, A. 2014b. "Reflections on an Urban Majority in Cities of the South". In S. Parnell & S. Oldfield (eds) *The Routledge Handbook on Cities of the Global South*. London: Routledge, 322–36.

Urban Task Force 1999. *Towards an Urban Renaissance*. London: Department of the Environment, Transport and the Regions.

CHAPTER 24

TOWARDS A QUEER PHENOMENOLOGY OF SOCIAL REPRODUCTION: INSIGHTS FROM LIFE HISTORIES OF INFORMAL ECONOMY WORKERS IN URBAN INDIA

Priti Ramamurthy and Vinay Gidwani

INTRODUCTION

In *For Space* (2005) and elsewhere Massey urges that "space" must be thought of in the same open-ended manner that "time" has in continental philosophy post-1920s: namely, as multiple rather than unilinear, as therefore a terrain of "openings" that constitutes the very possibility of "politics". In this recasting of time, space has been (implicitly or explicitly) posited as "closure", hence – for Massey – as apolitical. In opening up "space" Massey reiterates its relational nature as process and becoming; this implies too that there is never one "space" that is in the process of becoming but a heterogeneity of spaces associated with a heterogeneity of relations that encompass humans and non-humans. Since the processes in question have different temporalities it follows that when we talk of politics we must meaningfully talk of "spatial politics" and, by implication, heterogeneous spatiotemporalities. It is through a relational conception of space that Massey avoids the pitfalls of objectivist (Cartesian) and subjectivist (individual-centred or perspectivist) approaches to space. A relational conception is congruent with dynamic approaches to space as a "production".

Our chapter seeks to extend Massey's dialectics of the spatiotemporal and the political through an engagement with "queer phenomenology" (Ahmed 2006a). There are two components to our argument here, both important: the aspect of phenomenology and the aspect of queering. By drawing on life histories of urban migrants in two Indian cities, we show how queer phenomenology compels us to comprehend *humans as embodied beings* made by orientations, disorientations and reorientations to social reproduction. The *disorientations of handed-down life scripts* powerfully reveal the queering

325

that opens up spaces of political possibility. A phenomenological stance also enriches Massey's conception of spatial production by engaging with practices of imagination that are vital elements of people as meaning-making beings. Finally, attentiveness to queer phenomenology jolts "us" (as theorists) from our inherited categories and suppositions as well. The tensions that permeate our life histories show how lifeworlds (and the desires that course through them) exceed the framework of social reproduction. And here we reiterate our principal claim: a queer phenomenological stance to social reproduction offers a way to think in considered ways about the dialectic of the spatiotemporal and the political: namely, how the spatiotemporal is political and the political is spatiotemporal.

"SOCIAL REPRODUCTION" IN THREE MOMENTS

We find provocative, Massey's argument, "not just that the spatial is political (which, after many years and much writing thereupon, can be taken as given), but rather that thinking the spatial in a particular way can shake up the manner in which certain political questions are formulated, can contribute to political arguments already under way, and – most deeply – can be an essential element in the imaginative structure which enables in the first place an opening up to the very sphere of the political" (Massey 2005: 9). We propose turning this attentiveness to think about the spatiotemporalities of "social reproduction". "Social reproduction" is a powerful analytic, for good reason, which is why feminist scholars have employed it to think about spatial politics. At the same time, it is an analytic that has evolved. We begin by tracing some key moments in this trajectory, as a way of underscoring what the analytic enables and, ultimately, where it needs to go.

The concept of "social reproduction" has long antecedents in feminist theory. Katz, for example, understands "social reproduction" as the repertoire of "social practices through which people reproduce themselves on a daily and generational basis and through which social relations and the material basis of capitalism are renewed" (2001: 709). Social reproduction has been a cornerstone for feminist politics, from first wave feminists such as Rosa Luxemburg, to its current resurgence as the grounds to forge a response to the contemporary contradictions of capitalism. Whereas, on the one hand, theorizations of social reproduction have been amenable to the kind of shake up Massey advocates – namely, how certain political questions are formulated, on the other hand, social reproduction as an analytic appears unable to contain the heterogeneity of its actually existing forms (cf. Collins 2016: 109). One simple argument (and plea) we issue here is that the political possibilities incipient in the concept of social reproduction lie in substantively recognizing the difference that difference makes. In so doing, we echo Massey's summons that to open up "space", thereby opening it up as "political", is to foreground its relational nature as process and becoming; and equally important, that there is never one "space" that is in the process of becoming but a heterogeneity of spaces. We contend that apprehending life histories through queer phenomenology may offer a way to free the analytic of social reproduction from the impasse it recurrently encounters.

A detailed charting of these various impasses is beyond the scope of this chapter, but a few key moments are worth recounting.[1] We identify three noteworthy moments. The *first* moment takes us back to the 1970s and 1980s, when feminists renovated the concept of social reproduction present in the writings of Marx and Engels to foreground women's work in the biological and social reproduction of labour-power. They powerfully showed how this work, although productive in renewing capitalism's material basis (living labour), was *devalorized* and *unremunerated*: constituting an unpaid subsidy to capital. To explain what the domestic, private sphere enabled for capitalism is, feminists tracked the emergence of the "domestic" as a distinct sphere to a process of primitive accumulation, consolidated by the privatization of property which led to the establishment of the patrilineal family, a process that Maria Mies evocatively termed "housewifization" (Mies 1986; also Federici 2003). The domestic labour debate and the movement it birthed ultimately foundered on the question of value: how should one value activities which are subjective, expressive and hard to quantify as exchange value? The dual systems approach straightjacketed both patriarchy and capitalism in space and time, construing these as static – and, as later feminists were to point out – Eurocentric formations. The incapacity, both theoretically and politically, to attend to the spatiotemporal heterogeneities of social reproduction led to an impasse. Inattention to difference, especially race, was symptomatic of this blindspot (Combahee River Collective 1986 [1977]).

In the *second* moment, in the 1980s and 1990s, the critique of social reproduction developed along two lines. Nakano Glenn (1992) exemplified a US Woman of Color critique by pointing out the privileging of white, middle-class women's experiences in earlier Marxist feminist analyses. Difference, especially of caste, was central to Indian feminist theorizations of social reproduction as well (Jain & Banerjee 1985). "Lower" caste women are channelled into the most demeaning and stigmatized forms of reproductive work, often

---

1. There has been a resurgence in thinking about social reproduction recently. For rich and detailed accounts of its genealogies see Collins (2016, 2017), Meehan and Strauss (2015), Bakker (2007), Bezanson and Luxton (2006) and Fraser (2014, 2016).

under poorly paid and only nominally free conditions. This sort of excavation of social reproduction as a space "in which distinct trajectories coexist, as the sphere therefore of coexisting heterogeneity" (to invoke Massey 2005: 9) founds a politics that raises a different set of questions. It is no longer whether domestic work can be brought into the realms of exchange value. It is there already, but in racialized or caste-stigmatized forms. This places middle-class women who benefit from the services of social reproduction workers (the majority of them, people of colour or "low" caste) in the uncomfortable position, as feminists, of rallying around a decent wage that directly impacts their disposable incomes. Glenn also leveraged this as an opening to question "the hierarchy of work" (1992: 37) which demeans the skills required for reproductive labour. Indian feminists questioned if paid work could ever be unquestionably "liberating", when women's withdrawal from "outside" work signals higher household and caste community status (Rao 2003).

A second intervention in the 1980s and 1990s was the poststructuralist critique of Marxism as a cross-culturally applicable theory (Scott 1986; Benhabib & Cornell 1987; Nicholson 1987). Marx's unacknowledged "philosophical anthropology" reinforced a limited understanding of gender and Engels's later exegesis in *The Family, State and Private Property*, was undergirded by a narrow understanding of the "economy" as separate from the "family". By the 1990s, inspired by Althusser and Gramsci, feminists began to theorize the domestic sphere as one of "relative autonomy". Gibson-Graham in *The End of Capitalism (As We Knew It): A Feminist Critique of Political Economy* challenged the tendency to think of capitalism as a totality. In their assessment, households were spaces of economic difference, sites from where progressive politics could reclaim equality, pleasure, and the end of exploitation. Although Gibson-Graham made a promising beginning in deconstructing the representation of capitalism as monolithic and all-encompassing, by resurrecting the household as a place of radical alterity they fell back on dualized representations of "female exploitation and male domination" (1996: 237) as soon as they entered "actually existing" homes in Queensland, Australia. Again, the spatiotemporal heterogeneities of social reproduction led to a theoretical and political impasse.

In its most recent iteration, what we call the *third* moment, a surge of feminist scholarship – Mitchell *et al.'s Life's Work* (2004) being an early exemplar – has highlighted the effects of capital's current spatiotemporal transformations on social reproduction. A rich and varied literature has emerged on phenomena such as transnational care chains, surrogacy, affective economies and emotional labour, precarity, and new materialist understandings of human–nature relations (Fraser 2014, 2016; Meehan and Strauss 2015; Federici 2012; Weeks 2011; Butler 2009). In a recent and important edited volume Meehan and Strauss contend "work that seeks to directly link social reproduction, labor, and the body is ... surprisingly rare" (2015: 12). Although sympathetic to this summons, we are obliged to note that such statements elide Women of Color, Third World, and transnational feminist scholarship, which has insistently done so. Congruently, the spatiotemporal heterogeneities of contemporary capitalism – as a system and social formation – and its relationship to social reproduction continue to confound feminist theory and politics. In response, our argument here underscores the embodied, normative character of social reproduction, even as it tries to identify betrayals of normative scripts. In highlighting these disorientations, which entail heterogeneous habitations of space and time, we reveal how social reproduction is queered.

FIELDWORK AND RESEARCH DESIGN

Our chapter draws on oral histories of rural-to-urban migrants, specifically their experiences of living and working in urban informal economies in two Indian cities: Delhi in north India and Hyderabad in south India. With the aid of our institutional collaborators in Hyderabad (Hyderabad Urban Lab) and Delhi (Centre for Policy Research) we took (roughly NW to SE) transects through each city with the aim of identifying five working-class neighbourhoods that house migrants. These transects covered the cities' core (older) areas as well its peripheral (newer) areas. Our expectation was that we would find earlier generations of migrants to the cities in older settlements, and the more recent migrants in the newer settlements. We also expected to capture occupational variability and trajectories in migrants' livelihoods by canvassing older as well as newer settlements.

We found it necessary early in the research process to develop a keen understanding of not only the details of our chosen settlements but also of the broader processes that have shaped both cities in very different ways. For this, we undertook key informant interviews with people who have a particularly acute understanding of either a settlement or forces that have shaped one of the cities, or both. The authors gathered oral histories between September 2015 and August 2016, aided by two RAs in each city. We were able to conduct a total of 135 oral history interviews (54 in Delhi, 81 in Hyderabad, comprising 58 women and 77 men). Predictably, the depth and richness of oral histories varied considerably (cf. Rogaly & Qureshi 2017). While we intend to pursue follow-up meetings with a subset of interviewees over the coming months, for this chapter we rely on a set of completed oral histories (real names altered) that illuminate the varied dynamics of social reproduction among urban migrants who work in the informal economy.

ORIENTATIONS, DISORIENTATIONS AND REORIENTATIONS TO SOCIAL REPRODUCTION

Petite and attractive, Putli bears herself with grace. Though 29 she looks much younger. A contract worker, she cleans the toilets at a girls' hostel on the University of Delhi campus. She lives in a nearby settlement, in an immaculately clean and tidy one-room, with a shared water pipe and toilet. One of the millions of rural migrant women who do commodified social reproduction work in the city, Putli is relatively well-paid (compared to women who work in private households) though her job is insecure. A Rajput (a relatively "high" caste) she is from a village in Bihar. At the age of seventeen, her mother arranged Putli's marriage to Mahesh, also a Rajput, from a village at some distance from Putli's. Mahesh was visiting his aunt in Putli's natal village, her *maike*, the aunt thought it was a good idea to get them married. Putli's mother, landless, impoverished and with a drunken, ailing husband, agreed. "A girl is caste" (*ladki jaat hai*), Putli explains. It is through marriage and kinship that caste is reproduced. If her father had died, and her mother had brought along Putli, "a single, unmarried daughter" to the city, reasons Putli, she would have been accused of "living by prostituting me". "It was due to my *majboori* (utter helplessness), that my mother [not an elder male] negotiated my marriage." After a short, simple ceremony, in a temple, Putli went straight on to live with her in-laws. "I'd have liked a wedding with pomp and glitter, just like everybody else. I'm still sad."

Sara Ahmed in *Queer Phenomenology* (2006a) offers a model of how bodies become oriented by the ways they take up time and space. When we are orientated, we follow familiar lines. "Considering the politics of the straight line", Ahmed writes, "helps us rethink the relationship between inheritance (the lines that are given as our point of arrival into familial and social space) and reproduction (the demand that we return the gift of the line by extending that line)" (2006b: 555). Putli's mother followed the straight lines of caste and patriarchy in getting her daughter married. If she hadn't, the social consequences could have been dire: "If you don't have anyone, people throw sludge at you", Putli told us. The heterogeneous spatiotemporalities of social reproduction are discernable in Putli's mother's anticipation of her husband's (untimely) death, the expectation that society will talk when she moves out of the space of the extended family and caste community, and from the village to the city, and her desire to "extend the line" through a normative marriage of her daughter into a Rajput Bihari village family. This palimpsest of sociotemporalities reveals it is not just the dire economic fate that awaited her in the village as a widow, but the putative affective trauma of being cast out of her village and out of her caste, that propelled Putli's mother to arrange her marriage.

On arrival in her marital home (*sasuraal*), Putli too followed the straight lines of patriarchy and caste. "Rajput women don't work outside the house", Putli tells us.

> Women work only inside, and men work outside. In my in-laws' house, I always wore a sari. In front of the men, and whenever I went out, I always covered my head with my *ghunghat* (sari). I've never spoken straight to my eldest brother-in-law. Or sat on a bed in front of my in-laws.

In following these practices so dutifully, Putli conveys how her bodily comportment was tacitly oriented by the lines of caste and patriarchy. Putli also remembers her marital house as a hellhole of never-ending physical reproductive work. "It was a big house. Lots of people." She spent from sun up to sun down every single day doing all the cleaning, cooking, washing the dirty clothes, dirty cooking pots and everyone's dirty plates, with absolutely no respite. The house was a prison of discord, she was taunted continuously, and her in-laws were uncaring. "No one as much as asked if I'd eaten." Putli escaped, finally convincing her migrating husband to take her along, and has not been back for a while.

Yet her orientation to inhabiting the world as a high caste Rajput, a woman, a married woman, still powerfully moves her. The physical impress of space on bodies is palpable in Putli's hatred for her paid job of cleaning toilets. It is not what high caste Rajput women do. The unspoken here is that this is "outside" work that Balmikis (formerly untouchable caste people) do. So unclean is the work and so strong the feelings it elicits in Putli that she has not told anyone in her natal or marital villages what her job is. Putli's own orientations to the straight lines of caste and patriarchy – the ways she tacitly "knows" how to follow the lines – lead her to knowing when they are not being followed. She complains,

> My brother's wife doesn't know how to behave properly. Here [in Delhi] it is OK but in the village it will be [considered] disrespectful. I've tried to teach her – don't call your mother-in-law or husband *tum* (the informal version of "you"), don't sit on the bed in front of your mother-in-law, you can sit on the bed with your younger brother-in-law's wife or husband's younger sister, but not with elders. When my own mother-in-law or sisters-in-law returned from their natal homes, I used to wash their feet before serving them food.

Thereby, Putli disciplines her sister-in-law into high-caste gendered respectability by trying to make her embody the comportment befitting a Rajput, something the girl has not been properly socialized into by her own family and is unlikely to get in the city. The spatiotemporal pathways of social reproduction Putli traverses draw our attention to the variegated social ontologies of reproduction. We learn what constitutes "caring" through and beyond labouring and how comporting oneself inside and outside the home matters.

But orientations that follow the straight lines of caste and patriarchy are neither Putli's or her mother's full stories. Again, Sara Ahmed is provocative when she suggests that when we are orientated we may not even notice that we are. We may not even "think" to think about this point. Consequently, when we experience disorientation, we may notice orientation as something we do not have. Disorientations could be emotional, psychological and physical, perceived in the body's exterior movements and in the interior movements of thought which move bodies in certain ways, down certain paths.

Putli's lifeworld after her marriage to Mahesh has been one of unrelenting physical and mental abuse. He drinks and beats her up. Time and time again. So violent is he that their son, now 10, has often tried to save his mother, banging on neighbours' doors for help before his drunken father killed her, counselling her to run away before she gets dragged into the street by her hair and her clothes torn asunder once again, pleading with Putli to please kill him – the son – before she is murdered or commits suicide herself. Putli has run away from Mahesh multiple times. Often, she has returned to her *maike*, her natal village, for succour, economic support and regeneration.[2] Once she left to live in Hyderabad for six months, supported by "a friend". The male friend *(dost)* and Putli shared "*lagaav* (attachment)". "He wasn't that attractive to look at, but he had good qualities, he was thoughtful and loving. I'll always wish my husband was like him."

Queering social reproduction makes us aware just how far Putli went from the point zero of orientation – her marital house in village Bihar – from where her world unfolds. Literally and metaphorically blinkered by her head-covering "there" – "I didn't know my way around that village because I never went out" – she moved to a completely different universe, a "here" of her own dwelling, to which she travelled, where she loved another man, and lived in a faraway state. Putli never hid her intimate relationship with her *dost* from her mother or brothers. Her mother told her, "If you like someone, go ahead and marry him." But Putli was reoriented to normative social reproduction. She recalls,

> He wanted to marry me, but I said no. My mother's twice married
> [the second time to an already married Dalit man Putli disapproves

---

2. In many of our oral histories, women's natal homes continued to be places of refuge, even in north India, where village endogamous marriage leads to the attenuation of these ties.

of]. If I had married twice too what would people in my (maternal) village say? The loss of respectability would be crippling for my brothers and paternal uncles. With him [the *dost*], there was no problem with his caste [he's Rajput too]. But he wasn't prepared to marry me with my child. At times he'd say, give the boy to Mahesh.

The multiple straight lines of patriliny – her uncle's, her brother's, her male friends' – reorient Putli away from marrying her *dost*. Nevertheless, she savours the warm memory of that slice of time, a plenitude she feels fondly.

Putli's husband found her at the *dost*'s and dragged her back to live with him, on the outskirts of Hyderabad.[3] His abusive behaviour got worse. She ran away again, this time to Delhi and lived there, working for the first time in her life as a domestic in private houses to feed her son and herself. Two years went by when one day word came from the village that her husband had died. Hearing this, Putli's mother made Putli go through the rituals of widowhood, physically stripping off her *bindi*, her *sindhur*, her bangles, her toe rings. Engulfed by the darkness of widowhood the press of grief on Putli's body, as she tells it, guided Putli's subsequent lines of action. On learning that her husband had staged his death, and was alive after all, she allowed him back. The disorientation, the feeling of being so out of place, pushed Putli back to the familiar. But it's also a conjugal relationship, improbably, filled with love and desire. For Putli considers her husband "beautiful, fair, articulate, capable ... even though I have never lived peacefully with him". She adds, "Now I have a husband. I'm married to him. We may fight but I have respectability in the village; in the city." Whenever he gets drunk he still calls her a "dirty slut" (*gandi aurat*). But she doesn't let the insult stick. She knows "people in her *maike* will tell him she is a good woman". She believes it in her own bones. She's had an extramarital relationship with a male who is not her husband, everyone knows about it, and yet, remarkably, she is not defined nor is her body defiled by physical intimacy with another man. Thereby, Putli queers desire and the regulation of female sexuality through marriage.

A queer phenomenology of social reproduction pays keen attention to bodies as they move through and produce space, transforming the "power geometries" (Massey 1991a) which orientate them to norms, sometimes resulting in *durable* disorientations that endure and sometimes finding themselves reorientated to those same norms, albeit as a repetition with difference. By tracing processes of embodiment as people constitute and

---

3. She's quite foggy about exactly where these places are; in her mental map they are simply the places where the husband and *dost* worked, where people who speak a strange language live, and where they put salt, not sugar, in *sooji*.

are constituted as social beings, heterogeneous spatiotemporalities of social reproduction come into view. Recognizing this heterogeneity begins to address the impasses previous theorizations of social reproduction have stumbled on. Putli has overcome the drudge and "sludge" of social reproduction she was "born" into and that she inherited. She laughs at the memory of her first train journey, accompanying her husband for the first time to Siliguri from Bihar when she didn't know where or how to pee nor how to ask him, a relative stranger at the time, what to do. Now she confidently takes trains and moves through space. Once, Putli broke the patriarchal phenomenology of touch and the regulatory phenomenology of desire within marriage by living intimately with a man who was not her husband. Putli has broken the caste phenomenology of touching shit and working outside the home as well but is repulsed by her stigmatized job even as it provides her food on the table, a future for her son, and a modicum of protection from her husband. In these spatiotemporal dialectics lie the openings to found a politics.

Embodiments of imagination and reorientations

Embodiments of imagination – the ways in which aspirations are felt and acted upon – to create meaningful life-worlds *now* are important aspects of social reproduction that have been relatively neglected in the literature. For many women and men we interviewed, their children are their joy, their life. Women, in particular, visualize radically better futures for their children, vastly different to their own lives. Those imaginations inform their presents in ways that are more important today than what may eventually transpire in the future. Futurity fills the present with meaning through acts of imagination. This, here and now, is what shapes the could be.[4] Imagining fundamentally better futures for their children inform social reproduction and move them, sometimes reorienting women to normative patriarchy and caste, as in the case of Putli, and in others, queering hierarchical kinship to transform the spaces of social reproduction

Malika, around 35 years old, is self-possessed and confident; on the day we meet her in Hyderabad, she is dressed in a bright green, brassily embroidered polyester sari, complimented with bangles, a watch, and a big handbag. Malika is a beautician with a difference: she used to work in a beauty parlour

---

4. Queer of colour scholar Muñoz (2009) has theorized the importance of imagining futurity for re-thinking the present in the here and now. We draw on his thinking. With queer scholars like Edelman (2004), we critique the temporality of reproductive time or lifecycle time for violently excluding people of non-normative genders and sexualities; however, our interviews powerfully demonstrated how imagining different futures for their children enables women to actions now.

but now gives women facials, manicures, pedicures, plucks eyebrows, waxes arms and legs, dyes and shampoos hair, all in the comfort of their own homes. She works long hours, providing personalized care to working women on weekends, and stay-at-home women during the week. A couple of years ago she bought herself a scooter, and recently learned to drive it, so she no longer depends on her husband to take her to clients. Men in the settlement Malika lives in may still belittle her as they watch her driving by, but they can no longer demean her family, or taunt: "What do *these* people know? They lack worldliness." Malika recollects, "They used to be so disparaging about us, but now that I am so successful we get shown respect."

Malika's raison d'être for working is her children's futures. Time and again in her story she returns to the imperative for her to work so they could "go to private elementary schools", and then go to "model" (exorbitant, private) high schools, and "will do engineering" in college. Malika has had to fight her mother-in-law and brothers-in-law every step of the way to work. Malika is Mangali, a "backward" caste; barbers, nail-cutters, masseuse and midwives, Mangali men and women have both provided (ritually "dirty") personal care services to higher castes in Telangana villages for generations. However, on migrating to the city, Mangalis signal their rising caste status by proscriptions on women's movement and work "outside" the home. Yet, Malika was determined. She started by watching TV shows to teach herself embroidery and taking orders for embellishing saris at home. Not satisfied, she borrowed money from her brothers and enrolled in a beauty technician's class, eventually convincing her husband that this was in line with their Mangali caste. Virulent objections and jibes by her in-laws about her wearing churidar-kurta "suits", not saris, to work in a beauty parlour made Malika quit. Eventually, by providing personalized beauty care in other women's homes *dressed in a sari* ("even though it shows more of the body than suits", she wryly notes) she has reoriented the traditional caste occupation of Mangalis. Noticing men were demanding beauty treatments too, Malika has taught her husband beauty skills, thereby reorienting the pedagogical script, and, on occasion, she massages and bathes babies, a social reproduction task her female forebears were highly skilled at and that which Malika claims her body tacitly knows. Malika earns far more than her husband, but maintains the pretense of his being the breadwinner by having a joint bank account. It is she, however, who decided to invest in a house and a plot in Hyderabad and in real estate along the main drag leading to a temple in her village – a village experiencing a boom as the new state of Telangana promotes it as an answer to Tirupati, one of the richest temples in the world. Females in India are supposed to "belong" to their marital households on marriage, yet Malika's ties to her brothers remains strong and she is queering the script of her relationship to their wives from a hierarchical relationship to a lateral one. Malika has trained her sisters-in-law (brother's wives) in tonsuring hair, and has now set them up on the main drag to shave the heads of women pilgrims. One could argue that caste continues to orient Malika but that would be missing the remarkable changes in professionalizing personal care-work, in the relations of gendered work, in household dynamics, and kinship which she has wrought. By reorienting vertical male–female relationships of power within households to lateral relationships of care between female in-laws who are kin, Malika queers social reproduction.

Malika's changing relationship to her mother-in-law is a good illustration of queering social reproduction by transforming a hierarchical, vicious kin relationship into a lateral relationship of care and intimacy. Early on, Malika's mother-in-law wanted her to have a third child but refused to pay for a hospital when Malika miscarried in the fifth month of her third pregnancy and nearly bled to death. Malika reasoned that "to take care of her existing two children well, with so little financial support [from her in-laws], I needed to stop having kids". Without telling her husband or mother-in-law she did just that. "Even if we had to eat rice with pickle, because we couldn't afford to buy vegetables, they hung on to their rule that women shouldn't work." Her mother-in-law also objected to Malika sending her kids to private schools (when government schools were good enough for her other grandchildren) and she objected to Malika buying "modern" kitchen appliances like a gas stove, a refrigerator, a blender and so on, which would have lessened the burdens of daily reproduction, as "a waste of money". Yet, towards the end of her life, Malika's mother-in-law, now waning in household authority, came to realize that she has more control over her own food, space and TV watching if her daughter-in-law goes out to work. Her mother-in-law also realized that *she* can learn how to operate newfangled gadgets – a gas stove, for example – to make tea whenever she wanted. And, that the kids of women who work are not less well behaved. Malika's mother-in-law, for the first time, started helping out with daily reproduction tasks in Malika's house, braiding her granddaughters' hair and telling the kids not to bother Malika when she came home tired after a long day at work. Malika regrets, "Just as my mother-in-law started understanding me, she died." She would have liked to reset her relationship with her mother-in-law more fully. In both their lives the relationship between the two women was finally something other than the hierarchical, normative relationship between daughter-in-law and mothers-in-law: it was caring not obligatory or oppressive. Intimate relations of laterality between kin reorient normatively hierarchical relationships, thereby queering understandings of social reproduction.

Malika reorients the caste phenomenology of touch by rescripting the occupation of a Mangali woman in the city. Simultaneously, she remakes space in her own home, in the settlement, and in her clients' homes. Social reproduction itself is dynamically transformed by her embodiment of imagination; anticipating a future for her children opens up the space to act now. She overcomes the drudge of daily reproductive work and discourse and, in the process, transforms her affiliation with her sisters-in-law and her mother-in-law from normatively hierarchical relations to perceptibly horizontal ones. The queer-studies scholars Patricia Clough (2013) and Nayan Shah (2011) have argued that intimate relations of laterality in non-heterosexual, non-reproductive relationships between siblings and relative strangers are vital to the infrastructure of existence. Malika, by creating lateralities out of normatively hierarchical kin relations, invites us to queer understandings of social reproduction within households.

CONCLUSION

Inspired by Doreen Massey, we have theorized social reproduction as a space of heterogeneous spatiotemporalities. Queer phenomenology, with its attention to bodies and intimacy, orientations and disorientations, is a vital approach for charting the dialectics of the spatiotemporal and the political. A dialectical approach, which conceives of social reproduction as always in process, a becoming, that transforms an existing "power-geometry" addresses the impasse of earlier feminist theorizations of social reproduction.

The life histories of the women and men who participate in informal economies demonstrate how bodies are aligned to the material, affective and intimate activities that let human beings to be reproduced as labourers and social beings. But frequently it is by queering or disorienting norms of gender, marriage, kinship and caste that social reproduction is made meaningful and can continue. Our observations of how bodies move, respond and comport through spaces of social reproduction brought home how often people rewrite normative scripts or refill the scripts with meaning. Social reproduction emerges as a space to imagine a future and create the conditions of possibility to realize it. In these processes of becoming, politics in a minor key is palpable in the fullness of stolen moments, small pleasures. Space itself is transformed and borders of gender, caste, sexuality, marriage and kinship redrawn. Massey, of course, recognizes these border struggles are political.[5]

---

5. Our foremost thanks go to the generous people in Delhi and Hyderabad who shared their life histories with us, often over multiple sittings. Their kindness continues to humble us. Enduring appreciation also for the staff at Hyderabad Urban Lab and at the Centre for Policy Research, New Delhi, for their guidance and insights. Our research would have faltered without the talents of our matchless research assistants: Sunil Kumar and Lokesh in Delhi,

REFERENCES

For works authored and co-authored by Doreen Massey, please see Select Bibliography of Doreen Massey, beginning on p. 371.

Ahmed, S. 2006a. *Queer Phenomenology: Orientations, Objects, Others*. Durham, NC: Duke University Press.

Ahmed, S. 2006b. "Orientations: Toward a Queer Phenomenology". *GLQ: A Journal of Lesbian and Gay Studies* 12 (4): 543–74.

Bakker, I. 2007. "Social Reproduction and the Constitution of a Gendered Political Economy". *New Political Economy* 12 (4): 541–56.

Benhabib, S. & D. Cornell 1987. *Feminism as Critique: Essays on the Politics of Gender in Late-capitalist Societies*. Cambridge: Polity.

Benston, M. 1969. "The Political Economy of Women's Liberation". *Monthly Review* (September): 1–14.

Bezanson, K. & M. Luxton (eds) 2006. *Social Reproduction: Feminist Political Economy Challenges Neoliberalism*. Montreal: McGill University Press.

Butler, J. 2009. "Performativity, Precarity and Sexual Politics". *AIBR: Revista de Antropologia Iberoamericana* 4 (3): i–xiii.

Clough, P. T. 2013. "Rethinking the Social and the Psyche". *S&F Online* 11.1–11.2 (Fall 2012 – Spring 2013). http://sfonline.barnard.edu/gender-justice-and-neoliberal-transformations/rethinking-the-social-and-the-psyche/ (accessed 18 January 2018).

Collins, J. 2016. "Expanding the Labor Theory of Value". *Dialectical Anthropology* 40 (2): 103–23.

Collins, J. 2017. *The Politics of Value: Three Movements to Change How We Think about the Economy*. Chicago, IL: University of Chicago Press.

Combahee River Collective Statement 1986 [1977]. *The Combahee River Collective Statement: Black Feminist Organizing in the Seventies and Eighties*. Albany: Kitchen Table: Women of Color Press.

Dalla Costa, M. & S. James 1975. *The Power of Women and the Subversion of Community*. 3rd edition. Montpelier, VT: Falling Wall Press.

Edelman, L. 2004. *No Future: Queer Theory and the Death Drive*. Durham, NC: Duke University Press.

Federici, S. 2003. *Caliban and the Witch: Women, the Body and Primitive Accumulation*. Brooklyn: Autonomedia.

Federici, S. 2012. *Revolution at Point Zero: Housework, Reproduction, and Feminist Struggle*. Brooklyn, NY: Common Notions, PM Press.

Fraser, N. 2014. "Behind Marx's Hidden Abode". *New Left Review* 86: 65–72.

Fraser, N. 2016. "Contradictions of Capital and Care". *New Left Review* 100: 99–117.

Gibson-Graham, J.-K. 1996. *The End of Capitalism (As We Knew It): A Feminist Critique of Political Economy*. Oxford: Blackwell Publishers.

and Akash Barman and Srujana Boddu in Hyderabad. We are grateful to the American Council of Learned Societies (ACLS) and the American Institute of Indian Studies (AIIS) for underwriting our research collaboration.

Glenn, E. N. 1992. "From Servitude to Service Work: Historical Continuities in the Racial Division of Paid Reproductive Labor". *Signs* 18 (1): 1–43.

Jain, D. & N. Banerjee 1985. *Tyranny of the Household: Investigative Essays on Women's Work*. New Delhi: Shakti Books.

Katz, C. 2001. "Vagabond Capitalism and the Necessity of Social Reproduction". *Antipode* 33 (4): 709–28.

Meehan, K. & K. Strauss (eds) 2015. *Precarious Worlds: Contested Geographies of Social Reproduction*. Athens, GA: University of Georgia Press.

Mies, M. 1986. *Patriarchy and Accumulation on a World Scale: Women in the International Division of Labour*. London: Zed.

Mitchell, K., S. A. Marston & C. Katz (eds) 2004. *Life's Work: Geographies of Social Reproduction*. Malden, MA and Oxford: Blackwell.

Muñoz, J. E. 2009. *Cruising Utopia: The Then and There of Queer Futurity*. New York: NYU Press.

Nicholson, L. 1987. "Feminism and Marx: Integrating Kinship with the Economic". In S. Benhabib & D. Cornell (eds) *Feminism as Critique: Essays on the Politics of Gender in Late-Capitalist Societies*. Cambridge: Polity Press.

Rogaly, B. & K. Qureshi 2017. "'That's Where my Perception of it all was Shattered': Oral Histories and Moral Geographies of Food Sector Workers in an English City Region". *Geoforum* 78: 189–98.

Rao, A. (ed.) 2003. *Gender and Caste*. New Delhi: Kali for Women.

Scott, J. 1986. "Gender: A Useful Category of Historical Analysis". *The American Historical Review* 91 (5): 1053–75.

Shah, N. 2011. *Stranger Intimacy: Contesting Race, Sexuality, and the Law in the North American West*. Berkeley, CA: University of California Press.

Weeks, K. 2011. *The Problem with Work: Feminism, Marxism, Antiwork Politics, and Postwork Imaginaries*. Durham, NC: Duke University Press.

CHAPTER 25

GLOBAL FACTORY, SUPPLY CHAINS AND SPATIAL DIVISIONS OF LABOUR AT THE MEXICO–US BORDER

Christian Berndt

The debate around the North American Free Trade Agreement (NAFTA) continues to be dominated by familiar one-sided positions. On the one hand a direct line can be drawn from Ross Perot's "giant sucking sound" (*New York Times* 1992) to Donald Trump's (2016a) representation of NAFTA as "worst trade deal in history" benefiting Mexico at the expense of the US. Early accounts of NAFTA as a win–win-scenario, on the other hand, are continued today by representations of an ever-deeper global division of labour characterized by interlocking supply chains. Both sides mobilize their trivial geographies: a deeply entrenched methodological nationalism turned political chauvinism and contrasting imaginations of globalization as "the economic equivalent of a force of nature" (Clinton 2000). But the NAFTA debate demonstrates also that the zeitgeist has changed during the last 25 years or so. Tales of economic laws of nature are increasingly eclipsed by a mix of frustration and anger that is predominantly inward-looking and driven by diffuse longings for the good old Fordist times.

Doreen Massey would have been critical of this development, but almost certainly not too surprised. It seems to be increasingly difficult to avoid the impulse to either succumb to a globalizing neoliberal vision or fall back to methodological nationalism. As different as they seem to be, both positions are connected to models of development that are part of the same trajectory of modernization. This is why Bruno Latour has referred to this view as "modernizing the modernization" (Latour 1998: 1). One way to escape this dilemma is to recall Massey's repeated reminder of the "structured divides, the necessary ruptures and inequalities, the exclusions, on which the successful prosecution of [capitalist modernity] itself depends" (Massey 2005: 84) and her earlier insistence that "globalization of social relations is yet another source of (the reproduction of) geographical uneven development, and thus of the uniqueness of place" (Massey 1991a: 29).

The Mexico–US border has long been a paradigmatic site on which to study these inequalities. It is here that political borders get decentred and turn into borderlands, "a vague and undeterminate place (…) in a constant state of transition" (Anzaldúa (2007 [1987]: 25). Choosing with Ciudad Juárez a paradigmatic site for this condition as an example, my contribution has two aims. The first is to map the city's entanglement with the global factory as a deeply contradictory process defying easy narratives of linear modernization and the trivial geographies still informing the globalization debate. This is the purpose of the first two sections of this chapter. The second aim is to reflect why it has become so problematic to formulate progressive "left" alternatives. I attempt to do this mainly in the final, third section.

THE GLOBAL FACTORY IN A REARTICULATED SPATIAL DIVISION OF LABOUR

> It used to be cars were made in Flint and you couldn't drink the water in Mexico. Now, the cars are made in Mexico and you can't drink the water in Flint. (Trump 2016b)

Approaching "Naftaland" from the Mexico–US borderlands, we enter the territory of the mythical "global factory", the ever-changing social and spatial arrangement of production sites, capital flows, workers that is not only "a site of investment and exploitation" but also "a producer of meaning, one that frames manufacturing activity as a stage of development" (Werner 2016: 47). It is exactly this frame that stylized accounts of the arrival of the global factory mobilize in cities such as Ciudad Juárez. The region's more recent history is represented as one of (economic) modernization and a linear evolution of development stages: the implementation of the so-called Border Industrialization Program (BIP) in 1965 marked a turning point from the import substitution area, followed by the opening of the national economy during the 1980s and reaching a temporary high point with NAFTA from 1994 onwards. At a closer look, however, Mexico's apparent journey northwards has not been nearly as straightforward.

In the collective memory of the border region, BIP is synonymous with the maquiladora industry (MI). What apparently began as a peripheral exception to import-substituting industrialization, eventually turned out to be an export-oriented Trojan horse, skillfully inserted and subsequently made bigger with the help of powerful players on both sides of the border (see Berndt 2004, ch. 2). Around the "*maquiladora*" a flexible discourse emerged that combined logics of territorial difference (e.g. wage gap, regulatory difference) with tropes of connectivity and common interests. These representations connect with a larger linear modernization discourse that has a strong regional presence. One manifestation is a stylized three-generation model of maquila plants popularized by the Mexican economist Jorge Carrillo in the 1990s (from "manual labour intensive" to "rationalization of manual labour" to "highly competent skilled labour" companies; see Carrillo & Hualde 1998; Carrillo & Lara 2005). Rather than an exact description of the reality the role of the narrative is a legitimizing one, providing a normative frame with which to make sense of developments "on the ground" that veils what are in fact deeply contradictory and ambivalent processes. Let me illustrate this with an exemplary case study.

My example is Delco, an unassuming US American maquiladora plant located in Ciudad Juárez whose twisted biography I was able to follow over a time period of seven years from 1999 to 2006, with a follow-up visit in 2011.[1] From the outside Delco appears to be a good example for the modernization narrative. When the plant began its life in 1986 as the first and only international production facility of a Massachusetts based producer of mechanical and electronic switches, it may have looked like a typical first generation manual assembly plant. This apparently changed dramatically in 2000, when a production unit for a new generation of switches (membrane switches) and the respective jobs were relocated from the US headquarters. Together with the new product came more sophisticated production processes that had stringent "production hygienic" requirements. This led to the installation of a cleanroom, a hermetically sealed unit that quickly became the symbol for progress amongst plant workers.

At a closer look, however, this seemingly straightforward story of upgrading has to be qualified. There are four reasons for this. First, the Juárez plant was not "modernized" across the board. The majority of the workers continued to perform relatively simple manual tasks, predominantly handling mechanical switches. Second, the work regime remained surprisingly similar despite the different technological conditions of the work tasks in question. All workers continued to be subject to the sticks and carrots regime so characteristic for this type of activities. Third, while the addition of new production units created about 60 jobs, the plant's workforce as a whole shrank from 650 to 400 during the period of 2000 to 2002. This had mainly to do with the simultaneous relocation of production to Delco's new facility in Dongguan, China. This was true for Juárez as a whole, the city losing over 70,000 jobs roughly a quarter of its maquiladora employment in the same period. China had just entered the world stage as a serious player in the global division of labour after becoming a member of the WTO.

---

1. The findings of this research are published in Berndt (2004). Unless referenced otherwise, the data presented in this essay are taken from this book.

Fourth, Delco's history took a somewhat unexpected twist in 2005 and 2006. When I returned to the plant in 2006, the modern *linea membrana* was gone, leaving a glaring hole on the factory floor. A US company had bought the production line. This company opened a new plant in Ciudad Juárez and simply relocated machinery, equipment and workers. Only a year later, Delco as a whole was sold to another US company. The new owner moved more of the high-volume product to Dongguan, bringing in additional production from the US. But the internal balance of power had changed irrevocably. In 2006 the Dongguan plant handled more product than the factory in Juárez for the first time since production had started in China.

This is an example of how the linear modernization discourse is translated geographically. In this discourse of the "not yet" the US American present is Mexico's future, the latter chasing a goalpost that is constantly moving away. In contexts such as the maquiladora industry corporate upgrading does neither mean an end to the subordinate position in global supply chains nor a marked improvement for a large proportion of the workforce and also the cities that emerged around them. A look at the development of so-called "labour compensation costs" supports this judgement (see Figure 25.1). The

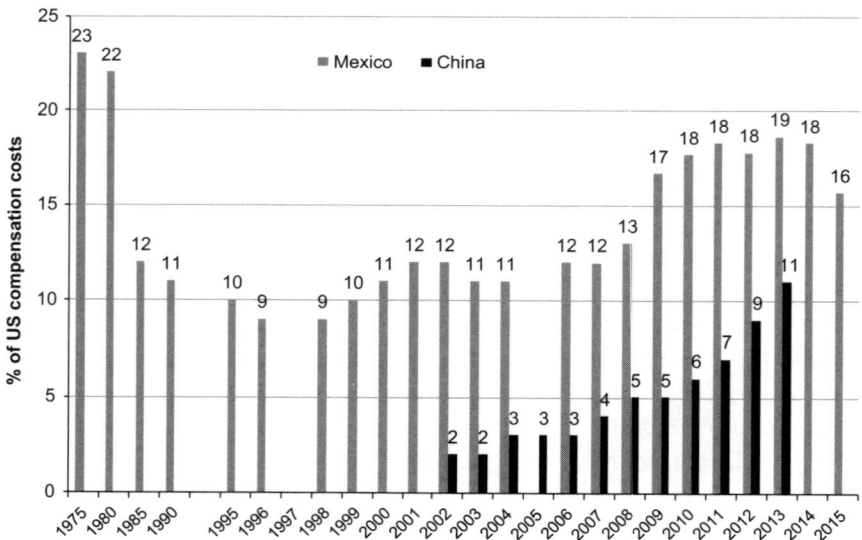

Figure 25.1 Mexican hourly compensation costs in manufacturing, as percentage share of costs in the United States, 1975–2015

Data sources: US Bureau of Labor Statistics International Comparisons of Hourly Compensation Costs for Production Workers in Manufacturing (various issues); The Conference Board, International Labor Comparisons program, April 2016 (www.conference-board.org/ilcprogram/index.cfm?id=38269#Table1; 25.6.17).

gap had narrowed as Mexican wages rose from 10 per cent in 1995 to 16 per cent of the US equivalent in 2015. It is worth noting, however, that China reduced the gap much faster than Mexico. In addition to this, it should be added that the 19 per cent reached in 2013 are still below the figures for the mid-1970s. And finally, there is widespread agreement amongst observers of everyday life at the Mexico–US border that maquiladora salaries continue to be a far cry from being living wages (Animal Político 2016).

Wage differentials with China play an important role for developments in Juárez. Downturns in the 2000s coincided with increasing competition from China and the recent revival with evidence for rising wages for Chinese workers (see Figure 25.2). There is of course a host of other reasons explaining these developments, not least economic crises in the US, exchange rate adjustments, the narco-related violence in Mexico and a re-regionalization of supply chains more recently (for the latter point, see Peck 2017: 145). It is difficult to conclude from all this that cities such as Juárez have significantly improved their position in the global division of labour under NAFTA.

Another continuity concerns the way labour is disciplined in the plants. Rhetorical lip service to skilling and learning notwithstanding, the labour

Figure 25.2 Maquiladora employment, Ciudad Juárez, 1980–2017

Note: The Mexican government changed regulations for state programmes that support export plants; the numbers from June 2007 onwards can therefore not easily be compared to previous figures.

Data sources: INEGI 2012 – Estadística de la Industria Maquiladora de Exportación (numbers for 1980–2006); INEGI 2017 – Estadística Mensual del Programa de la Industria Manufacturera, Maquiladora y de Servicios de Exportación (2007–2017; subset manufacturing plants, direct and subcontracted workers); monthly figures.

regime is authoritarian and paternalistic, *operadores* and *operadoras* having to work long hours at often frenetic pace. This has been the case in the 1970s and 1980s (e.g. Fernández-Kelly 1983) as well as the 1990s and 2000s (Wright 1997; Berndt 2004). And this continues until today:

> "In the factory speed is everything – you have to be fast. But no matter how much you work you barely have enough to eat", a female maquila worker is reported of saying, a single mother, aged 37, who has packed ink cartridges at the Lexmark maquila 48 hours a week for six years earning 670 pesos ($35) a week – or $7 a day. (Carroll 2016)

What is more, the maquiladora labour regime continues to be deeply gendered, male supervisors and managers regularly exerting dominance over female bodies. A particularly disturbing aspect of this is the continuing discrimination of pregnant women.

A recent wave of unprecedented labour unrest in Ciudad Juárez, involving strikes and sit-ins at a number of plants – amongst them Foxconn and Lexmark – is evidence for the frustration of workers after more than twenty years of standstill under NAFTA. So, to return to the quote at the beginning of this section, while cars may perhaps have left Flint, they have never really arrived at the northern Mexican border. And while US industrial workers may have lost out to globalization, their Mexican counterparts did profit very little.

GEOGRAPHIES OF SUPPLY CHAIN CAPITALISM

Delco's corporate history is emblematic for global factory operations. There is a constant comparison of production costs and quality benchmarks, and relentless adjustment to an extent that renders claims of linear upgrading problematic. This hints at a more general problem in the current debate around NAFTA and free trade in the US: the shortsighted conflation of industrial restructuring in the US as a causal effect of NAFTA. This totally neglects what may be called the *longue durée* of the development of the spatial division of labour, going along with the emergence of complex outsourcing systems that increased social and spatial complexity (Silver and Arrighi 2011: 62–3). Anna Tsing (2009) has coined the term "supply chain capitalism" to account for this development.

This transformation went hand in hand with an emerging "global market for labor and for sites of production" (Silver and Arrighi 2011: 62) that profoundly reworked north–south relations. These are contradictory entanglements of "dependent development" (Sklair 1989: 16) that are necessary for global capitalism to thrive, but which both the market-radical free traders and national-populist protectionists try their best to veil. The former mobilize the modernizing tale of the "inevitability of market expansion" (Massey 2005: 82). This is the Western-centred belief in one history according to which the future and the past of the United States are the future and the past of Mexico. Bruno Latour (1998: 3, 7) associated this position with the "globalizing Right" and contrasted it with the perspective of a second "ethnicizing Right" which chimes well with the current advance of national-populism and protectionism. This is a relativist position according to which the US and Mexico have different pasts and futures, an essentialist reading that apparently justifies radical measures of separation in the name of the "national interest". The corresponding geographical imagination is the powerful idea of "defensible places, of the rights of 'local people' to their own 'local places'" (Massey 2005: 86), that Trump's reference to "bad hombres down there" epitomizes quite well (Associated Press 2017).

It becomes very obvious when looking more carefully at Mexico–US relations that both positions have their limits and have been adjusted all-too-readily to the "realities" on the ground. From very early on in the NAFTA debate the "globalizing Right" argued for a selective regime that prioritized certain cross-border movements over others. This has been most clearly visible when looking at Mexican migrant labour. Pro-trade US government representatives have made it repeatedly clear that strict immigration controls can only be eased when living conditions in Mexico have caught up with the ones north of the border (e.g. former Secretary of State Madeleine Albright in Reuters 2000).

But there is a similar gap between what is said and what is being done on the side of the "ethnicising Right". The rhetoric of "America first" has its obvious limits given the dependence of US producers and consumers on cross-border flows in the global division of labour. It is quite ironic therefore that both positions, as different as they are, lead to quite similar results: the figuration of a deeply asymmetric transnational market order. This is a market order so contradictory that it ought to be impossible to contain and reproduce it. But this is exactly what may be referred to as modernity's sleight of hand: As long as there is an appearance of order and the particular framing of the state of affairs remains uncontested, it is possible to do almost the exact opposite, entangling what should be disentangled, disconnecting what in fact is deeply intertwined (see Latour 2003: 38).

What gets evident therefore is how global capitalism always involves both – the erasure of territorial borders and growing importance of cross-border flows, and the stubborn persistence of bordered inequalities. For

Massey (2005: 86) it has been the spatial organization of labour in particular that is of paramount importance, involving as it does "a geography of borderlessness and mobility, and a geography of border discipline". There are few examples that illustrate these insights better than the border regime that emerged around NAFTA.

At the same time when there were attempts to accelerate the flows of goods and capital, freedom of movement for ordinary people was severely reduced. For NAFTA also marks the beginning of the so-called militarization of the border (Spener 2003). Another crucial moment in this transformation occurred in the wake of 9/11 and a public discourse in the US that centred on the question of territorial security and sovereignty. The US government pushed its NAFTA partners to agree to a border regime that promised the best of both worlds: secure territorial borders and unhindered cross-border trade. All this culminated in the idea of the "smart border" (see Sparke 2006). On the southern border George W. Bush used this term for the first time in public when he visited new high-tech border installations in El Paso in 2002 (Bush 2002). The "smart border" terminology is no coincidence. This was the time of the Iraq War and public representations of clinical, surgical air strikes. This is also the promise of the smart border: intelligent technology that enables desirable cross-border movements while blocking undesired ones – illegal substances and drugs, cheap "Chinese" counterfeit products, and of course Trump's "bad hombres".

This is a graphic example of how today's border regimes sort mobilities in a deeply asymmetric way. This may have always been a quality of borders, but has acquired a new quality with the increasing use of surveillance technologies. Today's borders give form to ambivalent border regimes that are a necessary condition for the construction of global markets and trade systems. Yet, in order for these markets to work, these ambivalences have to be hidden.

A good example for this is the figure of the "illegal migrant". The smart border does not of course keep out Mexican migrants. It turns into a kind of filter that produces a much-needed labour segment. It is mainly the able-bodied and strong who are able to cross. And at the very moment of crossing, the border transforms a heterogeneous group of human beings into an abstract category: the illegal alien that is tailor-made for particular labour market segments. Rather than talking about borders we should therefore better talk about b/ordering. Understood as ambivalent double play of entanglement and disentanglement, of territorialization and deterritorialization b/ordering is not limited to the political line demarcating the territories of supposedly sovereign nation-states. The Mexico–US border emerges in Mexican fields and greenhouses where all sorts of fruit and vegetable are produced for US consumers. The line demarcating export from lower grade produce is drawn in the field, at the sorting machines, in the packing houses – farmers scrutinize the collected produce rejecting substandard crop with considerable implications for farmworkers' salaries, US government inspectors patrol the installations, US officials perform fast-track custom procedures etc. (see Berndt & Boeckler 2011).

In the maquiladora factories similar things happen. At Delco the arrival of the new "advanced" production unit changed the power-geometries in the plant. In the representations of managers and workers a more conventional production space emerged as a symbol for "cheap" and "backward" Mexican assembly work against the new production unit as a sign for modernized skilled labour on the one hand and the English language-dominated administrative area on the other. There was also a small extraterritorial US American space, the packing and shipping unit that worked under fast-track regulations that effectively move border controls into the plants.

The mobile Mexico–US border poses a challenge to theories of regional integration. From the perspective of mainstream economics, the current situation is a result of the successful dismantling of tariffs with NAFTA and the increasing importance of non-tariff barriers to trade. Exerting their differentiating force not at the political border itself, the latter are also referred to as "behind-the-border barriers" (Haggard 1995: 2). These concern a potentially unlimited array of regulatory issues that were "once deemed wholly domestic" (*ibid.*) and have increasingly become the subject of international negotiations aiming at harmonization and standardization. In contrast to the "shallow integration" of classical free trade mainstream economists have called this process "deep integration", actively to be managed by political actors across borders (see Haggard 1995).

From the point of capital the b/ordered realities of global markets continue to provide the best of possible worlds. After the neoliberal moment of shallow integration intensified trade and capital flows across borders other dimensions of difference got accentuated and exploitable. These have a lot to do with labour (regulations, wage differentials, education), but include access to land, financial capital and also consumers. There is a constant tension between standardization and differentiation, not least since every move towards harmonization inevitably opens up new opportunities for arbitrage.

This can be observed at different spatial scales, from the fragmented space of "Naftaland" to the shopfloor. At Delco, for instance, standardization had been a result of the plant's increasing entanglement with global supply chains. The arrival of advanced switches coincided with the introduction of the latest quality management templates of the time, symbolized above all by Six Sigma, a global quality-benchmarking device that originated at Motorola in 1981. What the introduction of Six Sigma brought to the fore at Delco were differentials in quality performances, the Juárez plant struggling to meet the ambitious targets set by the corporate headquarters. The local plant management rationalized these differences in typical ways. On the one hand, decision-makers made sure that Mexico's progress is scaled against US history. "Mexico (...) is just back in time and that is how the US was back then" (Mexican manager, 27 February 2002). On the other hand, the factory in China was positioned in the accounts as producing both cheaper and at better quality.

When I visited the former plant manager in El Paso 2011, however, his opinion appeared to have changed: "I realized I should have done it the other way ... keep everything in Mexico. The quality sucks, China is too far away to control everything, wages are getting high, the culture." The point of this is not to discuss what is right or wrong. Rather, what this illustrates is the extent to which decision-makers at more peripheral nodes of global supply chains have to cope with a constantly changing landscape of shifting frames and dis/connections. These follow no obvious long-term logic and pose difficult challenges. The cultural stereotypes employed in these accounts serve as attempts to re-establish some sort of order. Although the arbitrary message of these statements unmasks them as purely representational, they have all too real, material effects. This holds in particular for the workers in Mexico and China, given that they legitimize what continue to be "third world" working conditions.

WHAT'S NEXT?

The preceding discussion illustrates how the geographical illusion of a national age has become increasingly difficult to maintain. The lines demarcating "core" and "periphery" move and multiply to an extent that acknowledgment of this complexity becomes unsettling, both for society at large and political decision-makers. Doreen Massey very early recognized the challenges that arise from this complexity. She sought inspiration from the poststructural insight that people have multiple identities and used this idea to conceptualize what she referred to as a more open "global sense of place" against a more inward-looking local one (e.g. Massey 1991a). The latter involves attempts of reterritorialization via processes of othering that negate and suppress difference on both sides. It is against this that Massey demanded recognition of these differences and inequalities and of the fact that these geographical identities and the borders demarcating them have nothing natural about them. Instead they have to be negotiated, there is a constant need for b/ordering. A reconfigured notion of the border is therefore key to this project. But this comes at a cost. It means that we have to take responsibility not only for our own lives and the lives of those in our immediate environment, but also "at a distance". Massey was adamant that this can obviously be demanded more easily from a privileged position. It is quite a different thing, if you happen to live at the margins of the neoliberal world order.

My analysis illustrates how an omnipresent, but malleable modernization dispositif has erased these contradictions from view. But the tide appears to have turned. The nationalist, populist projects that currently develop both on the right and left are attempts to turn back the clock, to undo modernization gone awry by redirecting development to the trajectory of methodological nationalism. The geographies of protectionism, whether represented in a more progressive or regressive tone, are inherently national-territorial ones. In order to be successful they crucially involve the establishment of a national consensus across axes of social differences. In the US and in Europe right-wing populist movements have captured the minds of what once was the working class and created a kind of reactionary bloc in which resistance against the globalizing elite is hopelessly entangled with xenophobia, homophobia and misogyny (represented most clearly by events in the US and post-Brexit UK, but also visible in Germany and France; with a view to France see, for instance, Eribon 2013). But it is equally important to ask what made these regressive social formations possible. After the radical market neoliberalism of the 1980s had polarized societies in the global North along the traditional axes of left and right, the subsequent reaction in the form of "roll out neoliberalization" (Peck & Tickell 2002) led to a disintegration of this established political order. On both sides of the political spectre the winners and losers of globalization had increasingly less in common and looked for new allies. The "upper" segments of society formed what Bruno Amable and Stefano Palombarini (2014) have termed a *"bloc bourgeois"* across old right–left divides (illustrated, for instance, by the rise of new Labour in the UK, Clinton's "new" democrats, Gerhart Schröder's reinvention of social democracy in Germany, and more recently the emergence of Macron's *"En Marche"* movement in France). The disenfranchised segments of the population in turn sought refuge in populist movements, all of them in some way or the other formulating exclusionary projects that appeal to some kind of "national consciousness". In most cases these new formations still have a fragile societal base. What we are currently witnessing are struggles over hegemony in many of the countries mentioned above.

These struggles are, however, neither solely fought around the question of whether there should be more or less market nor about less or more state involvement. This neglects the extent to which market and state agents have quite different interests, just like society at large. In different contexts these struggles play out differently. It is of crucial importance therefore to adopt a decentred understanding of history and geography. This includes the reminder that participating agents' histories may be interwoven, but cannot be projected on a single, universal plain modelled after the experiences of a handful of Western European countries. And it means to take seriously Massey's (2005: 82) demand not to deny the "essential multiplicities of the spatial". The acknowledgement of these multiplicities has two important consequences. First, the need to be wary of simplistic geographies that locate neoliberal capitalist pressure always outside at the global scale and that routinely delegate resistance against neoliberalization to the local level; second, the problem that political movements in principle are essentialist projects that strategically and/or violently suppress internal differences in order to bring external differences to the fore. At least in my view the response to the regressive developments along the US–Mexico border cannot be simply to "remodernize modernization". Rather it would demand a different tactic. Massey (2005: 83) argued that critique and struggle need to be directed at the relations that mutually construct both the global as an abstract spatial scale out there and the local as the realm to be defended. I read this as a plea to resist all-too-simple promises of a future that will be somewhat less complex, whether in the vision of the globalizing right (less regulation, more market, more mobility) or in "localist" dreams on the right and left (cutting global economic ties, reclaiming national sovereignty, regionalism, less mobility). This is possible because both projects have their obvious limits. They work only as long as their exclusionary character remains hidden or accepted as a quasi-natural fact. As soon as it gets obvious that (global) "market forces are imposed on some but not others" (Hall *et al.* 2013: 14), and that regionalist projects are produced with and against their various outsides, contradictions and tensions get visible that provide opportunities for alternatives.

REFERENCES

For works authored and co-authored by Doreen Massey, please see Select Bibliography of Doreen Massey, beginning on p. 371.

Amable, B. & S. Palombarini 2014. "The Bloc Bourgeois in France and Italy". In H. Magara (ed.) *Economic Crises and Policy Regimes: The Dynamics of Policy Innovation and Paradigmatic Change*. Cheltenham: Edward Elgar, 177–216.

Animal Político 2016. "Cambios al TLC que Quiere Lograr Trump Amenazan a la Industria Maquiladora de Juárez, Chihuahua". *Animal Politico* 11 December. www.animalpolitico.com/2016/12/tlc-trump-juarez-maquiladoras (accessed 9 July 2017).

Anzaldúa, G. 2007 [1987]. *Borderlands – La Frontera: The New Mestiza*. 3rd edition. San Francisco, CA: Aunt Lute.

Associated Press 2017. "Trump Reportedly Threatens to Send U.S. Military to Mexico in Call with Mexican President". *Los Angeles Times* 1 February. http://beta.latimes.com/nation/nationnow/la-na-pol-trump-mexico-call-20170201-story.html (accessed 9 January 2018).

Berndt, C. 2004. *Globalisierungs-Grenzen: Modernisierungsträume und Lebenswirklichkeiten in Nordmexiko*. Bielefeld: Transcript.

Berndt, C. & M. Boeckler 2011. "Performative Regional (Dis-)Integration: Transnational Markets, Mobile Commodities and Bordered North–South Differences". *Environment and Planning A* 43 (5): 1057–78.

Bush, G. W. 2002. "Radio Address by the President of the Nation". 23 March. https://georgewbush-whitehouse.archives.gov/news/releases/2002/03/20020323.html (accessed 18 January 2018).

Carrillo, J. & A. Hualde 1998. "Third Generation Maquiladoras? The Delphi-General Motors Case". *Journal of Borderlands Studies* 13 (1): 79–97.

Carrillo, J. & A. Lara 2005. "Mexican Maquiladoras: New Capabilities of Coordination and the Emergence of a New Generation of Companies". *Innovation: Management, Policy and Practice* 7 (2–3): 256–73.

Carroll, R. 2016. "Juárez Factory Workers are Protesting for their Rights with an 'Occupy-style' Sit-In". *The Guardian* 5 April. www.theguardian.com/world/2016/feb/17/juarez-factory-workers-protest-rights-occupy-style-sit-in-lexmark (accessed 9 January 2018).

Clinton, B. 2000. "Remarks at Vietnam National University in Hanoi, Vietnam". *The American Presidency Project* 17 November. www.presidency.ucsb.edu/ws/?pid=1038 (accessed 8 February 2017).

Eribon, D. 2013. *Returning to Reims*. Los Angeles, CA: The MIT Press.

Fernández-Kelly, M. P. 1983. *For We Are Sold, I and My People: Women and Industry in Mexico's Frontier*. Albany, NY: State University of New York Press.

Hall, S., D. Massey & M. Rustin 2013. "After Neoliberalism: Analysing the Present". *Soundings* 53: 8–22.

Haggard, S. 1995. *Developing Nations and the Politics of Global Integration*. Washington, DC: Brookings Institution.

Latour, B. 1998. "Ein Ding ist ein Thing – a (Philosophical) Platform for a Left (European) Party". *Concepts and Transformation* 3 (1–2): 97–112.

Latour, B. 2003. "Is Re-Modernization Occurring – And If So, How to Prove It?: A Commentary on Ulrich Beck". *Theory, Culture and Society* 20 (2): 35–48.

New York Times 1992. "The 1992 Campaign; Transcript of 2nd TV Debate Between Bush, Clinton and Perot". *New York Times* 16 October. www.nytimes.com/1992/10/16/us/the-1992-campaign-transcript-of-2d-tv-debate-between-bush-clinton-and-perot.html?pagewanted=all (accessed 9 January 2018).

Peck, J. 2017. *Offshore: Exploring the Worlds of Global Outsourcing*. Oxford: Oxford University Press.

Peck, J. & A. Tickell 2002. "Neoliberalizing Space". *Antipode* 34 (3): 380–404.

Reuters 2000. "Cuando en México Mejore el Nivel de Vida, EU Abrirá la Frontera: Albright". *La Jornada* 2 December. www.jornada.unam.mx/2000/12/02/028n1pol.html (accessed 9 January 2018).

Silver, B. J. & G. Arrighi 2011. "The End of the Long Twentieth Century". In C. Calhoun & G. Derluguian (eds) *Business As Usual: The Roots of the Global Financial Meltdown*. New York: New York University Press, 53–68.

Sklair, L. 1989. *Assembling for Development: The Maquila Industry in Mexico and the United States*. Boston, MA: Unwin Hyman.

Sparke, M. 2006. "A Neoliberal Nexus: Economy, Security and the Biopolitics of Citizenship on the Border". *Political Geography* 25 (2): 151–80.

Spener, D. 2003. "Controlling the Border in El Paso Del Norte: Operation Blockade or Operation Charade?" In P. Vila (ed.) *Ethnography of the Border*. Minneapolis, MN: University of Minnesota Press, 182–98.

Trump, D. 2016a. "Campaign speech in Monessen, Pennsylvania". *Time* 28 June. http://time.com/4386335/donald-trump-trade-speech-transcript (accessed 9 July 2017).

Trump, D. 2016b. "Speech at the New York Economic Club". *Time* 15 September. http://time.com/4495507/donald-trump-economy-speech-transcript (accessed 20 February 2017).

Tsing, A. 2009. "Supply Chains and the Human Condition". *Rethinking Marxism* 21 (2): 148–76.

Werner, M. 2016. *Global Displacements: The Making of Uneven Development in the Caribbean*. Malden, MA: Wiley-Blackwell.

Wright, M. 1997. *Third World Women and the Geography of Skill*. Unpublished PhD dissertation. Baltimore, MD: John Hopkins University.

CHAPTER 26

PLACE AND THE POWER-GEOMETRIES OF MIGRATION

Jennifer Hyndman and Alison Mountz

INTRODUCTION

Doreen Massey was an intellectual powerhouse and feminist inspiration in so many parts of the discipline of geography and to so many people (like us). As a feminist and a Marxist, she refused the unspecified grand narratives of theory, a journey that catalysed the writing of *Spatial Divisions of Labour* (1984b). In the book, her careful account of the historical and material (re)making of place, and the selection of production sites based on the features of a place, had a major impact on the industrial location literature (cf. Massey 1973). *Spatial Divisions of Labour* came out in second edition in 1995, and was a foundational text for one author's (Hyndman) graduate research, facilitating her "conversion" to a PhD in geography.

For the second author (Mountz), Doreen Massey's writing offered one of her earliest encounters with feminist geography. While she discovered Massey as an undergraduate student, Massey's geographical thinking about gender drew her into graduate study in geography. As a feminist, Massey critiqued David Harvey and Ed Soja's influential books (*The Condition of Postmodernity* and *Postmodern Geographies,* respectively*)* for their high theory that subsumed patriarchy to relations of capital, thus enacting "flexible sexism". The phrase was a reference to – and critique of – the shortcomings of emerging concepts of capitalist restructuring at the time, wedded to the term "flexibility" (e.g. Harvey's "flexible accumulation") but indifferent to the centrality of social difference in achieving its effects. In so doing, she reminded economic geographers who might otherwise ignore gender relations that "it is often forgotten to what extent women were the first labour-force of factory-based capitalism … it was in the cotton industry around Manchester that the challenge was first laid down" (Massey 1994: 195). She also positioned her work early on as scholarship influenced by political economy, poststructuralism and feminist thinking. In this sense, she was a hybrid, expansive thinker and this was exciting. Her conception of place as produced by and productive of social, cultural, political and of course economic relations is perhaps best known and illustrative of this hybrid conceptual framing.

When we reflect on Massey's work, her concepts richly captured the process of globalization; they have lingered with us over the course of decades. From her conceptualization of historicized regional economic change as "rounds of accumulation" in *Spatial Divisions of Labour* to the topological notion of "power-geometry", our geographical imaginations have been augmented by Massey's scholarship. This chapter begins with the importance of Massey's historical, spatial approach to regional change in *Spatial Divisions of Labour*, one of the strongest if not most pervasive theorizations of globalization in geography. We then turn to her later notion of the "power-geometry" of time-space compression and its influence on thinking around mobility and migration in geography, including in our own work.

HISTORICIZING REGIONAL CHANGE: "ROUNDS OF ACCUMULATION"

In the early 1980s, Fröbel *et al.* (1980) outlined a "new international division of labour" (NIDL) between First and Third World countries and traced capital's search for new greenfield, low-wage meccas. Shortly thereafter, Massey (1984b) scaled down this grand narrative, which lacked a geographical analysis despite its multiple "case studies". She analysed *intranational* divisions of labour by city, industry and region. Footloose capital and downsized workers were not a singular or inevitable shift between global North and global South. Specific sites of job loss could be analysed in the context of spatial divisions of labour. If Fordism was no longer the main mode of production, post-Fordism took many forms and was expressed very differently across space.

Drawing on thinking from the French Regulation School, Massey (1984b) was clear that regulation – loosely defined as the national, international and institutional norms that govern and facilitated particular regimes of accumulation – also had to be considered in terms of locational decisions (Aglietta 1979 [1976]; Lipietz 1987). *Spatial Divisions of Labour* was situated within a cluster of related works by economists, geographers, and others that addressed the crisis of profitability faced by firms and the acute decline of mass production in the global North. Massey (1984b) argued that uneven development drives the spatial division of labour. Focusing on the UK, she noted that

> In the mid-1960s a new spatial division of labour became dominant within the United Kingdom, in which control functions were concentrated, even more than before, in London, scientific

and technical functions were clustered in the south-east … and direct production, while present throughout the country was a higher proportion of economic activity in the regions outside of the south and east. That new spatial division of labour was the outcome of a whole series of changes affecting different parts of the economy in different ways. (Massey 1994: 90–1)

While subtle, Massey's focus on "rounds" rather than "regimes" of accumulation makes her work far more geographical and accountable to the historical and material conditions of place than her non-geographer counterparts.

While the NIDL became the prevailing "regime of accumulation" for some scholars, regional production clusters also attracted attention as spatial expressions of neo-Fordist "flexible specialization". For example, alongside Massey, Michael Piore and Charles Sabel's (1984) *The Second Industrial Divide* analysed the decline of Fordism and the rise of more "flexible" relations of capital in unexpected places, like the textile districts of Northern and Central Italy where craft-based production in small firms and small batches thrived for a time. Sociologists Scott Lash and John Urry (1987) published *The End of Organized Capitalism*, further tracing this trend and the erosion of the vertical integration in the large corporate firm. The decentralization of businesses and a range of contributions to the production process were well underway.

Hyndman drew on Massey's thesis to analyse the spatial divisions of labour in Cork in the Republic of Ireland, a city of 150,000 people in 1989 with an official unemployment of 18 per cent. While several multinational corporations in the electronics industry were setting up in the early 1980s, traditional staple industries were also closing down. Apple Computers opened its doors in October, 1980 and Western Digital, a US electronics assembly firm that made PCs, followed suit in September, 1983. From 1983 to 1984, however, Cork suffered a triple blow as the Ford assembly plant downsized, and 500 workers lost their jobs. The Dunlop tyre manufacturing plant closed altogether, and the Verolme dockworks also shut down. The Ford plant had employed a large workforce since the 1920s; Dunlop had provided employment in the area since the 1930s; and dockworkers at Verolme had worked in Cork since the early 1950s. The loss of these unionized jobs, held mostly by men, that resulted from these shutdowns was devastating to the local economy, and unemployment spiralled to over 27 per cent (Hyndman 1989).

The new spatial division of labour of the 1980s created assembly and light industrial jobs in Cork, but these paid less, were more likely to be temporary, and were taken up by a more feminized workforce. Almost three-quarters of the employees were women, and over half worked on week-to-week contracts. Layoffs were common, and narrated by the employer as merely part of the business cycle in a temperamental hi-tech industry. In Cork, employers' attempts at control shifted from the labour process to the employment contract (Hyndman 1989). While Western Digital employees became members of the Irish Transport and General Workers Union (ITGWU), they had very little job security. Shorter-term jobs in the assembly plant could be subcontracted to non-union businesses outside the company, a condition unheard of during earlier "rounds of regulation" on the shop floor.

Along with Pfizer – a multinational chemical and pharmaceutical company – Apple and Western Digital became the largest employers in the city, ushering in a new but less prosperous "round of accumulation", and without employing many of the men who lost their jobs at Ford, Dunlop, and Verolme. As Massey observed in the 1980s, and reflected upon a decade later, multinational capital successfully mobilized "flexible sexism", a form of post-Fordist flexible specialization and locational rationality with a gendered twist (Massey 1994: 212).

Yet, Ireland was still appealing as a site for assembly not only because of its high unemployment rates and supply of relatively inexpensive, feminized labour, but also because it was within the European Union which charged a 17 per cent tariff for good assembled outside the EU. At the time of the research, labour costs in Ireland remained extremely competitive by international standards averaging 43 per cent of US, 59 per cent of German and 74 per cent of French rates (Hyndman 1989).

Reflecting on analogous, and indeed, linked processes of disinvestment in the UK, Massey did not simply *illustrate* the logic of location selection and strategy. Her framework enabled Marxist geographers (and sociologists) to rethink the locus of power relations *in relation to* economic activity. In Cork, Ireland, for example, the shift from Fordism to neo-Fordism, as the dominant economic activity allowed the metaphor of sedimentary layers to extend from "rounds of accumulation" to a new "round of union formation", the conditions of the latter shaped by the region's historical labour relations. These new union members faced degraded terms in the employment contract, ultimately rooted in – and reproducing – Cork's peripheral position within the EU.

The notion of rounds of accumulation, and the layers of dominant economic activity over time, prove useful today to historicize the uneven legacies of capital accumulation and disinvestment that coexist in a given site as economic activities – and the social relations that form in relation to them – wax and wane over time and across space. In our view, this metaphor becomes a catalyst for Massey's theorization of place (1993b) as a power-geometry of time-space compression.

Reflecting on her approach to globalization much later, Massey observed that "Everything is invented in particular places, and neoliberalism and globalization were quite significantly invented in London" (cited in Chanan & Salter 2012). The meaning of every place is a product of its interrelations with other places. In this reflection, one can witness the shift in Massey's thinking about places as sites where production is located and clustered, to a deeper, more nuanced way of thinking about place as itself a site produced through global mobilities and the uneven distribution of resources. For us, she expresses this way of thinking most fully in her later work on power-geometries.

POWER-GEOMETRIES

Massey developed her idea of power-geometry in her highly influential essay, "A Global Sense of Place", first published in *Marxism Today* in 1991. Power-geometry was her way of grounding and testing the concept of time-space compression, advanced by David Harvey (1989) in *The Condition of Postmodernity*, where he fleshed out Marx's generative but somewhat opaque assertion that capital develops through the "annihilation of space through time". Time-space compression suggested that people and places around the globe were now in fact closer together because of developments in technology and organization which accelerated turnover time for accumulation in late capitalism. But Massey observed that not all people and places were equally closer or more mobile in this new global milieu. On the contrary, globalization unfolded unevenly, leaving some with hypermobility and transnational interconnectedness, and others with far less access to the riches of the global economy, including the ability to move. Massey turned her feminist questions on time-space compression thus:

> [W]e also need to ask about its causes: what is it that determines our degrees of mobility, that influences the sense we have of space and place? Time-space compression refers to movement and communication across space, to the geographical stretching-out of social relations, and to our experience of all this. The usual interpretation is that it results overwhelmingly from the actions of capital, and from its currently increasing internationalization. On this interpretation, then, it is time space and money which make the world go around, and us go around (or not) the world. It is capitalism and its developments which are argued to determine our understanding and our experience of space. But surely this is

insufficient. Among the many other things which clearly influence that experience, there are, for instance, "race" and gender.

(Massey 1991a: 25)

She went on to explain that time-space compression happens unevenly
and "needs differentiating socially" (1991a: 25). To accomplish this, Massey
proposed the "power-geometry of time-space compression" (1991a: 26).
Massey's power-geometries offered an important corrective to much early
scholarship on the global economy that placed capitalism as the main driver
determining the movement of people and resources around the globe as relatively passive pawns. Instead, Massey wanted to understand the role that
socially situated people play in relation to globalization, work and the constitution of place:

> For different social groups, and different individuals, are placed in
> very distinct ways in relation to these flows and interconnections.
> This point concerns not merely the issue of who moves and
> who doesn't, although that is an important element of it; it is
> also about power in relation to the flows and the movement.
> Different social groups have distinct relationships to this anyway
> differentiated mobility: some people are more in charge of it than
> others; some initiate flows and movement, others don't; some are
> more on the receiving-end of it than others; some are effectively
> imprisoned by it.   (1991a: 25–6)

In her essay, Massey grounded this idea empirically by providing examples
of different kinds of subjects engaged in a range of global activities such as
consumption and labour.

Power-geometries are thus about place and identity; they offer a way to
understand the global distribution of power and resources and their lived
intimacies. Massey raised two important points in her essay that have endured
over the decades since its publication: first, the production of space and one's
mobility in it are not simply acts of individual choice; and second, mobility is
inherently political, social, economic and racialized. Mobility, displacement
and migration are all constituted through politico-spatial relations. Indeed,
power-geometries can be understood as an early form of intersectionality to
understand people, identity, place and mobility as relational and contextual
phenomena – long before contemporary geographers explored these ideas
and their spatial dimensions (Valentine 2007). We can read Massey's approach
to understanding the ways that intersecting identities and oppressions influence differential power-geometries, retrospectively, as an approach related

to feminist intersectionality, developed by Kim Crenshaw (1991) as a way to understand identities as multiple, intersecting axes. Massey argued that one could never simply understand all people as subject in similar ways to time-space compression, but rather that late capitalism would have differential impacts on people and their mobility and subjectivities in place, depending on their intersecting identities and lived experiences.

The new mobilities paradigm v. power-geometry

Mimi Sheller and John Urry (2006) coined the "new mobilities paradigm" with reference to the scale and scope of human movement globally. They argued, "it is not a question of privileging a 'mobile subjectivity', but rather of tracking the power of discourses and practices of mobility in creating both movement and stasis" (Sheller & Urry 2006: 211). The authors draw specifically on transnational feminist studies to account for processes and politics of exile, migration and transnationalism. Caren Kaplan (1996), in a similar vein, uses the concept of "deterritorialized nomadism" as a way to marginalize the centre and unsettle white, masculinist, imperial cultures of the west. While the feminist angles aim to understand who moves, how and why – questions that no doubt interested Massey – she would likely have refused these notions because of key missing ingredients: uneven development across world regions and within the borders of a given state or region; and the wholly unequal distribution of resources – social, cultural, political and economic. As Massey wrote, "different social groups have distinct relationships to this ... differentiated mobility: some people are more in charge of it than others" (1991a: 26). Massey's work tacitly challenges the mobilities paradigm to theorize such disparities.

How did these authors of the "new mobilities paradigm" (Urry 2000; Sheller & Urry 2006), and geographers writing along similar lines (e.g., Cresswell 2006) miss Massey's theorization of the intersectional analysis of differential mobility? Obviously, there are theoretical differences here: Urry, Sheller and Cresswell might well be identified as poststructuralist in their orientation, rather than Marxist feminist. In fact, many would read Massey, especially from *For Space*, as a poststructuralist, but with a strong commitment to materialist analysis. Massey never equated poststructuralism as the abandonment of questions of class, labour, uneven development, etc., but instead drew on poststructuralism – and critiqued its limited understanding of space – to forge a more robust understanding of sociospatial difference. Still, we do not take these theoretical differences to be a sufficient reason to "miss" Massey's work, both on place as a dynamic and always "in the making" site of social, cultural and economic relations, and on power-geometry and the spatial expression of disparate power relations more generally, which the new mobilities paradigm largely ignores.

John Urry sadly died during the same week as Doreen Massey, in March 2016. His later work (Urry 2000: 35) demonstrated how mobilities transformed sociology as a discipline in his "brave manifesto", a book that argued "society" was the wrong object of enquiry for sociologists, and that "mobility" was its proper focus. Such thinking aligns well with Massey's relational understanding of economic activity, but also about the production of place. Networks, flows, technology and mobilities undermined the idea of a fixed society, and Urry, like Massey, focused on more relational notions of ordering: movement, mobility and contingent sequences, rather than sedentary people, structures and social order. Thus,

> If we rethink culture ... in terms of travel then the organic, naturalizing bias of the term culture – seen as a rooted body that grows, lives, dies, etc. – is questioned. Constructed and disputed historicities, sites of displacement, interference, and interaction come more sharply into view.
> (Clifford 1992 cited in Cresswell 2006: 43–4)

The shifts towards more poststructuralist perspectives start to become clear, but proceeded in directions that were divergent from those forged by Massey, who insisted on power-geometries as an abiding – if specific and complex – limit.

Echoing John Urry (2000), geographer Tim Cresswell observes that stillness is valorized over nomadism and movement in modern Western cultures: the "metaphysics of sedentarism is a way of thinking and acting that sees mobility as suspicious, as threatening, and as a problem. The mobility of others is captured, ordered, and emplaced in order to make it legible in a modern society" (Cresswell 2006: 55). Cresswell labels this sedentarist metaphysics as modern, a constellation of social, economic and political power relations, implying that the current epoch is thus, postmodern, wherein movement is the norm rather than the exception.

Geographer Matt Sparke also challenges this "everyone and everything is in motion" approach, by showing how a particular pattern or mapping effaces the multiple relations of power that *produce* that pattern:

> When geographers and whomever else set out to describe a particular geography, and even more so, when they invoke geography and space metaphorically, there is a metaphysics of presence at work – what might be called a metaphysics of geopresence – that fixates on the "geo" of a particular spatial pattern or a particular

poetics of location while simultaneously downplaying the geographic diversity of the constitutive processes that produced it.

(2005: xxix)

This metaphysics of presence can reduce multiple processes, contextual factors and power relations into a single map or explanation for a particular phenomenon.

POWER-GEOMETRY: UNEVEN ACCESS TO MOVEMENT

Perhaps because our research tends to focus on displacement, asylum, and refugees who are often denied *access* to places, legal processes and the material entitlements that they afford, we find that mobility is always constrained. All persons are subject to the calculus of Massey's (1993b) power-geometry, but conditions of highly restricted mobility, even containment, are more common for those bodies that are criminalized, displaced, and/or construed as a security threat to the state and its citizenry. The case of the "offshoring" of border enforcement to prevent asylum seekers from reaching sovereign territory and accessing the right to asylum once they land is instructive (Mountz 2011). Through this process, the state enacts border controls in a foreign airport, for example, screening visas and identity documents well before a person arrives on the sovereign territory of a country, where mere presence may imply basic rights. A second example involves the transfer of cash from countries of the global North in exchange for the containment of displaced persons by countries of the global South (Hyndman 1997; Castles 2008). This "cash for containment" approach is vividly illustrated in the 2016 EU–Turkey deal whereby the Government of Turkey agreed to keep all asylum seekers who want to travel to the EU in Turkey, providing education and basic health services to them, in return for 3 billion euro (in year 1). These EU "partnership" compacts have emerged with Jordan and Lebanon as well, and replace earlier "readmission return agreements" that enabled destination countries to return people who entered their sovereign territory to transit countries and countries of origin without authorization (reducing bureaucratic burdens of paperwork). The EU Emergency Trust Fund for Africa, established in 2015, offers massive development funds to tighten mobility with a registry system under development that will involve surveillance and registration of identity documents and human mobility across the African continent (Landau, forthcoming). The International Organization for Migration is an international organization that runs "Assisted Voluntary Return Programmes", which provide cash to asylum seekers who withdraw their asylum applications in countries such as the United Kingdom, Canada and elsewhere, and agree to return home.

Massey's power-geometries operate as an effective interpretive framework of these competing mobilities of cash, migration and development policy, and people. While the most wealthy global citizens can move quickly across vast distances, and government authorities can meet to share migration management policies and best practices, they may do so at the expense of other people's mobilities – the latter contained by the actions and policies of the former. Not only are African migrants from across the continent spending years in limbo and containment in Libyan and Tunisian coastal cities, but they are also transforming those sites as well, remaking place. Their experiences of place, violence and precarity along transnational migration routes spanning distances from Eritrea or Somalia to Italy or Greece bear little resemblance to the travel routes and infrastructures of more elite global workers and tourists who travel to nearby locations.

In Mountz's (2011) research on islands as in-between locations where migrants and asylum-seekers end up in prolonged periods of spatial, temporal, legal and psychological limbo, power-geometries also offer an incisive interpretive framing. While migrants negotiate islands through their own power-geometry calculus, seeking more permissive legal jurisdictions and hosts over less generous ones, they face a wall of exclusion policies as well. When migrants land on islands, they are often contained in detention facilities, and reconstitute the economies and meanings of place on small islands. They find themselves in limbo so often that Mountz has reconceptualized the geographical metaphor of the border as island, as migrant encounters have also shifted spatial metaphor from crossing to containment. In this sense, islands introduce new dimensions to power-geometries – and migrants' subversion of these relations – into what otherwise might appear as a linear journey.

CONCLUSION

As scholars who study human migration, mobility and geopolitics, Massey's ideas about globalization, spatial distribution, and her ways of conceptualizing power and movement pervade our thinking. Global migration is all about power-geometries: who can move, who must move, who can stay, and who cannot move. In this brief chapter, we have explored Doreen Massey's early ideas, and their remarkable foreshadowing of and resonance within the present. Nowhere is this more apparent than the pervasive ways that Massey's concepts and approaches to labour, gender and place have been so thoroughly taken up that they are now simply taken for granted and often no longer attributed to her through citation. New spatial divisions of labour and enduring power-geometries characterize the global refugee regime and the cities in which we live. Thanks in no small part to Doreen Massey and the legacy of her scholarship, geographers have analytical tools to make sense of place, who moves, who does not, and why.

REFERENCES

For works authored and co-authored by Doreen Massey, please see Select Bibliography of Doreen Massey, beginning on p. 371.

Aglietta, M. 1979 [1976]. *A Theory of Capitalist Regulation: The U.S. Experience.* Translated from the French by David Fernbach. London and New York: Verso.

Castles, S. 2008. "The Politics of Exclusion: Asylum and the Global Order". *Metropolis World Bulletin* 8 (October): 3–6.

Chanan, M. & L. Salter 2012. "Doreen Massey on London – Extracts from the Secret City". www.youtube.com/watch?v=zhHeelvwEN0 (accessed 9 January 2018).

Crenshaw, K. 1991. "Mapping the Margins: Intersectionality, Identity Politics, and Violence Against Women of Color. *Stanford Law Review* 43 (6): 1241–99.

Cresswell, T. 2006. *On the Move: Mobility in the Modern Western World.* London: Routledge.

Fröbel, F., J. Heinrichs & O. Kreye 1980. *The New International Division of Labour: Structural Unemployment in Industrialised Countries and Industrialisation in Developing Countries.* Translated by Peter Burgess. Cambridge: Cambridge University Press.

Harvey, D. 1989. *The Condition of Postmodernity: An Enquiry into the Origins of Cultural Change.* Malden, MA: Blackwell.

Hyndman, J. 1989. *Sweatshops in the Rain? Electronics Assembly in Cork, Ireland.* MA thesis. Lancaster University, unpublished.

Hyndman, J. 1997. "Border Crossings". *Antipode* 29 (2): 149–76.

Kaplan, C. 1996. *Questions of Travel: Postmodern Discourses of Displacement.* Durham, NC: Duke University Press.

Landau, L. n.d. "Shunning Solidarity: Durable Solutions in a Fluid Era". In M. Bradley, J. Milner & B. Peruniak (eds) *Shaping the Struggles of their Times: Refugees, Peacebuilding and Resolving Displacement.* Forthcoming.

Lash, S. & J. Urry 1987. *The End of Organized Capitalism.* Madison, WI: University of Wisconsin Press.

Lipietz, A. 1987. *Mirages and Miracles: The Crisis in Global Fordism.* Translated by David Macey. London: Verso.

Martin, L. & A. Secor 2014. "Towards a Post-Mathematical Topology". *Progress in Human Geography* 38 (3): 420–38.

Mountz, A. 2011. "The Enforcement Archipelago: Detention, Haunting, and Asylum on Islands". *Political Geography* 30 (3): 118–28.

Piore, M. & C. Sabel 1984. *The Second Industrial Divide: Possibilities for Prosperity*. New York: Basic Books.

Rose, G. 1996. "As if the Mirrors Had Bled: Masculine Dwelling, Masculinist Theory and Feminist Masquerade". In N. Duncan (ed.) *Bodyspace: Destablizing Geographies of Gender and Sexuality*. London and New York: Routledge, 57–74.

Sheller, M. & J. Urry (eds) 2006. *Mobile Technologies of the City*. London and New York: Routledge.

Smith, N. & C. Katz 1993. "Grounding Metaphor: Towards a Spatialized Politics". In M. Keith & S. Pile (eds) *Place and the Politics of Identity*. London: Routledge.

Sparke, M. 2005. *In the Space of Theory: Postfoundational Geographies of the Nation-State*. Minneapolis, MN: University of Minnesota Press.

Urry, J. 2000. *Sociology Beyond Societies: Mobilities for the Twenty-First Century*. London and New York: Routledge.

Valentine, G. 2007. "Theorizing and Researching Intersectionality: A Challenge for Feminist Geography". *The Professional Geographer* 59 (1): 10–21.

EPILOGUE: "HOW WE WILL MISS THAT CHUCKLE": MY FRIEND, DOREEN MASSEY

Hilary Wainwright

Doreen Massey, socialist feminist, engaged geographer and influential public intellectual, died during the night of 11 March 2016, aged 72. Tributes poured in through social media. "Now we understand what everyone else felt about David Bowie", was one attempt to convey the scale of our loss. Doreen's body could sometimes be frail but her mind was brilliant and tough. As a character she was strong, passionate, curious and also imbued with modesty and kindness. She radiated political energy and humanity, sparkling with a cheeky wit.

With a steady flow of books, such as *For Space, The Anatomy of Job Loss*, and *Space, Place and Gender* and contributions to the books of others such as Huw Beynon's *Digging Deeper*, a collection on the politics of the 1984/85 miners strike, she worked alongside others such as David Harvey, to establish geography as the intellectual source of a powerful, integrated critique of predatory capitalism in the age of climate change and the corporate driven global market.

Her academic base was the Open University to which she was strongly loyal because of its openness and accessibility to all who wanted to learn. She turned down professorships from elsewhere, including from Oxford, which she considered too exclusive and elitist for her far-reaching educational mission. She became a mentor, both through her writings and through a tireless round of talks and personal conversation, to generations of young geographers; their appreciation was evident in the Twitter storm that followed the shock of her death.

She was proud of winning the Prix Vautrin Lud ("Nobel de Geographie") and she organized a celebration with her large and varied circle of friends in typically convivial Doreen style. On the other hand, she was appalled to hear that the vehemently hated establishment wanted to award her an Order of the British Empire (OBE). The result was an urgent phone call to get contact details of other refuseniks of royal awards Ken Loach, John Palmer and more in order to plan how to ensure that her refusal had a maximally antiroyalist impact.

Her strong sense of class underpinned all her politics; it drove her energetic support for the women in the mining communities during the 1984/85 strike (we went together to Blidworth in Nottinghamshire to give moral and practical support to the local pickets and Women Against Pit Closures); and it was one reason for her enthusiasm for Jeremy Corbyn's leadership of the Labour Party and the possibilities it opened up. (Indeed the last time I was with her was when together with Chantal Mouffe, another close Kilburn friend, we discussed what we could *do* to support Corbyn's leadership.) She wrote a heartfelt call to intellectual arms in her last *Soundings* blog:

> Corbyn's commitment to democratic engagement and openness, and to doing politics in a different way, as well as his rejection of individual celebrity status, is a real strength ... We must do everything we can to keep this initiative growing and to play our part in the wider movement that keeps bubbling up.

But she did not just make editorial calls from the platform of *Soundings*, the thoughtful and imaginative quarterly journal that she founded with Mike Rustin and the late Stuart Hall (published by Lawrence & Wishart's Sally Davison). She and fellow Kilburn residents, Mike Rustin and Stuart Hall, worked hard and with an impressive sense of urgency to produce the pathbreaking and remaking Kilburn Manifesto. Path-breaking because of the way it challenged the dominant terms of debate and deconstructed the language of neoliberalism. As she wrote:

> Think, for instance, of that bundle of words "investment/expenditure/speculation". Each term bears moral connotations: investment good thing; expenditure a cost, possibly a burden; speculation in a financial sense a bit dodgy. Now think about how we use these terms in popular and political discourse. The couplet investment/expenditure for example. Investment, in the national accounts, is money laid out for things such as buildings and infrastructure. Expenditure on the other hand is money laid out, for instance, on the wages of people operating the services for which the investment provides the infrastructure. So building a new school is investment (moral connotation: a good thing) but paying for teachers and dinner ladies is expenditure (a pure cost, a burden). (Ponder, immediately, the gender implications of this.)

The radical, socialist and labour movement think tank CLASS (Centre for Labour Studies) was another important platform for Doreen and her work for them was greatly appreciated by its staff.

Her upbringing in Wythenshawe, one of the largest council estates in Manchester, was a constant reference point in our political discussions and I think it fired her fury at class exploitation and the exercise of ruling class power, intertwined in complex ways with the inequities of economic geography and of gender and race. The northwest remained a focus for several of her lifelong passions. There were constant train journeys from Euston to Liverpool to cheer on her football team, to which she was fiercely loyal and then on to Ulveston, to stay with her much loved sister Hilary, go walking with her and do a bit of birdwatching (another of Doreen's enduring loves) in the inspiring fells of the Lake District.

She was also passionate about London, her city of residence and often of study. She was an active architect and engineer of the Greater London Council under the leadership of her friend and Kilburn neighbour, Ken Livingstone. She became a member of the Greater London Enterprise Board on which fellow board member, John Palmer, remembers how she "would ask searching questions on issues surrounding the advancement of the rights of women and ethnic minorities in the preparation of development strategies for GLEB investments". And Robin Murray, also a member of the board as Economic Advisor to the GLC, describes how she "insisted that space was social not just physical [but also] gendered space, class space". She brought these perspectives, which enlightened so many, to the deliberations of the GLEB Board as they considered how the GLC principles could be translated into the concrete projects that came before it.

Several of us, including Robin Murray and one of her co-founders of *Soundings*, Mike Rustin, also remember the Ariel Road group that met with Ken Livingstone in her front room after Mrs Thatcher's abolition of the GLC, and how, in Robin's words, "a great theorist was equally at home in discussing the key political issues of the day".

She applied this ability to combine theoretical innovation and political engagement on an international scale, especially in Latin America. She spoke fluent Spanish and spent a year in Nicaragua, writing a book about it. Her continuing interest in space and power led her to a longstanding engagement with political change in Venezuela. I remember how chuffed she was that her concept of power-geometries was taken up as part of the effort to extend grassroots democracy and participatory democracy. She was also a member of the Editorial Board of the journal *Revista Pos*, the School of Architecture and Urbanism of Sao Paulo, Brazil.

In addition to her long involvement with Latin America, she also worked with South African activists. During the transitional government, she worked with Frene Ginwala, later the first person of colour to become Speaker of the South African Parliament, in a workshop on gender and unpaid labour, at a time when such issues were sidelined in economic debate. "It was an extraordinary event", remembers another friend, Maureen Mackintosh, also involved with the struggles in Southern Africa.

On a personal note, I came to know Doreen through both of us becoming trustees of a charity, the Lipman-Miliband Trust, whose mission was to fund education and research about socialism. And it was through the Trust that we discovered we shared the same birth date. Our two older trustees at the time, doughty historian John Savile, and magisterial Marxist political theorist Ralph Miliband, suggested a date for the winter meeting: 3 January. We both muttered somewhat grumpily, "but that's our birthday". But as good soldiers to John Savile's military command, we went along with it. After a birthday drink we arrived at the meeting at the Miliband/Kosack home, to be greeted in the most chivalrous manner, by John and Ralph presenting each of us with a large bouquet! It was a fond memory of these two comrades over which we often had a good chuckle.

How we will miss that chuckle! But we are fortunate that through the stunning (a favourite word of hers) range and scope of her writing, her ideas are still with us, though we are able to reinforce her influence. So she would want me to end this salute and make sure it resounds, with a sharp political remark. The opening of her recent *Soundings* blog would do the job: ever since Corbyn was elected, pundits have been predicting doom for Labour. In fact Labour's doom is more likely to be sealed if the party does not rally round and work to make his leadership successful.[1] Whatever the party – and in particular the Parliamentary Party – does, friends of Doreen – a large and influential network – I know will rally round to remember Doreen and keep alive her determined spirit and ensure the continuing influence of her transformational and deeply radical politics. Viva Doreen! Viva the transformation of geometries of power! Viva![2]

1. See www.lwbooks.co.uk/soundings/blog/why-corbyn-leadership-still-best-hope-labour.
2. This appreciation was first published in *Open Democracy* on 15 March 2016. www.opendemocracy.net/uk/hilary-wainwright/how-we-will-miss-that-chuckle-my-friend-doreen-massey.

SELECT BIBLIOGRAPHY OF DOREEN MASSEY

Allen, J. & D. Massey (eds) 1988. *Restructuring Britain: The Economy in Question*. London: Sage.

Allen, J., D. Massey & A. Cochrane 1998. *Rethinking the Region*. London: Routledge.

Amin, A., D. Massey & N. Thrift 2000. *Cities for the Many Not the Few*. Bristol: Policy Press.

Amin, A., D. Massey & N. Thrift 2003. *Decentering the Nation: A Radical Approach to Region Inequality*. London: Catalyst.

Cordey Hayes, M., T. A. Broadbent & D. Massey 1970. "Towards Operational Urban Development Models". Working Paper 60, Centre for Environmental Studies, London.

Hall, S. & D. Massey 2010. "Interpreting the Crisis". In R. Grayson & J. Rutherford (eds) *After the Crash: Re-inventing the Left in Britain*. London: Soundings, Social Liberal Forum and Compass, 37–46.

Hall, S. & D. Massey 2012. "Interpreting the Crisis". In S. Davison & K. Harris (eds) *The Neo-liberal Crisis*. London: Soundings, 55–69.

Hall, S., D. Massey & M. Rustin 1995. "Editorial: Uncomfortable Times". *Soundings* 1: 5–18.

Hall, S., D. Massey & M. Rustin (eds) 2012. *After Neoliberalism? The Kilburn Manifesto*. London: Lawrence and Wishart.

Hall, S., D. Massey & M. Rustin 2013. "After Neoliberalism: Analysing the Present". *Soundings* 53: 8–22.

Harrison, S., D. Massey & K. Richards 2006. "Complexity and Emergence (Another Conversation)". *Area* 38 (4): 465–71.

Harrison, S., D. Massey & K. Richards 2008. "Conversations across the Divide". *Geoforum* 39 (2): 549–51.

Harrison, S., D. Massey, K. Richards, *et al.* 2004. "Thinking across the Divide: Perspectives on the Conversations between Physical and Human Geography". *Area* 36 (4): 435–42.

Henry, N. & D. Massey 1995. "Competitive Time-Space in High Technology". *Geoforum* 26 (1): 49–64.

Lury, K. & D. Massey 1999. "Making Connections". *Screen* 40 (3): 229–38.

Martin, R., A. Markusen & D. Massey 1993. "Classics in Human Geography Revisited: Spatial Divisions of Labour". *Progress in Human Geography* 17 (1): 69–72.

Massey, D. 1968a. "Problems of Location: Linear Programming". *Working Paper* 14, Centre for Environmental Studies, London.

Massey, D. 1968b. "Problems of Location: Game Theory and Gaming Simulation". *Working Paper* 15, Centre for Environmental Studies, London.

Massey, D. 1969. "Some Simple Models for Distributing Changes in Employment within Regions". *Working Paper* 24, Centre for Environmental Studies, London.

Massey, D. 1971. "The Basic: Service Categorization in Planning". *Working Paper* 63, Centre for Environmental Studies, London.

Massey, D. 1973. "Towards a Critique of Industrial Location Theory". *Antipode* 5 (3): 33–9.

Massey, D. 1974. "Social Justice and the City: A Review". *Environment and Planning A* 6 (2): 229–35.

Massey, D. 1978. "Regionalism: Some Current Issues". *Capital and Class* 2 (3): 105–25.

Massey, D. 1979. "In What Sense a Regional Problem?" *Regional Studies* 13 (2): 233–43.

Massey, D. 1983a. "The Shape of Things to Come". *Marxism Today* April: 18–27.

Massey, D. 1983b. "The Contours of Victory – Dimensions of Defeat". *Marxism Today* July: 16–19.

Massey, D. 1983c. "Industrial Restructuring as Class Restructuring: Production Decentralization and Local Uniqueness". *Regional Studies* 17 (2): 73–89.

Massey, D. 1984a. "New Directions in Space". In J. Urry & D. Gregory (eds) *Social Relations and Spatial Structures*. London: Macmillan.

Massey, D. 1984b. *Spatial Divisions of Labour: Social Structures and the Geography of Production*. Basingstoke: Macmillan.

Massey, D. 1984c. "Introduction: Geography Matters". In D. Massey & J. Allen (eds) *Geography Matters! A Reader*. Cambridge: Cambridge University Press, 1–11.

Massey, D. 1985. "New Directions in Space". In D. Gregory & J. Urry (eds) *Social Relations and Spatial Structures*. London: Macmillan, 9–19.

Massey, D. 1986. "Nicaragua: Some Reflections on Socio-Spatial Issues in a Society in Transition". *Antipode* 18 (3): 322–31.

Massey, D. 1987a. *Nicaragua: Some Urban and Regional Issues in a Society in Transition*. Milton Keynes: Open University Press.

Massey, D. 1987b. "Spatial Labour Markets in an International Context". *Tijdschrift voor Economische en Sociale Geografie* 78 (5): 374–9.

Massey, D. 1988a. "A New Class of Geography". *Marxism Today* May: 12–17.

Massey, D. 1988b. "Uneven Development: Social Change and Spatial Divisions of Labour". In D. Massey & J. Allen (eds) *Uneven Re-development: Cities and Regions in Transition*. London: Hodder and Stoughton, 250–76.

Massey, D. 1991a. "A Global Sense of Place". *Marxism Today* June: 24–9.

Massey, D. 1991b. "Flexible Sexism". *Environment and Planning D: Society and Space* 9 (1): 31–57.

Massey, D. 1991c. "The Political Place of Locality Studies". *Environment and Planning A* 23 (2): 267–81.

Massey, D. 1992a. "A Place Called Home?" *New Formations* 17: 3–15.

Massey, D. 1992b. "Politics and Space/Time". *New Left Review* 196: 65–84.

Massey, D. 1992c. "Space, Place and Gender". *London School of Economics Magazine* Spring: 32–4.

Massey, D. 1993a. "The Different Side of the 'Sixties'". *Environment and Planning A* Anniversary Issue: 10–13.

Massey, D. 1993b. "Power-Geometry and a Progressive Sense of Place". In J. Bird, B. Curtis, T. Putnam, G. Robertson & L. Tuckner (eds) *Mapping the Futures: Local Cultures, Global Chance*. Abingdon: Routledge, 59–69.

Massey, D. 1994. *Space, Place and Gender*. Cambridge: Polity Press.

Massey, D. 1995a. "Masculinity, Dualisms and High Technology". *Transactions of the Institute of British Geographers* 20 (4): 487–99.

Massey, D. 1995b. *Spatial Divisions of Labour: Social Structures and the Geography of Production*. 2nd edition. Basingstoke: Macmillan.

Massey, D. 1995c. "Thinking Radical Democracy Spatially". *Environment and Planning D: Society and Space* 13 (3): 283–8.

Massey, D. 1997a. "Editorial: Problems with Globalization". *Soundings* 7: 7–12.

Massey, D. 1997b. "Spatial Disruptions". In S. Golding (ed.) *Eight Technologies of Otherness*. London and New York: Routledge, 218–25.

Massey, D. 1999a. Curriculum Vitae. Papers of Doreen Massey.

Massey, D. 1999b. "Imagining Globalization: Power-Geometries of Time-Space". In A. Brah, M. J. Hickman & M. Mac an Ghaill (eds) *Global Futures: Migration, Environment and Globalization*. Basingstoke: Macmillan, 27–44.

Massey, D. 1999c. "Negotiating Disciplinary Boundaries". *Current Sociology* 47 (5): 5–12.

Massey, D. 1999d. "Philosophy and Politics of Spatiality: Some Considerations". *Geographische Zeitschrift* 87 (1): 1–12.

Massey, D. 1999e. *Power-Geometries and the Politics of Space-Time: Hettner Lecture 1998*. Heidelberg: Department of Geography, University of Heidelberg.

Massey, D. 1999f. "Spaces for Co-Existence?" *Soundings* 12 (Summer): 7–11.

Massey, D. 1999g. "Space-Time, 'Science' and the Relationship between Physical Geography and Human Geography". *Transactions of the Institute of British Geographers* 24 (3): 261–76.

Massey, D. 2000a. "Bankside: International Local". In I. Blazwick (ed.) *Tate Modern: The Handbook*. London: Tate Publishing, 24–7.

Massey, D. 2000b. "Travelling Thoughts". In P. Gilroy, L. Grossberg & A. McRobbie (eds) *Without Guarantees: In Honour of Stuart Hall*. London: Verso, 225–32.

Massey, D. 2001a. "Geography on the Agenda". *Progress in Human Geography* 25: 5–17.

Massey, D. 2001b. "Living in Wythenshawe". In I. Borden, J. Kerr, J. Rendell & A. Pivaro (eds) *The Unknown City: Contesting Architecture and Social Space*. Cambridge, MA: MIT Press, 458–75.

Massey, D. 2001c. "Talking of Space-Time". *Transactions of the Institute of British Geographers* 26: 257–61.

Massey, D. 2002a. "Don't Let's Counterpose Place and Space". *Development* 45 (2): 24–5.

Massey, D. 2002b. "Globalisation: What Does It Mean for Geography? *Geography* 87 (4): 293–6.

Massey, D. 2002c. "Geography, Policy and Politics: A Response to Dorling and Shaw". *Progress in Human Geography* 26 (5): 645–6.

Massey, D. 2003a. "No More, No Less". *Building Design* July 25: 36.

Massey, D. 2003b. "Some Times of Space". In S. May (ed.) *Olafur Eliasson: The Weather Project*. London: Tate Publishing, 107–18.

Massey, D. 2004a. "Geographies of Responsibility". *Geografiska Annaler B: Human Geography* 86 (1): 5–18.

Massey, D. 2004b. "The Responsibilities of Place". *Local Economy* 19 (2): 97–101.

Massey, D. 2005. *For Space*. London: Sage Publications.

Massey, D. 2006a. "Space, Time and Political Responsibility in the Midst of Global Inequality". *Erdkunde* 60 (2): 89–95.

Massey, D. 2006b. "Landscape as a Provocation: Reflections on Moving Mountains". *Journal of Material Culture* 11 (1–2): 33–48.

Massey, D. 2006c. "London Inside-Out". *Soundings* 32: 62–71.

Massey, D. 2007. *World City*. Cambridge: Polity.

Massey, D. 2008. "When Theory Meets Politics". *Antipode* 40 (3): 492–7.

Massey, D. 2009a. "Invention and Hard Work". In J. Pugh (ed.) *What is Radical Politics Today?* Basingstoke: Palgrave Macmillan, 136–42.

Massey, D. 2009b. "Concepts of Space and Power in Theory and in Political Practice". *Documents d'Anàlisi Geogràfica* 55: 15–26.

Massey, D. 2010. "The Political Struggle Ahead". *Soundings* 45: 6–18.

Massey, D. 2011a. "Ideology and Economics in the Present Moment". *Soundings* 48: 29–39.

Massey, D. 2011b. "A Counterhegemonic Relationality of Place". In E. McCann & K. Ward (eds) *Mobile Urbanism: Cities and Policy Making in the Global Age*. Minneapolis: University of Minnesota Press, 1–14.

Massey, D. 2011c. "Espacio y Sociedad: Experimentos con la Espacialidad del Poder y de la Democracia". In A. G. González (ed.) *Latinoamérica: Laboratorio Mundial*. Madrid: La Oficina de Arte y Ediciones, 29–46.

Massey, D. 2011d. "Landscape/Space/Politics: An Essay". The Future of Landscape and the Moving Image Research Project website. https://thefutureoflandscape.wordpress.com/landscapespacepolitics-an-essay/ (accessed 17 December 2017).

Massey, D. 2012a. "Landscape/Space/Politics: An Essay". In R. Tyszczuk, J. Smith, N. Clark & M. Butcher (eds) *Atlas: Geography, Architecture and Change in an Interdependent World*. London: Black Dog Publishing, 90–5.

Massey, D. 2012b. "Learning from Latin America". *Soundings* 50: 131–41.

Massey, D. 2012c. "Los Significados de la Multiplicidad". *Debates y Combates* 27 March. www.youtube.com/watch?v=9wn4QphR4Yw (accessed 9 January 2018).

Massey, D. 2013a. "Vocabularies of the Economy". *Soundings* 54 (Summer): 9–22.

Massey, D. 2013b. "Neoliberalism has Hijacked our Vocabulary". *The Guardian*, 11 June. www.theguardian.com/commentisfree/2013/jun/11/neoliberalism-hijacked-vocabulary (accessed 17 January 2018).

Massey, D. 2014. "The Kilburn Manifesto: After Neoliberalism?" *Environment and Planning A* 46 (9): 2034–41.

Massey, D. 2015a. "Globalización, Espacio y Poder". In CEPAL *Memoria del Primer Encuentro de Expertos Gubernamentales en Políticas de Desarrollo Territorial en América Latina y el Caribe*. Santiago de Chile: CEPAL, Naciones Unidas, 9–14.

Massey, D. 2015b. "Why the Corbyn Leadership is Still the Best Hope for Labour". *Soundings blog*, 18 November. www.lwbooks.co.uk/soundings/blog/why-corbyn-leadership-still-best-hope-labour (accessed 18 December 2017).

Massey, D. 2015c. "Vocabularies of the Economy". In S. Hall, D. Massey & M. Rustin (eds) *After Neoliberalism? The Kilburn Manifesto*. London: Lawrence & Wishart, 24–36.

Massey, D. & J. Allen (eds) 1984. *Geography Matters: A Reader*. Cambridge: Cambridge University Press.

Massey, D. & P. Batey 1977. "Introduction". In D. Massey & P. Batey (eds) *Alternative Frameworks for Analysis*. London: Pion, 1–5.

Massey, D., S. Bond & D. Featherstone 2009. "The Possibilities of a Politics of Place Beyond Place? A Conversation with Doreen Massey". *Scottish Geographical Journal* 125 (3–4): 401–20.

Massey, D. & A. Catalano 1978. *Capital and Land: Landownership by Capital in Great Britain*. London: Edward Arnold.

Massey, D. with the HGRG [Human Geography Research Group] 2009. "The Possibilities of a Politics of Place Beyond Place? A Conversation with Doreen Massey". *Scottish Geographical Journal* 125 (3–4): 401–20.

Massey, D. & J. Jacobs 2016. "Significant Geographies: Reflections from a Paper Age". Unpublished proposal. Papers of Jessica Jacobs.

Massey, D. & K. Livingstone 2007. "The World We're In: Interview with Ken Livingstone". *Soundings* 36: 11–25.

Massey, D. & R. Meegan 1978. "Industrial Restructuring versus the Cities". *Urban Studies* 15 (3): 273–88.

Massey, D. & R. Meegan 1979. "The Geography of Industrial Reorganisation: The Spatial Effects of the Restructuring of the Electrical Engineering Sector Under the Industrial Reorganisation Corporation". *Progress in Planning* 10 (3): 155–237.

Massey, D. & R. Meegan 1982. *The Anatomy of Job Loss: The How, Why and Where of Employment Decline*. London: Methuen.

Massey, D. & R. Meegan (eds) 1985. *Politics and Method: Contrasting Studies in Industrial Geography*. London: Methuen.

Massey, D. & N. Miles 1984. "Mapping Out the Unions". *Marxism Today* May: 19–22.

Massey, D., P. Quintas & D. Wield 1992. *High-Tech Fantasies: Science Parks in Society, Science and Space*. London: Routledge.

Massey, D. & M. Rustin 2015. "Displacing Neoliberalism". In S. Hall, D. Massey & M. Rustin (eds) *After Neoliberalism? The Kilburn Manifesto*. London: Lawrence & Wishart, 191–221.

Massey, D. & A. Stevens 2010. "The Future of Landscape: Doreen Massey: An Interview by Andrew Stevens". *3:AM Magazine* 29 September. www.3ammagazine.com/3am/the-future-of-landscape-doreen-massey (accessed 8 January 2018).

Massey, D. & H. Wainwright 1985. "Beyond the Coalfields: The Work of the Miners' Support Groups". In H. Beynon (ed.) *Digging Deeper: Issues in the Miners' Strike*. London: Verso, 149–68.

Massey, D. & N. Warburton 2013. "Doreen Massey on Space". *Social Science Bites* 8 May. www.youtube.com/watch?v=Quj4tjbTPxw (last accessed 8 January 2018).

McDowell, L. & D. Massey 1984. "A Woman's Place". In D. Massey & J. Allen (eds) *Geography Matters! A Reader*. Cambridge: Cambridge University Press, 128–47.

Peck, J., D. Massey, K. Gibson & V. Lawson 2014. "The Kilburn Manifesto: After Neoliberalism?" *Environment and Planning A* 46 (9): 2033–49.

INDEX